D0142929

STUDIES IN COMPUTER SCIENCE

MAA STUDIES IN MATHEMATICS

Published by
THE MATHEMATICAL ASSOCIATION OF AMERICA

Studies in Mathematics

The Mathematical Association of America

William E. Ball
Washington University in St. Louis

Ronald V. Book
University of California at Santa Barbara

E. R. Buley
General Research Corporation, Santa Barbara

Mark Franklin
Washington University in St. Louis

R. H. Pennington
General Research Corporation, Santa Barbara

Seymour V. Pollack
Washington University in St. Louis

Terrence W. Pratt
University of Virginia

Franco P. Preparata
University of Illinois, Urbana-Champaign

James R. Slagle
Naval Research Laboratory, Washington, D.C.

C. F. Starmer
Duke University

Studies in Mathematics

Volume 22

STUDIES IN COMPUTER SCIENCE

Seymour V. Pollack, editor
Washington University in St. Louis

Published and distributed by
The Mathematical Association of America

Library of Congress Catalog Card Number 82-062390

Complete Set ISBN 0-88385-100-8
Vol. 22 ISBN 0-88385-124-5

Printed in the United States of America

Current printing (last digit):
10 9 8 7 6 5 4 3 2 1

INTRODUCTION

The Association of Computing Machinery's Administrative Directory for 1979 lists 207 American computer science departments granting bachelor's degrees, 127 granting master's degrees, and 73 offering Ph.D. or D.Sc. degrees. In addition, the directory includes computer science programs at all levels embedded in 163 mathematics departments, 56 business schools, 29 electrical engineering departments, and 40 other schools or departments, including such diverse areas as physics, industrial engineering, and economics. What makes these figures remarkable is the fact that the first computer science department appeared less than two decades earlier. To me, this rapid growth is but one of several factors that combine to place computer science in an exceptional position vis-à-vis other areas of inquiry. A brief exploration of these points will be helpful in providing some general perspectives within which the articles in this study can be considered.

First, it is important to note that the burgeoning of computer science programs cannot be equated with the maturation of computer science. There still is no "standard" (i.e., universally inoffensive) definition of computer science.* In fact, the existence of such a

---

* *The Encyclopedia of Computer Science* (A. Ralston and C. Meek, eds., Petrocelli/Charter, 1976) defines computer science as follows: "Computer science is

discipline continues to be a debatable point for a substantial number of people. (A prominent educator, though himself a chairman of a computer science department, cautions his audiences to regard with suspicion any discipline with "science" in its name—urban science, consumer science, economic science, social science, computer science.) Some people think of computer science's "universe" as a relatively restricted one, limited by definition to electronic digital information processing systems. Definitions at the other end of the spectrum perceive an arena consisting of an arbitrarily wide range of information processing systems, including biological ones. Despite this diversity, the digital computer system is clearly the dominant vehicle for study. Mildly stated, this is a very unusual situation: Instead of exploring the behavior of a cell, a fluid, an organism, or a galaxy, the computer scientist seeks basic observable phenomena from an artifact (i.e., the computer itself or the program processed therein). Thus, the quest for "natural laws" carries little meaning here. There is no ultimate and final reality against which explanatory structures are to be assessed. The "reality," represented by the computing hardware, software, and information, is arbitrarily alterable. Though it may sound almost facetious, the fact remains that, if an attempt (regardless of its degree of formalism) falls short of explaining observed events, those events (i.e., reality) can be changed to meet the explanation halfway. Inevitably, this has a profound effect on the phenomena computer scientists seek to describe and the ways in which such descriptions are voiced.

A second major peculiarity lies in computer science's inherent invisibility. End products of computer science, i.e., information processing systems, generally are used (and motivated) by people with little interest in computer science. A major objective in the implementation of such products is to obscure their inner workings so that the user's attention remains focused on the externally perceived behavior. For instance, a well-designed translating system for a high-level programming language (such as FORTRAN) suc-

---

concerned with information processes, with the information structures and procedures that enter into representations of such processes and with their implementation and information processing systems. It is also concerned with relationships between information processes and classes of tasks that give rise to them."

cessfully promotes the illusion that the user's programs execute directly on a FORTRAN computer, with no apparent intervention between the program (as written) and the machinery. (This is emphasized by the jargon, which terms such a system "transparent" to its users.) Conversely, implementors of such systems generally are less concerned with the ultimate uses than they are with the technical issues pertaining to a system's design and behavior. As system objectives became more ambitious, their requirements grew more intricate, thereby generating increasingly complex design and organizational problems. Not surprisingly, many of these problems, when abstracted, began to take on a separate existence, atrracting many people interested in pursuing them within their own context. The resulting dichotomy produced a situation in which a considerable amount of creative and intellectually exciting work has been done with little or no connection to events in the field. Although it is labeled "computer science theory," it has not been motivated by actual computing problems, nor has it exerted any substantial influence on computing practice. Rather, the breakneck pace that characterizes the advance of computer applications has been fueled for the most part by a loosely interwoven fabric of empirical technologies.

The extent to which formal computer science and applied computer technology will continue on essentially independent paths remains a matter of speculation. There are strong signs that certain areas of formal inquiry (such as computational complexity, formal languages, and automata theory) already are beginning to have an impact on the design of a variety of information processing procedures. (This, in part, motivates some of the emphasis on these areas in this Study.) Moreover, the realities of continued growth in computer applications militate for more concerted efforts to promote such interactions. The conceptual demands imposed by many of these new systems are beginning to fall beyond the range of complexity that can be accommodated without more generalized models and applicable formal structures. However, an emerging pattern of bridge-building between abstract computer science and applied computing technology remains to be defined.

Another important aspect of computer science relates to the computer's pervasiveness. The indoor record for platitudes may well be held by variants of the statement that computers have touched every aspect of human endeavor. Overworked as this chestnut may be, it has long stopped being a hyperbole. While a considerable amount of this proliferation is reasonably attributable to a sustained, impressive selling effort, a significant impetus has come from computer science, even as the discipline was seeking its identity. Despite the almost chaotic diversity that characterized early perceptions of computer science, one fundamental concept emerged whose profound effect on the breadth of computer applications was felt almost immediately: This was nothing more (more precisely, we should say "nothing less") than the realization that the computer is primarily a symbol manipulator, with numerical calculation being but one relatively specialized activity. The primitive manipulations that are specifiable on a computer receive their context from the algorithm designer. Thus we can define any system we choose to define, representing its components via an arbitrary set of symbols and imbuing that system with a set of procedural characteristics consistent with our definition. When these symbols are processed, the "meaning" of the operations performed on them derives strictly from our perception. The staggering result, then, is that the computer presents itself as a vehicle for controlling (managing) any procedure that we are able to describe precisely.

Self-evident as this recognition may appear, it all but escaped any notice by many early computer users. However, once substantial interest was aroused in examining computer procedures and languages for specifying such procedures, the revolution began in earnest and it persists unabated a quarter of a century later. Computer usage continues to proliferate, exerting a feedback effect on the directions of hardware and software development.

It is this mutual interaction between computer usage and computer-oriented research and engineering* that has helped engender the enormous collection of applicable technology. At the same time, this process has catalyzed (to a substantial degree) the identity problems that continue to complicate computer science's movement toward adulthood. When computer applications are developed in a particular area, the resulting effect on that field may

* This is a community distinct from more abstractly oriented computer scientists, mentioned earlier, who have explicitly exempted themselves from such interactions.

be quite profound. For example, basic directions in thermodynamics have been altered by making "exotic" computations routine and previously "impossible" computations plausible. In a number of disciplines, these effects have been so fundamental that it seemed natural to associate computers and computing with those disciplines. Thus, computer-oriented workers in, say, numerical analysis found it reasonable to treat studies in computing as being inexorably linked with those in numerical analysis. Then, when it came to define an academic program, its natural habitat (for those workers) was clearly in that area. This situation replicated itself to a sufficient extent so that as recently as 1974 the National Science Foundation deemed it necessary to affirm computer science as being distinct from other disciplines.

These perplexing characteristics make it all the more difficult to identify a group of concerns that indisputably are "central" to computer science and therefore represent inevitable candidates for inclusion in a compendium such as this one. Consequently, centrality (by whatever criteria) was not the only motivation in the selection of these articles. Considered also was the desirability of emphasizing topics that would be particularly interesting to mathematicians and would underscore the wide range of areas deeply and continually affected by computer science.

The collection resulting from these deliberations consists of nine articles. After a historical overview of the major factors motivating the development of computer sicence, there are three articles dealing with aspects of programs and the programming process. Regardless of the objectives of a given computerized information handling process, the inescapable fact is that the generation of any observable phenomena involves the execution of a program. Consequently, any study claiming to pay attention to the computer science's mainstream issues must examine an interlocking chain of concepts and technologies that includes the characteristics of programming languages, design and implementation of algorithmic translators for these languages, and the analysis of programs written in these languages. Accordingly, an article by William E. Ball provides a systematic examination of programming languages as communication vehicles for expressing algorithms. Corresponding language processors are then examined as means for analyzing and translating these programs, stated in terms convenient to the user, into the more primitive forms required for execution on a computer.

The pivotal role of programming languages has stimulated a growing interest in the study of their features and properties within the broader context of formal language theory. Mathematical models developed for the exploration of natural languages are serving as effective bases for describing and characterizing computer languages as well. Results stemming from this work have provided invaluable insights that are exerting considerable influence on the direction of new language designs and implementations. The scope and impact of these inquiries, together with the underlying theoretical framework, are discussed in an article by R. V. Book on the specification of formal languages.

Work on programs and programming has engendered another major avenue of formal study, this one dealing with the programs themselves. While there are many people who still are convinced that the process of designing and implementating computable algorithms is characteristically resistant to the imposition of any substantial formalism, a growing number of investigators holding the opposite conviction have made important strides toward defining basic analytical tools for the formal description of a program and its behavioral properties. In an article on formal analysis of programs, T. W. Pratt discusses the major directions this work has taken and defines realistically the prospects for a theoretical basis that will make possible the systematic generation of optimally constructed, probably correct programs.

Another focal point in the study of computational processes addresses the computations themselves. Closely intertwined with the properties of programming languages and the behavior of algorithms, but still quite distinct, is a set of concerns regarding the characterization of algorithms in terms of their complexity. The development of theoretical structures for assessing precisely the relative difficulty of competing solution methods, well beyond the contemplative stage, is examined in an aritcle by F. P. Preparata.

The last four articles deal with areas that are of great interest and serious concern to computer science but whose position relative to the "core" of the discipline continues to be debated. For some, areas such as artificial intelligence, numerical analysis, statistics, and simulation are important applications of computer science. Others take a diametrically opposite view, perceiving these fields as separate disciplines in which computer science plays a significant but supportive role (as electron microscopy might be perceived in relation to genetics or metallurgy). Still others may think of some or all of these areas as branches of computer science itself. Irrespective of one's orientation, there is an undeniably pervasive attribute shared by such fields: All of them have been altered profoundly by their interaction with computing and computer science.

Perhaps the most conspicuous example of a branch of knowledge in which computers have provided primary impetus is that of artificial intelligence. Machine-oriented work has produced dramatic results in pattern recognition, theorem proving, game playing, and other cognitive processes that have affected the structure and capabilities of information processing systems in several important ways. More fundamentally, achievements in artificial intelligence have prompted some unsettling questions about the nature and constancy of the boundaries between human and machine intelligence. Accordingly, this area is addressed in a comprehensive article by James R. Slagle.

Both numerical analysis and statistics have been revolutionized by the sheer mass of computational power that can be marshaled to attack problems in these areas.* Methodologies which in the past could only be described are now part of the standard repertory; classes of algorithms whose potential computational requirements had placed them beyond contemplation are now actively pursued, designed, implemented, and used. One basic effect of this available computational power has produced a most provocative consequence in both these areas: Traditionally, many of the assumptions underlying a wide range of statistical and numerical algorithms were accepted implicitly because of the extensive amount of computation required to test them and the even more formidable work required to introduce corrections (or alternative methodologies) should the tests reveal a violation.

---

* Numerical analysis, in particular, has emerged with renewed prominence and continues to experience vigorous growth. In response, a separate volume in this series is being planned to characterize this field.

In an article on the impact of computers on numerical analysis, E. R. Buley and R. H. Pennington explore these effects as related to a number of major areas, including function evaluation, numerical quadrature, and systems of equations. Particular emphasis is placed on ways of exploiting more sophisticated numerical algorithms whose use for improving approximations and reducing error build-up no longer can be eschewed based on computational difficulty.

Digital simulation exemplifies an area whose basic spectrum has been widened dramatically by the introduction of digital computers. Mark Franklin discusses the range of analytically inaccessible systems whose exploration has been made possible by the development of models based on continuous and discrete simulation. In addition, the article presents a discussion of pertinent modeling and validation processes, along with a discussion of algorithms designed to specify these models and special languages for implementing such algorithms.

In his treatment of computational tools for statistical data analysis, C. F. Starmer examines a basically analogous set of effects resulting from simplifying assumptions that tend (artificially) to coalesce inherently different entities into groups defined as being "the same" for purposes of computational convenience. At a more intrinsic level, there are numerous applications in which the combination of computationally efficient algorithms and economical automatic data collection techniques bring into serious question the continued use of estimated parameters (forced on the statistician because of sampling) instead of the exactness provided by gathering data from an entire population.

In a study such as this one, it is futile to hope for completeness. Accordingly, there are inevitable gaps, several of them quite conspicuous. For example, the study of functional organization of computing systems known as computer architecture, is an important computer science subject. Moreover, its prominence is increasing because dramatic advances in equipment technology have broadened the range of practical configurations so that the practicing computer scientist is faced with an almost arbitrary range of feasible options for a system configuration. These options include architectures in which a number of computers are linked together, to allow the subdivision of an application into several simultaneously executing components. However, opportunities to exploit this parallel processing are severely restricted because there is only a rudimentary understanding of the underlying phenomena. Consequently, there are growing areas of inquiry aimed at finding formal vehicles for characterizing such processes, languages for specifying them, and algorithms for generating them from equivalent sequential processes.

These and other aspects of computer science may be the subjects of additional studies in the future. Meanwhile, it is hoped that this study provides an interesting look at an explosive field still in the process of becoming.

SEYMOUR V. POLLACK

CONTENTS

THE DEVELOPMENT OF COMPUTER SCIENCE

Seymour V. Pollack

Anytime one undertakes to chronicle the development of any human endeavor, it is tempting to climb on Time's ascending balloon so that the receding past appears more orderly. The license granted by this "perspective" offers the convenience of reinterpreting certain occurrences, or perhaps laying others aside as the teeming jumble fades into a systematic patchwork of related but well-bounded entities. Events, one can claim, foreshadow events, and we develop a case study in inevitability.

When it comes to computing and computer science, in their present state, it is easy to stand there, eyes ahead, and proclaim successful resistance: The temptation to make things orderly simply is not there. Computer science, after all, deals with our desire to understand the transformations occurring in an information processing system (more precisely, an electronic digital computing system). And this amazing contrivance, though now ubiquitous, has been with us barely a third of a century. (The "amazing" has not yet worn off.) Moreover, its arrival was not accompanied immediately by universal recognition that we have here a device whose study warrants (much less demands) a full-blown "science," complete with university departments, journals, esoterica, international conventions, and other legitimizing accoutrements. Consequently, we are looking at a very new area of inquiry, with direct forerunners limited to a handful of strikingly prophetic individuals. The imaginative people produced ideas for machines that were far more than automatic calculators: Not only would they follow a sequence of prescribed instructions, but those instructions could be altered by the machine itself. It was not until a century after one man's initial insights that a second wave of visionaries, abetted by a new technology, a greater urgency, and a more responsive bureaucracy, revived these concepts, brought them to fruition, and set the stage for the emergence of a distinct area of study.

Additional background, less direct, must be pulled together from a variety of responses to man's search for efficient means of generating data to alleviate a shortage and, paradoxically, his growing need to organize and handle an overabundance of data. While not direct antecedents of computer science, these diverse sources have provided motivation for some of the equipment methodologies that have been crucial to its development. This is an important point, for it underscores the central notion that computer science deals, inevitably, both with the machines themselves and with the vehicles for expressing, implementing, analyzing, and explaining processes meant to operate on those machines. (Many people in the field emphasize this duality by using the name "computing science" in preference to "computer science"; persistent use of the latter term here does not imply any lesser emphasis.)*

THE SEARCH FOR DATA

It is no surprise that the widening interest in computers and computing has prompted detailed inquiry into the history and development of the entire range of calculating machinery. Accordingly, these studies are turning up a kaleidoscope of devices and mechanisms to add to the already fascinating array of better-

---

* In some views, computer science is more aptly termed "information science" because its domain is seen to encompass all information processing systems, including biological ones.

known machines. Several excellent historical accounts are cited at the end of this article. Of particular note is B. Randell's *The Origin of Digital Computers* [1]; a superb chronicle in which a feeling of drama is intensified by including reproductions of many of the original papers and accounts.

Many of these machines were designed in response to an emerging need for reliable data, spurred by the growth of commerce and the awakening of European science and technology. Napier's invention of logarithms (1614), crucial as it was in revolutionizing the ways of dealing with numerical calculations, required the production of logarithmic tables before it could be exploited. Such tables, laboriously built by hand, bristled with errors. Better celestial observations began to provide data which, in conjunction with the improved theories thus engendered, promised major advances in astronomy and navigation. Here again, these advances would remain as promises without the extensive sets of computed tables to help fuel them. Even the simple additions and subtractions used in bills of lading and other commercial documents began to place unacceptable burdens on the time and endurance of the human arithmeticians trying to meet growing demands while maintaining accuracy.

Thus there was no lack of motivation for reliable, tireless calculating machines. (The abacus, though used extensively and routinely in the East, was known in Europe but never went beyond the "interesting toy" stage there.) In 1642 the 19-year old Blaise Pascal built the prototype of a machine which added and subtracted, seeking to provide computational relief for his father's customs work. This was the first in a long and varied process of digital calculators invented (and reinvented) throughout the ensuing three centuries. Excellent summaries are to be found in Bowden [2], Goldstine [3], and Eames [4], with more detailed chronicles being cited therein.

By the early nineteenth century, the use of mathematical tables had grown considerably, but the few mechanical calculators then available had little impact on the preparation of these tables. (Neither their speed nor reliability caused sustained excitement.) Consequently, the promise of accurate tables still was unfulfilled at that time. One of the people angered by these deficiencies was Charles Babbage. In 1812 this young English mathematician (then 20) decided to seek a mechanized vehicle for computing function values.

He proposed to simplify the automation process by using the method of differences, an approach that would allow the evaluation of a polynomial function at systematic intervals without involving anything more complicated than addition. This is illustrated below for $Y = X^2 + X + 41$ by evaluating Y for $X = 0, 1, \ldots, 8$. This probably is the function Babbage first used to present his ideas:

X	Y	D1	D2
0	41		
		2	
1	43		2
		4	
2	47		2
		6	
3	53		2
		8	
4	61		2
		10	
5	71		2
		12	
6	83		2
		14	
7	97		2
		16	
8	113		

$$\left\{\begin{matrix} D1_i = Y_{i+1} - Y_i \\ D2_i = D1_{i+1} - D1_i \end{matrix}\right\}$$

It is clear from the table that, by starting with the constant second-order difference (nth order for a polynomial of degree n), one could apply successful addition and come up with corresponding function values at successive integer values of the variable. Moreover, the constant nth order difference can serve as a verifier. For example, a Y value of 175 for $X = 11$ would produce a corresponding D2 of 4, indicating an error. Procedures already had been laid out whereby similar tables were mass-produced by squadrons of computers (the term by which such *people* were known then) who were set to work, each person performing a particular addition in the cycle. Verification did not seem to be a particularly widespread custom, as manifested by the numerous errors to be found in such tables. This is likely to have provided additional motivation for Babbage, who planned a mechanism for handling up to sixth-order differences with an accuracy of 32 decimal places. A responsive government provided Babbage with support for the project in 1824. Work proceeded in fits and starts until support ultimately was withdrawn in 1842. Despite all the setbacks, a working difference engine was built by Babbage's son, demonstrated, and used for many years by the British government and by their insurance industry. Moreover, it provided the inspiration for a more modest difference engine (capable of producing fourth-order differences to thirteen places) produced by Scheutz in Sweden in 1854 and used productively for many years. The method of differences, and "engines" based on it, remained useful for table preparation well into the twentieth century. In 1928, Leslie J. Comrie, then Deputy Superintendent of the Royal Naval College's Nautical Almanac Office, used a variety of mechanical calculators as difference engines to produce tables ranging from Bessel functions to nautical tables of lunar data. Thus we find in these difference engines, and in the quest for reliable data, something approaching a continuous thread through the "progression" from calculating to computing machines. Moreover, it served as the impetus for the surge of interest in digital calculations that eventually produced the first automatic electronic computers. Interestingly, this need for rapid, voluminous, accurate numerical computations also helped lay the groundwork for the early and persistent misconception that computer science is an intriguing but relatively limited offshoot of numerical analysis.

THE ANALYTICAL ENGINE

While providing valuable new insights into the ingenuity and diversity of early calculating machines and difference engines, renewed inquiries also have served to reconfirm the uniqueness of Charles Babbage's position as the unmistakable "father" of the automatic digital computer. As he pursued the development of his difference engine, Babbage already began to be bothered by its inherent dedication to a very specific task, i.e., computation of a table of values for a polynomial function. With the engine still incomplete (1833), he envisioned its replacement by a more general "analytical engine" which, by following appropriate instructions, would produce values for any function. To an amazingly accurate extent, the fundamental concepts governing the organization and behavior of today's digital systems are embodied in Babbage's plans for his analytical engine: His design specified a machine that would follow a sequence of instructions submitted to it. Those instructions would activate computational components, accept input data, and produce human compatible (i.e., printed) output. More significantly, results produced at some interim point in the process would dictate the nature of the machine's subsequent activity by selecting an alternative pathway of instructions for the machine to follow.

Of course the idea of self-regulating devices did not originate with Babbage. By the time he was thinking through some of the ideas for his analytical engine, James Watt's steam engine, regulated by a ball governor, already was an established device and the principle of the thermostat had been worked out. Scientific and technological histories of the nineteenth century are abundantly dotted with instances of such devices in which a portion of the output is fed back to the input side, thereby regulating the system's behavior. In some instances, this "self-guiding" property has prompted the efforts to represent these mechanisms as forerunners of computers. However, tempting as this analogy might be, it is a wide and serendipitous jump from these single-purpose mechanisms to Babbage's dazzling idea of a general vehicle in which the operation (as perceived by the users) changes with each new set of automatically sequenced instructions.

The great frustration of Babbage's life was that he never saw his analytical engine design move from the drawings to reality. The device itself was to be a very ambitious one: Its storage unit, to be implemented by means of pegged cylinders, would accommodate up to 1,000 50-digit decimal numbers. Motivated by instructions supplied on punched cards (more about these later), selected numbers would be brought automatically to the central computational mechanism (which Babbage called the mill) where arbitrarily specified combinations of arithmetic operations would be performed. Results, transferred from the mill to storage, eventually would find their way to a printing device which, again automatically, would produce human readable output directly or prepare a plate from which copies could be made. All of this activity was to be purely mechanical, driven by steam power. The accompanying tolerance requirements were so stringent that they fell beyond the manufacturing capabilities then current. It was not until decades later that his son, General Henry Babbage, was able to demonstrate the soundness of his father's theories by successfully assembling and operating a subset of the mill, equipped with a printing mechanism.

There is no (discernible) continuous thread from Babbage to the early pioneers in electronic digital computing. In fact, Babbage himself left relatively little in the way of documentation. Recognition of his colossal intellectual achievement in this regard stems primarily from an annotated translation by his associate Countess Lovelace (Lord Byron's daughter) of a detailed description written by an Italian engineer named Menebrea.* These two individuals seem to comprise the population of Babbage's contemporaries who understood the impact of his invention. Several subsequent designs for analytical engines, independently conceived, have surfaced in response to renewed historical interest and are described in [1] and [2].

THE PROBLEM OF ABUNDANT INFORMATION

While the scientific and commercial communities of the nineteenth century industrial world began to deal with their growing need for reliable data, another facet of this growth produced a somewhat contrasting problem—the prospect of a deluge of information that could not be digested for meaningful use. For instance, the required decennial census of the United States was beginning to produce so much data that there was barely enough time to prepare and disseminate basic summaries before the next census. By the 1880's the problem became sufficiently acute to prompt the Bureau of the Census to seek ways of mechanizing the tabulation for the forthcoming 1890 census. (Some mechanical aids were introduced, to a limited extent, as far back as the 1870 census, but only rudimentary assistance was provided, a far cry from the automated tabulation being sought.) At the urging of Dr. John S. Billings, head of planning for the 1880 census, a young bureau employee named Herman Hollerith developed and patented an electric tabu-

* The paper is included as Appendix 1 in [2].

lating system in which data were represented as a series of holes on a continuous length of perforated paper, or on individual punched cards. Both Hollerith and Babbage may have drawn their inspiration for punched card usage from a common source—the Jacquard loom of 1804, in which punched paper cards controlled the selection of strands used to weave patterns. After successful applications to regional data in Maryland, New Jersey, and New York (1887–1889), Hollerith's machinery for punching and counting data was used successfully for the 1890 census, during which well over 50 million cards were punched.

This success was repeated in a number of European census operations, including an Austrian census in 1890 and a Russian census in 1897. Improved keypunching equipment, among other innovations, heightened the success of the 1900 U. S. census, after which the Bureau of the Census began producing its own equipment under the leadership of James Powers. Hollerith's original company (the Tabulating Machine Company) became part of the Computing-Tabulating-Recording Company, the direct forerunner of the International Business Machines Corporation. Powers eventually left the census bureau to form his own company. This enterprise merged with Remington Rand, which ultimately merged with Sperry Gyroscope. Thus, the patents granted to these two men provided the basis for competing punched card systems, with Remington Rand (by then the Univac Division of Sperry Rand) giving up the ghost on its card design in the late 1960's.

The success of punched card equipment in handling masses of census data was recognized quickly and exploited in a variety of commercial enterprises with abundant data problems (e.g., insurance companies, railroad companies, public utilities, large retailers). By 1910 the possibilities of Hollerith's machines (which by then were able to add as well as sort and tabulate) already had prompted the systematization of many financial and managerial procedures. In addition to automating what has been done manually by groups of clerks, this equipment engendered applications which had no equivalent manual predecessors. Thus were born such processes as cost analysis and sales analysis. More importantly, there emerged a recognition that these information gathering and processing devices were agents of change—catalysts for reexamining and redefining ways of doing things.

CONVERGENCE OF TECHNOLOGIES

As usage of tabulating equipment spread, calculating machinery experienced a parallel growth. By the time the twentieth century entered its second decade, calculators such as Steiger's "millionaire" and Odhner's Brunsviga numbered in the thousands and were to be found in routine use in widely varying business and scientific contexts. At the same time, machines like Felt's Comptometer and Burroughs' Adding and Listing Machine were making appearances outside of their customary accounting environs. However, only a small amount of computational ability crept into the design of Hollerith's and Powers' tabulating equipment. What now appears (with 20-20 hindsight) to be a "natural" marriage between machines able to generate sizable amounts of data and those capable of organizing and handling such data was not so apparent in 1910. Thus we find Karl Pearson, who was a gigantic factor in the spread of statistical applications, preparing his massive tables with a simple Brunsviga calculator.

It was not until the 1920's that L. J. Comrie, an official of the British Nautical Office (who had learned about calculators from Pearson), began his systematic exploitation of tabulating equipment for scientific purposes. Departing from "normal" usage for such equipment, Comrie applied Babbage's different engine techniques on several commercial machines, among them Hollerith's tabulators. The results, which included a set of greatly improved nautical tables, underscored Comrie's important insight: By seeking approaches designed specifically to exploit the properties of these machines, he emphasized a quest for newly achievable applications in contrast to those representing a direct carry-over of manual procedures.

A similar synthesis occurred at Columbia University under the guidance of Benjamin D. Wood, an educational psychologist. Sensing the importance of statistical analysis in educational applications, he persuaded IBM's Thomas J. Watson (in 1928) to lend the university a number of tabulating machines, thereby providing the basis for Columbia's Statistical Bureau, the first of its kind devoted to education. Working closely with IBM, Wood was instrumental in helping revolutionize educational testing by making it economically feasible on a large scale and greatly expanding pertinent methodology.

In 1931 Columbia's statistical laboratory attracted the attention of Wallace J. Eckert, an astronomer, who began using the computational facilities for his work. By 1933 this usage had developed to a sufficient extent to prompt the establishment of a separate computational laboratory for astronomy. This facility, later to become the Astronomical Computing Bureau, was equipped with IBM's latest tabulating and accounting equipment, which Eckert tied together by means of a "mechanical programmer" that allowed the execution of a succession of steps automatically. Later on, the Astronomical Computing Bureau was to play a significant supporting role in several major projects during World War II, including the Manhattan Project. Eckert, meanwhile, left Columbia to assume the directorship of the Nautical Almanac Office, where he continued to apply his techniques for adapting punched card machinery to scientific computing. The numerous tables thus produced were soon to be applied to a variety of wartime uses.

While it is impossible to pinpoint all the specific dates or events, it is clear that the combined use of data processing equipment and electromechanical calculators in a single context marks the beginning of a steadily accelerating movement toward electronic digital computers. In the specific case of Columbia University's Astronomical Computing Bureau, this facility served as a catalyst that helped move IBM decisively in the direction of scientific computing. With continued corporate support, the laboratory eventually (1937) became the Thomas J. Watson Astronomical Computing Bureau. While its primary interest was focused on astronomy and astronomical applications, the bureau served as a cradle for ideas in scientific computing which were to have substantive effects on other computing projects. Some appreciation of the bureau's central role can be gained from Herman H. Goldstine's excellent account of those days [3]. One quickly builds a perception of a rich and turbulent atmosphere in which people having some contact with the bureau keep showing up in various other seminal computing projects.

The bureau's original Board of Managers included T. H. Brown of Harvard. Consequently, when Howard Aiken, then a Harvard graduate student, expressed a strong active interest in digital computing, Brown sent him to spend some time with Eckert at the Watson Computing Bureau. The result was a collaborative project between IBM and Harvard begun in 1939 and culminating in 1944 with an electromechanical digital computer known as the IBM Automatic Sequence Controlled Calculator and eventually termed the Mark I. The computer was capable of storing 72 signed 23-digit decimal numbers as well as 60 manually set constant values. Its machinery enabled it to perform a multiplication in about six seconds. More significantly, a string of instructions, supplied to Mark I on perforated paper tape, permitted the execution of an arbitrary sequence of operations without intervention. L. J. Comrie hailed the computer as a realization of Babbage's dream.

While the Harvard–IBM project was taking shape, an independent undertaking with basically similar intent was going on at Bell Telephone Laboratories under the leadership of George R. Stibitz. Using telephone relays for storage, Stibitz's group developed a machine for performing arithmetic on complex numbers (1940). Instructions and data were introduced via teletype, either through direct connection or by long distance telephone. Stibitz astutely recognized the advantages of using binary arithmetic and designed the Complex Number Computer (as it was called) to use a binary coded decimal system very similar to that still employed. The machine, used routinely till 1949, served as a basis for more ambitious digital computer projects at Bell Laboratories, establishing that organization as an early and continuing contributor to the new computer technology.

Under different circumstances, these electromechanical computers would have generated considerable excitement. After all, here were untiring workers that could compute reliably at rates many times faster than possible heretofore. However, the timing was unfortunate in that these devices appeared just as electronic technology was beginning to mature. As a result, much of the impact was neutralized and the electromechanical computer's major role became that of predecessor.

A number of other individuals and organizations recognized the great potential inherent in combining data processing and computational technologies and made the appropriate intellectual leap. There is no intent here to obscure their work and deny them due credit. The purpose, rather, is to emphasize the importance of this technological marriage in providing an essential stepping stone into the computer age.

BALLISTICS AND BALLISTICS RESEARCH

In outlining the convergence of major forces to produce electronic digital computers, we must go back in time to pick up another important source of motivation. Unfortunately, this source stems from our seemingly unrelenting desire to hurl lethal projectiles at one another. In pursuing this somewhat bizarre approach to population control, there always has been a need to develop methods for calculating conveniently and accurately the landing points of various bodies dropped on, flung, fired, or launched at an enemy. As weapons became more sophisticated, such information began to take the form of elaborate tables with thousands of trajectories to account for a correspondingly bewildering array of projectiles, launching conditions, and external factors. Accordingly, a substantial segment of scientific computing effort throughout the centuries has been devoted to the development of theoretical models and practical methodologies that would permit such tables to be produced with reasonable effort. Thus it is not surprising to find prominent mathematicians throughout history associated with artillery boards and other ordnance agencies.

World War I saw a concerted attempt to place the American ballistics effort on a sound scientific basis. This included the establishment of proving grounds staffed with highly competent individuals drawn together from a variety of disciplines, who, through a combination of incisive theoretical work and well-planned experiments, produced notable improvements in the accuracy of ballistic calculations. Instrumental in one major program was Forest Ray Moulton, who modified the finite difference methods used in astronomy and applied them successfully to ballistics computations. Of perhaps equal significance is the fact that he persuaded the armed forces to sponsor a program wherein talented officers could pursue graduate work in mathematics and physics with specialties in ballistics. This helped reinforce the idea of an ongoing ballistics research effort.

Concurrently, a second ballistics program was set up under the leadership of Oswald Veblen, who helped place the newly established Institute for Advanced Study at Princeton University. Moreover, he formed the association between that body and the American ballistics research effort, thereby setting the stage for an extremely fruitful collaboration. Specifically, it was Veblen who brought John Von Neumann to the institute and later placed him (as well as many others) in contact with the problems surrounding the ballistics work at the Aberdeen Proving Grounds. This, of course, was of profound importance in the subsequent development of computers and computer science.

The definition and verification of improved ballistic theories prompted an increased desire for machinery on which the new solutions could be implemented. Since much of this work involved the solution of differential equations, the researchers' attention was caught by Vannevar Bush's differential analyzer, an elaborate analog device designed to solve complex sets of differential equations encountered in the analysis of electrical network flow. While similar devices had been built to analyze specific problems (in effect, by serving as scale models of the particular systems being investigated), Bush's machine, built at the Massachusetts Institute of Technology, was much more general, being constructed to handle a wide variety of problems expressible in terms of differential equations.

The analyzer's effectiveness also caught the eye of the University of Pennsylvania's Moore School of Electrical Engineering. As a result, arrangements were made for Bush and his colleagues to build an analyzer for each of these two organizations. Installation occurred in 1935. Thus the sphere of associations between the American ballistics research efforts and university scientists continued to expand.

Although the Bush analyzer served as a very useful vehicle in solving ballistics differential equations, its effectiveness was seen to lessen as problems grew more and more demanding. This became especially noticeable with increases in required computational precision. Since an analog instrument is an embodiment of a model based on continuous mathematics, any increase in the model's precision requires a corresponding increase in the instrument's precision as well. For example, if some continuous variable in the model is represented by a voltage in the instrument, another significant figure in the variable would mean a tenfold improvement in the voltage's accuracy. The limited ability of an analog system to accommodate these new requirements prompted a shift in attention back to digital solutions and digital technologies. Thus, early controversy as to whether the analog or digital approach would predominate in machine-oriented scientific computing gradually gave way to a realization that the arbitrary level of precision made possible by using discrete symbols to represent physical values would force the dominance of digital approaches by "natural selection." Hence digital computations became the focus around which computing and computer science eventually developed. There no longer is an "analog versus digital controversy." The former approach, now implemented with contemporary electronic technology, still finds important use where its particular strengths can be fruitfully exploited.

Aberdeen's quest for effective digital computing machinery led it to ask Bell Laboratories for an expanded and improved version of its Complex Number Computer. The result was the Ballistics Computer. Installed in 1944, it had roughly three times the capacity of its predecessor and could operate automatically for extended periods (up to 24 hours) on instructions submitted via paper tape. This was superseded by a much larger computer (the Model V) designed for more general use. In addition to supporting the work at Aberdeen (where it was installed in 1947), this machine was used for a variety of ballistics applications both for the Navy and Air Force. Bell continued to develop and improve relay computers, predominantly for its own internal use, until they were superannuated by the onrush of electronics.

IBM's interest in scientific computing, nurtured in part through its collaboration with Harvard and Columbia universities, played a part in American ballistics work as well. Thus, by 1944, numerous IBM standard and specially designed digital devices were to be found in several ordnance installations including Aberdeen, where they were used in computing bombing and firing tables.

THE EMERGENCE OF THE ELECTRONIC DIGITAL COMPUTER

The onset of World War II intensified greatly the need for ballistics tables. Briefly stated, the gunner (normally) knows the location of his target, the characteristics of his gun and projectile, and certain weather conditions; a firing table takes these into account and, for a given set of conditions, specifies angles of deflection and elevation. Typically, each entry in a firing table describes a particular trajectory, and a firing table for a particular piece of artillery may contain on the order of 3,000 such trajectories. At that time, the Bush analyzer (which still could meet precision requirements for most cases) and Stibitz's relay computer operated at about the same speed: On either system it took somewhere between 10 to 20 minutes to perform the several hundred multiplications required to produce a single one-minute trajectory. At that rate, each system was capable of turning out a firing table in about a month. Since a steady supply of these tables had to reach the men at the fronts, the Ballistics Research Laboratory mobilized for the effort. The analyzer at the University of Pennsylvania was absorbed into the overall project, and scores of people were trained to operate the systems. Under these circumstances, it is no surprise that laboratory personnel were constantly on the lookout for faster machines, and research efforts continued to improve the body of ballistics theory itself.

By mid-1942 a number of people already had concluded that many electromechanical calculating circuits could realistically be supplanted by functionally equivalent electronic components. Among the strongest advocates were John W. Mauchly, a physicist at the Moore School, and J. Presper Eckert, Jr., a graduate student at that institution. Their strong arguments in favor of electronics as a practical means for increasing computational speed helped convince Aberdeen's Ballistics Research Laboratory that the time was right to embark on the development of an electronic calculator and, in June, 1943, the Moore School received a contract to produce such a device for the Laboratory. The machine was to be called Electronic Numerical Integrator And Computer (ENIAC). Enthusiasm for the project was not universal: Proponents of electromechanical computers maintained that the desired increase in speed was obtainable via electromechanical means, with much greater reliability, by partitioning the work among several units operating concurrently. Others felt that similar improvements could be realized by applying electronic technology to analog equipment. Nor were ENIAC's principals under any illusion: They fully recognized the enormous reliability problem in producing such a complex device (plans called for about 18,000 vacuum tubes alone). Yet the need for higher computational speed overrode these apprehensions and the project was approved.

ENIAC was not completed until 1946, but it worked, typically reducing trajectory computation time by a factor of 30.

The impact of ENIAC extended well beyond the computational improvements effected at the Ballistics Laboratory. For example, the need to provide ENIAC with appropriate input/output facilities without undergoing another major development effort prompted Mauchly to contact IBM and engage their help in adapting some of their tabulating equipment for this purpose. As a result, ENIAC was equipped to read from punched cards and to punch new cards compatible in format with available IBM printing devices. Thus IBM, already heavily committed to scientific computing through its association with the laboratories at Harvard and Columbia, became an early participant in a major move toward electronic computing.

Another major impact is well worth discussing: Although the ENIAC was being developed for ballistics work, it was fully intended to make the machine available for other applications. In fact, the first full problem actually run on the new computer was one whose solution was needed by the Los Alamos Laboratory. (Part of this run was incorporated subsequently into the official demonstration.) In addition, a variety of runs were implemented, including some in aerodynamics, hydrodynamics, pure mathematics, and weather prediction. This was done even while ENIAC still was installed at the Moore School. When the transfer to Aberdeen was complete and the machine was operating again (in the summer of 1947), general usage continued, with the machine serving a widening community well into the 1950's. In spite of the maintenance difficulties and other complications, the machine generally performed successfully, thereby leaving a very deep impression among its users regarding the future of electronic digital computers in scientific work.

Thus the historical importance of ENIAC is well established. Questions remain, however, regarding its status as the *first* electronic computer. Despite the fact that many of the principal figures still are alive and have been interviewed numerous times (a collection of such taped interviews has been given to the Smithsonian

Institution and the the Museum of Science in Kent), the picture remains clouded and chaotic. A good deal of the confusion was resolved in 1972 with the appearance of Herman H. Goldstine's detailed chronicle [3]. This is a particularly important work in that it was written by one of the principals, the author having served as the Army's chief direct technical participant in the ENIAC project and its major successors. But ambiguities still remain regarding the "correct" chronology and connectivity. While ENIAC was in progress, Konrad Zuse, having started with a home-built relay computer in the 1930's, steadily improved his machinery to a point where the German Air Ministry in 1943 ordered several of his general-purpose calculators for aerodynamic work. Zuse went on to establish his own successful computer company. A major effort also was carried on in England, but much of the work in the early 1940's was kept secret, dealing as it did with special purpose computers for cryptographic use. A number of working electronic digital devices were produced under these auspices (Randell has reported on some of these [1]), but the more general work appears to have received its crucial impetus from the ENIAC project. Three major British computer projects were the National Physical Laboratory's Automatic Computing Engine (ACE), Cambridge University's Electronic Delay Storage Automatic Calculator (EDSAC), and Manchester University's Manchester Automatic Digital Machine (MADM). Principals associated with these projects visited the Moore School in 1945 to see the ENIAC nearing completion and to exchange ideas about improvements in its design. Bowden [2] provides an excellent account from the British perspective.

At the same time John V. Atanasoff, a physicist and mathematician at Iowa State, acted on his conviction that electronic digital computation was the proper approach toward automatic computers and began building such components. By 1941 Atanasoff's work had attracted sufficient attention so that an interested John Mauchly (prior to his association with the Moore School) visited Atanasoff and held extended discussions with him. Numerous other related projects have been identified in various searches for a complete and accurate log of computer development; however, the intensive scientific and technological activity characteristic of the World War II years make this already complex task all the more difficult. Ironically, even at this writing litigation still is pending regarding which computers preceded which other ones and who invented what. Historians seeking to link these early electronic computer efforts to Babbage's work in some substantive way have been generally unconvincing. The fact that some of the crucial documentation remains classified adds further plot thickener. Also, the Russians still have not published the authoritative history of computers, in which the first electronic digital machine will be proved to be the work of a physicist in Odessa following principles defined by a Ukrainian mathematician 152 years earlier. Thus such controversy is likely to continue.

Beyond controversy is ENIAC's role as a major point of departure in the evolution of electronic digital computers. Once the continued development of ENIAC was established as being a matter of engineering and not of basic feasibility, plans were initiated for a direct successor. Primary orientation toward ballistics calculations, together with the imposition of a tight development schedule, left the ENIAC design with a number of recognized shortcomings. Consequently, the ENIAC group began an exploration of the kinds of techniques and components that could be used to produce a machine that would store a lot more information using substantially fewer vacuum tubes and would be easier to convert from one application to another. (The ENIAC had to be rewired each time.) A new research and development effort was recommended, the resulting computer to be called EDVAC (Electronic Discrete Variable Automatic Computer). The EDVAC is of particular interest in our context because of its role as a focal point around which many basic structural concepts first were articulated [3].

By 1944, John Von Neumann, already a consultant to several government laboratories, including Aberdeen, had become deeply interested in the ENIAC project, having perceived the enormous value of electronic digital computers in a wide range of applications on which he was working. Consequently, it was natural for him to become an active participant (with Goldstine's encouragement) in the EDVAC project. His role in that capacity was pivotal, both in terms of the project itself and with regard to its impact on subsequent thinking about computers: Fundamentally, Von Neumann began looking at computers as logical (rather than electrical) devices. Once this all-important premise was established, he organized and defined the logical functions which became a basis for specifying and analyzing computer operations irrespective of how those operations are implemented electrically. These precepts were expressed in the first draft of a report written by Von Neumann and issued by the Moore School in 1945 [5]. Somehow that draft never was revised and ultimately it became the nucleus of an ongoing dispute (still unresolved in many people's minds) about the exact origins of its contents. Regardless of who contributed what, the draft still stands as perhaps the most definitive single document about computers. One of its most crucial aspects is the enunciation of the idea that the EDVAC (and computers in general) should be equipped with electronically alterable storage components capacious enough to accommodate the data used by the computer in solving a given problem and the instructions guiding the computer in effecting that solution. (The ENIAC's memory held twenty values and instructions were prewired.) Moreover, each type of instruction was to be expressed as a particular numerical code, with the computer's control mechanism equipped to recognize these codes and to determine when it was dealing with such codes and not with data. The implication, then, was that the same memory could be used to hold both types of information and the instructions could be modified as easily as the data. Unlike the ENIAC, in which replicate components were provided to allow certain calculations to proceed concurrently (this was tailored especially for ballistics work), Von Neumann perceived the computer as a completely serial logical machine in which each individual instruction would be accessed from the machine's memory, analyzed (i.e., decoded), and executed in distinct sequence. Unless signaled to do otherwise, the machine's control unit automatically would execute these instructions in the order in which they were stored. (Incidentally, this last principle was one of the many points of contention between Von Neumann and others in the EDVAC project which contributed to the eventual breakup of the original group. Contrary to Von Neumann and Goldstine's strong recommendation, each EDVAC instruction was designed to specify explicitly the address in storage containing the next instruction to be executed.)

Significantly, the formulation articulated in Von Neumann's draft continues to serve as the standard model for digital computers. Even now, despite drastic changes in physical circuitry and overall performance, the overwhelming majority of computers are Von Neumann machines.

ENIAC spawned a number of major computer projects besides the EDVAC. Cambridge University's EDSAC, though it drew on the ENIAC work and emulated many of EDVAC's concepts, actually was completed (in 1948) before EDVAC and is acknowledged to be the first stored program electronic computer. Eckert and Mauchly formed their own computer company in 1946 and produced the first U. S.-made stored program machine, the BINAC (Binary Automatic Computer) for Northrop Aviation in 1950. (EDVAC itself was delivered to the Ballistics Research Laboratory later that year.) They went on to build the first completely general-purpose electronic computer, UNIVAC (Universal Automatic Computer), which was turned over to the U. S. Census Bureau in 1951. Not long thereafter, the company became part of Sperry Rand.

One of the most prominent visitors to the Moore School was Jay W. Forrester, a co-founder of MIT's servomechanisms laboratory. At work on a project to build a computer for a flight simulator, Forrestor's concepts were influenced profoundly by his observations in Pennsylvania and the flight simulator evolved into MIT's first major digital computer, the Whirlwind. Forrester went on to found that institution's Digital Computer Laboratory and to invent ferromagnetic core storage, a medium for computers' main memory components that lasted without serious challenge from alternative technologies until the 1970's.

ENIAC engendered another project, in its way perhaps more significant than the EDVAC. Even as EDVAC's concepts were taking shape, concerns were raised about the future of computer research and development once the impetus of war receded. Although the Moore School came to mind as the logical site for subsequent efforts, commitments to ENIAC and EDVAC, coupled with a variety of other factors, prompted other centers to be considered. After much deliberation, Princeton's Institute for Advanced Study was selected to house a major computer research project, to be conducted under peacetime auspices. Von Neumann returned to the Institute (1945) as project director and he was joined by Goldstine in the next year. A number of others came to the project from the Moore School, government, industry, and from the Institute itself.

Like its predecessors and contemporaries, the Institute's project was centered around the projection of a machine (the IAS computer). However, the machine was not earmarked for any single application, nor was it destined for some specific organization. Instead, funded from a variety of sources, it was envisioned as a research instrument available to support an arbitrary range of scientific and mathematical inquiries, including those relating to its own construction and behavior. As a result, the Institute issued detailed reports about the machine's organization and construction. These were widely disseminated, thereby making available an abundance of fundamentally useful information well in advance of the machine's actual completion. As a result, the IAS project catalyzed a number of others and many of these were well under way prior to completion of the IAS computer itself. These included MANIAC at Los Alamos, ORDVAC and ILLIAC at the University of Illinois, and major systems at RCA and IBM.

Thus, by 1947, the concepts allowing a unified functional perception of computers were well established and fairly widely recognized, so that a substantial number of projects already were in progress by 1950. Moreover, most of the systems around which these projects were oriented were intended for general use, with increasing emphasis on facilitating the changes from one use to another. Dramatic speed advantages over electromechanical predecessors, amply demonstrated in a variety of specific contexts, provided an additional contribution to a climate which (with the wisdom of hindsight) was suitable for a technological and commercial revolution.

BASIC COMPUTER ARCHITECTURE

Before attempting to trace the growth in awareness that the logical and procedural phenomena in a computing system warrant serious study, it will be useful to elaborate a little on the basic functional components of a computing system. This will provide a frame of reference against which one can mirror the concepts and insights developed in the other articles. Toward this end we shall develop a discussion around a model that (basically) characterizes

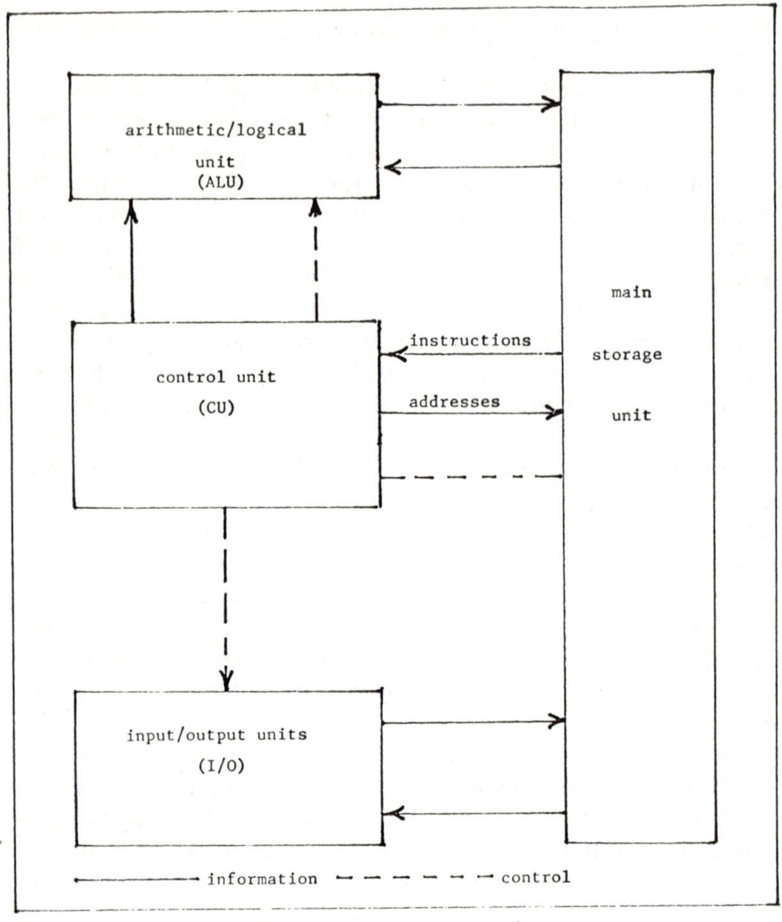

FIG. 1. Major components of a computing system.

most of today's configurations and is very similar to the one put forth for EDVAC and the IAS computer.

As shown schematically in Figure 1, a computing system is comprised of four fundamental components:

1. An arithmetic/logical unit (ALU) in which is embodied the circuitry designed to perform certain elemental operations (e.g., addition, negation, comparison for equality, internal movement of data). These operational circuits are supplemen-

ted by special memory units (registers) used to hold the data on which the ALU operates. The diversity and complexity of these operations can vary widely among computer systems.

2. A main storage unit (memory) subdivided into a prescribed number of equally sized elements (words), each with its own unique permanent numerical address. Depending on the type of computer, there may be tremendous variation in memory size (number of words,) word size (the amount of information accommodated by each word), speed (the time required to obtain and reproduce the contents of a specified location), and physical construction. However, the memory's functional aspects transcend these individual differences: it is a passive component, initiating no action of its own. Instead, it receives or delivers information on demand. Except for very special cases, this information is not characterized by its contents but rather by its location in storage. For example, the machine is not organized to print a specified value; instead, it is designed to print the contents of a specified storage element.

   For example, the process C ← A-B (in which the contents of location C are replaced with a value obtained by subtracting the contents of location B from those of location A) would require a sequence such as the following:

   (a) Reproduce A's contents in the ALU's register (A's contents are unchanged).
   (b) Subtract from the value in the ALU's register an amount equal to that found in location B (B's contents are unchanged).
   (c) Reproduce in location C the value from the ALU's register (C has a new value in it; the ALU's register is unchanged by step (c).

3. A set of input/output mechanisms to handle the transmission of data between the computing system and the outside world. Regardless of the number and diversity of input/output units, it is the function of this component to manage the flow of information to and from the processing system.
4. A control unit which directs the activities of the entire system. In the basic Von Neumann machine, instructions are taken from memory, examined, decoded, and executed one at a time.

Accordingly, the control unit is equipped with a counter that keeps track of the main storage location whose contents represent the next instruction, and a register into which that instruction is copied for subsequent decoding. In this context, the essential function of the control unit can be represented as an iterative cycle of the sequence shown below. For convenience, let us assume that the next instruction to be executed is stored in memory address n:

(a) The instruction stored in location n is copied into the appropriate control unit register.
(b) The control unit's counter is incremented to n + 1 (the location of the next sequential instruction).
(c) The control unit decodes the instruction, thereby determining the operation to be performed and the word of main storage involved in that operation. The repertoire of operation types, each represented by its own unique numerical code, comprise the machine language for a given type of computer. As an integral part of the decoding process, the control unit activates those circuits in the arithmetic/logical unit required to perform the activities corresponding to that operation.
(d) The ALU is activated, thereby executing the specified operation.
(e) Once the operation has been completed, the control unit resumes its basic activity beginning again with step (a) above.

Thus, the instructions are executed in strict sequential order. Changes in sequence are handled by an instruction whose execution causes the number (address) in the control unit's counter to be changed. Then, without any alteration at all in the unit's basic cycle, the sequential execution of instructions continues from that new point. This makes it possible to select an alternative sequence of events dynamically, based on prevailing conditions at that instant.

The Von Neumann architecture makes it possible to fulfill another basic function without compromising the fundamental simplicity of the control unit's cycle: Since a sequence of instructions can be used to obtain the contents of any storage location, to process those contents (i.e., change the value) and to place the new result in the location from whence the original value came, the author of such a sequence could contrive to specify a location whose contents happen to be one of the instructions in the sequence. With properly specified manipulations, the results could produce a situation in which a particular instruction is executed and then, through a normal sequence of activities, that instruction is replaced with a modified operation. A change in sequencing (as outlined above) forces the control unit back to execute again from that location whose contents underwent the change. Thus, in effect, the storage of instructions and data in the same memory immediately implies the availability of a self-modifying mechanism whose flexibility is limited only by the user's perception of how to exploit it.

EDUCATIONAL BEGINNINGS

Regardless of the "true" chronology of concepts or the ambiguity surrounding the proper assignment of credit for "firsts" to the right individuals, organizations, or countries, electronic digital computers clearly were production items in the United States by the mid-1950's. Well over a thousand systems were in use and a number of companies were seriously committed to the positional scramble within the burgeoning industry. (IBM was just beginning to assume preeminence.) This growth, anything but systematic, produced an increasingly acute shortage of personnel prepared to deal with computers and their use. Manufacturers tried to provide support for their customers by implementing their own training programs, but the effort fell short as growth continued to accelerate.

A possible solution was seen in the university computer laboratories. Although the major design and manufacturing work was shifting to industrial settings, activity in the laboratories did not halt. Some maintained their research/design programs (with several laboratories under manufacturers' subsidy); others redirected their emphasis toward developing and supporting new applications. This latter orientation brought into sharp focus the desirability of providing a computing resource for use across an arbitrary spectrum of university research and administrative activities. Prompted by the increasing availability of "ready-made" computers, more and more universities established such laboratories (primarily) as service facilities, inevitably including instruction in computing and programming among their functions. While much of this instruction was informal, intended to support internal users, many universities were quick to recognize an intrinsic educational responsibility to the rapidly growing population involved with computers.

By 1956 well over a dozen American universities had computation laboratories equipped with electronic digital computers and organized attempts to characterize these responsibilities were well under way [6]. As would be expected in any new pursuit, there was great diversity in people's perception of the nature and extent of the educational needs: For example, some saw a relatively clear dichotomy in which electronics engineers design, build, and maintain computers, and experts in their respective fields (i.e., astronomers, mathematicians, accountants, meteorologists, and so on) identify, design, and implement applications. Accordingly, the laboratory's educational duty, in this context, would be to provide training in how to write programs and use the university's facility. Other views were predicated on the need for a new type of person ranging from an "analyst" schooled in mathematics, electrical engineering, or business and trained in computers on the job, to a "computer expert" produced by a vocational institution, to a "mathematical engineer," the latter resulting from a graduate level curriculum within mathematics. Still another view held that every student headed for a career in science or business should receive a course in computers (either from the mathematics department, electrical engineering, or the computer laboratory).

Some institutions began to provide a framework for students interested in computer work, with the tendency being to provide specialized courses in computing and programming on top of a prescribed group of standard offerings in applied mathematics. By 1958 several universities, through their numerical analysis centers, electrical engineering departments, or computer laboratories, were offering fellowships specifically earmarked for concentration in computer work, and many others routinely included one type of computer-related course or project in science, engineering, mathematics, and business curricula. The center of activity, though, still was the computer laboratory, which continued to become an increasingly recognized source of trained personnel for industry and commerce.

As a result, computer manufacturers stimulated the growth of university computing with the same zeal and determination that characterized their campaigns for nonacademic customers. A number of companies instituted liberal discount policies, thereby facilitating the establishment of centers throughout the country. At the same time, university people interested in the structure, administration, and impact of such centers were encouraged to band together for cooperative study and discussion of these issues. (In many cases the "encouragement" was more substantial, taking the form of a *quid pro quo* in which the implementation of computer courses was a precondition for installation of a heavily discounted machine.) The federal government also encouraged formation of university computer laboratories via institutional grants.

By 1960 about 200 colleges and universities were equipped with digital computers, and the general infusion of computer usage into the educational process was well established. Processes for developing computer programs were facilitated greatly by the introduction of high-level languages, which brought the coding of programs closer to human terms. The use of a program (compiler) to mediate between the high-level language and that required by the machine was pioneered by Grace M. Hopper. As a naval officer she played a significant role in Harvard's Mark I project. After World War II, she implemented many of her ideas about language processors in her position as senior mathematician with the Eckert-Mauchly Computer Corporation.

Use of computing facilities was simplified further by a variety of convenient software products for creating, exploiting and maintaining program libraries. Many of these supportive programs operated beyond the users' perception and were, in effect, "invisible." For example, a numerical integration program coded in the FORTRAN language could be submitted in that form (along with data) to a computing system. Without further human intervention, the system would deliver final results (i.e., integral values). Unless explicitly informed otherwise, the user easily could sustain the illusion that the numerical integrator and the computer were the sole participants in that process.

As these procedures and services developed, realization grew that here was a sizable body of knowledge closely related to, but distinct from, the applications themselves. The problem addressed by these concepts and techniques (e.g., analysis and translation of languages, automatic generation of programs, creation and management of effective user-oriented working environments on computers) were far from trivial and they grew increasingly important with the advent of larger, more complex, and more versatile computing equipment. It would be unrealistic to assign any degree of unity to this realization or to associate it with some kind of coordinated reaction to it. While the fruits of this technology were indispensable components of any successful computing facility, there was nothing near a consensus regarding the characteristics of this technology; opinions differed widely as to whether there was a discernible "body of knowledge" here, and, if so, where its proper home was. Some argued that it was part of numerical analysis; others saw it as a newly emerging area subsumed, respectively, under mathematics, electrical engineering, industrial engineering, linguistics, library science, and business (among others). Advocacy for a new and distinct discipline still was very sparse, owing in part to the difficulties in ascribing some type of specific identity to this field. Some things were fairly well settled: Notably, the complexities inherent in the design and implementation of effective general purpose systems components such as language translators and other software products were sufficiently well appreciated to allow the general abandonment of the idea of vocational settings for such instruction. Nevertheless, there were fundamental questions regarding who teaches what to whom.

A number of institutions, feeling that detailed operational and structural knowledge of systems software was central to the preparation of an effective user of computers, sought to include such instruction by expanding the offerings in the department currently teaching programming and computing. The results, though disappointing pedagogically, helped underscore the fact that many (probably most) users had little or no interest in the science and technology of developing software. (Users' similar lack of enthusiasm for equipment details already was clear.) Consequently, alternatives were considered wherein software specialists, now to be distinct from application programmers, computer designers, and computer users, were to align themselves with computer laboratories, under whose auspices such training could be formalized.

Another model was based on reorganization of computer-related studies into components of a new administrative unit (e.g., an interdisciplinary program or department in which computer design/engineering, numerical analysis, computer applications, etc., could be divisions, or even a school in which they could be departments). These were implemented, in one form or another, by some universities. For example, by 1961 Carnegie-Mellon University (then Carnegie Institute of Technology) had an interdepartmental doctoral program in computer systems and communications; Stanford University had a computer science division within the mathematics department and the University of Wisconsin set up a numerical analysis department. However, most institutions remained cautious, adding individual courses as the market dictated.

Matters were complicated by individual identity problems, even among major practitioners. The following (hypothetical) situation was typical: A pharmacologist needing some machine computations found that, for his situation, it would be most expedient for him to implement the applications himself. After completing the computer laboratory's informal programming course, he developed and coded a program which, eventually, did what he wanted and the project was concluded successfully. In the process of testing and refining the program, the pharmacologist had learned much about the laboratory's operation. (He even may have written a utility program in support of his project and then generalized it for incorporation into the laboratory's resources.) As a result, he found himself spending more and more time helping others with their applications and with the logistics of laboratory use. After a while this became his major activity and the laboratory formalized the process by adding him to its staff. Yet he continued to think of himself as a pharmacologist.

Other, more fundamental factors also impeded crystallization of academic frameworks for computer-related instruction. Despite the mushrooming growth of computer usage, many people (among them those with heavy involvement in programming and applications) still held a vague and narrow perception of what a computer is. While there was a rapidly expanding literature on equipment design, applications, and computational techniques, relatively little of it addressed itself to more reflective aspects of this new area. This, coupled with the primary motivations for early computer development, made it very easy to propagate and reinforce the contention that a computer is a fast, powerful, mathematical tool, a "super slide-rule." Accordingly, it was argued, pertinent instruction should emphasize the design and maintenance of these machines, the techniques for programming them, and the selection of computational processes best suited for implementation on computers. (An indication of the roots for this perspective can be seen by the fact that it was not until the UNIVAC was developed that a digital computer was equipped to store and display nonnumeric characters. Moreover, numerous machines that were built subsequent to the UNIVAC were restricted to handling numeric data.) Further encouragement for this outlook came from computer manufacturers who actively propagated the myth that "scientific computing" and "business computing" were inherently different activities requiring different approaches and different types of machines. (The former endeavor was characterized by arbitrarily complex computations on relatively small amounts of data; the latter involved trivial computations on arbitrarily large amounts of data.) As user sophistication grew and the spectrum of applications broadened, it became increasingly difficult to defend this oversimplification, and one rarely hears it nowadays.

A second factor which retarded movement toward an educational identity was a notable dearth of "science" to go along with the rapidly growing technology. It was easy enough to identify a collection of very useful and clever techniques for improving various aspects of a computer system's effectiveness. In certain contexts, it was possible to point to sets of general precepts which were emerging as foundations for areas of applied work. For example, understanding of high-level programming languages and their translators in the early 1960's enabled such objects to be designed with a considerable amount of determinism, in contrast to their weed-like predecessors. However, there were no major phenomenological frameworks that served to unify and organize the tremendous amount of empirical information that already had accumulated. The people who were claiming existence of a separate discipline called "computer science" or "information science" were hard pressed to identify the characteristics or cornerstones of such a science. (One finds references to such puzzling concepts as the "theory of applications.")

Certainly, such difficulties were understandable. Unlike most other areas of inquiry, there was no natural arena such as an atom, tissue, or crystal lattice to serve as a source of observations. Instead, the "universe" of interest was an artifact barely a decade old; there was no direct heritage of contemplative structure to which the newly acquired observations could be reconciled. In these circumstances it was natural for seekers of academic respectability to fill the void with existing material that was strongly related to the emphases at a particular institution. Thus, at a university with strong commitments to engineering computer applications and a desire to formulate an educational program, the scientific aspect of such a program would be organized around relatively extensive offerings in numerical analysis. The more adventurous institutions introduced organized efforts to analyze and compare numerical algorithms with regard to their suitability for computer implementation; others taught numerical analysis unaffected by computers, leaving it to the old hands in the laboratory to provide interested students with the tricks of the trade. Similarly, an institution emphasizing computer design and construction would place numerical analysis in a less dominant role in favor of Boolean algebra, switching theory, and mathematical logic. Here again, depending on the individual program, the material was augmented with explicit connectors to computer technology, or questions of applicability were left for the student to discern. Regardless of the extent to which such dovetailing was attempted, the "pure" subjects generally were not treated as "borrowed" disciplines which would serve until the "real" computer science matured. Rather, there was genuine conviction that switching theory, or numerical analysis, or documentation theory (or whatever), was the proper nucleus around which computer science would develop and grow. As it turned out, this made it a little easier to provide a still new and amorphous area with some semblance of "tradition." A closer look at this notion will be of interest.

As discussed earlier, one can point to a collection of computational and logical devices that span over three centuries and connect such devices conceptually to electronic digital computers.

In the same general sense it is possible to identify certain contributions in mathematics, logic, and philosophy as "spiritual forerunners" of computer science. While this may provide a comfortable feeling of continuity, it really does not contribute substantially to our current perception of computer science and its central issues. That does not deny all connections; far from it. For instance, the importance of Boolean algebra to computer technology and, subsequently, to computer science is beyond dispute. But Boolean algebra has not become an aspect of computer science (or vice versa). Rather, the precepts and techniques of Boolean algebra have become part of the collection of indispensable vehicles used in computer science in the same way that partial differential equations have been crucial to the study of astronomy or ballistics. As separate computer science programs began to form in the early sixties, the perceived need for academic underpinnings made it especially easy for many to adopt these disciplines, construing them as central issues. Many of the subsequent evolutionary changes in computer science education can be related to an increasing awareness of the conceptual distinction between the science itself and the expressive and manipulative vehicle required for its pursuit.

Despite this jockeying for tradition, it would be misleading to say that computer usage was developing with no science at all. Considerable theoretical work was in progress and some of it provided the roots for today's substantial and relevant body of knowledge. For example, the complexity of 1964's state-of-the-art computing systems necessitated the implementation of powerful executive software structures (operating systems) whose cost began to rival (and soon would exceed) that of the hardware. Effective use of the equipment called for a multiprogrammed organization, wherein a number of independent (and probably unrelated) programs would be in the system at a given instant, each on its way to completion and each contending for some subset of the system's resources. Proper management of the resources and mediation among the contenders without inflicting excessive overhead called for an integrated system organization far more sophisticated than earlier executive programs. In response to these demands, theories began to emerge for characterizing such systems, analyzing their performance, and predicting behavioral effects wrought by specified changes in operating parameters. Such exploration draws heavily on the techniques of operations research and statistics, but it is the study of operating systems, reinforced by operating systems theory, that is peculiar to computer science.

At about the same time, there was a growing realization that the process of preparing, developing, and testing programs was subject to some degree of systematization that could raise such endeavors to a more consistent level. Eindhoven Technological University's E. W. Dijkstra began to identify coherent principles which characterize sound program structure and form a basis for eventual formal verification of a program's correctness. Work stemming from these important insights has resulted in new programming languages (and extensions to existing ones) designed to facilitate the application of these principles of structured programming. The effectiveness of these precepts has been demonstrated repeatedly in terms of less costly, more reliable software products. More significantly, these ideas have precipitated a fundamental change in attitudes toward program design and assessment such that the entire cycle of implementing algorithms, a process central to computer science, has lost a good deal of its craftlike character and is approached in a much more disciplined manner [7].

Other areas of concern to computer science began to coalesce in the same general way, driven by operational problems stemming from new equipment technologies and more challenging applications. The resulting information-handling processes began to overtax the ad hoc methodologies used to analyze simpler sequences of events, and more comprehensive models began to appear.

THEORY VERSUS PRACTICE

At the same time another trend was taking shape. The rapid growth of computer science education stimulated increased interest in theoretical areas (such as automata theory and formal languages) whose pursuit predated computers. Now, these areas were seen potentially to impinge on questions raised by the design and use of computer systems. Consequently, there appeared to be a prospect of concurrent and mutually nourishing development in computer science theory and practice. Curiously, this did not happen. The newly intensified effort generally maintained its own paths, interacting very little with the application-motivated problems that were helping to spur headlong advances in hardware and software technology.

A brief look at the early role of automata theory provides an interesting example of this situation. Mathematical logicians had been concerned for some time with classes of computable numbers and exact procedures for obtaining them. In 1936 these studies received tremendous impetus from Alan Turing and Emil Post. Working independently, each devised an abstract machine (an automaton) in which the outcome (a computable number) could be represented as a sequence of ones and zeros, and the exact procedure for producing that outcome as a sequence of well-defined primitive actions. Moreover, Turing was able to show that it was possible to specify a "universal" machine of this character such that it could duplicate the results of any particular automaton, even those producing arbitrarily complicated sequences. Thus, if a number was computable, it could be computed on a universal Turing machine. Both Turing and Von Neumann were acutely aware of the applicability of these results to the description, characterization, and analysis of automatic computers. (Turing declined an offer to become Von Neumann's assistant and went to head England's Automatic Computing Engine project.)

Once the idea of teaching the "science of computing" began to gain momentum, many people felt that the ability to formulate automata as abstractions of computers would provide the basis for an increasing flow of ideas and results between theory and practice. Accordingly, automata theory became a mainstay of many early graduate programs in computer science.

However, the crossflow did not occur. Workers engaged in devising coherent structures for the growing mass of observations obtained from computer-related processes hoped eventually to exploit the models and insights that would flow from automata theory. Meanwhile, it was important to teach the theory on its own for its rich cultural value (and it was respectable too). However, work in automata continued within its own context, basically undisturbed by newly focused attention from computer science programs. It was not until well into the 1970's that the pathways of automata theory and computer science began to converge, and now the impact of one on the other is notable and useful.

THE EMERGENCE OF COMPUTER SCIENCE

This metamorphosis of computing/computer technology/ computer science was observed closely by a number of organizations besides the academic institutions themselves. (We already have mentioned the government and computer industry.) In addition, this process was of great interest to professional societies, prominent among whom was the Association for Computing Machinery (ACM). Formed in 1947, the ACM had assumed a central role in encouraging the evolution of computer science as a distinct field of study. By 1963 this concern took form via ACM's Curriculum Committee whose first recommended program in 1965 [8] represents one of the earliest attempts to produce a coherent definition of computer science's major concerns. In a direct predecessor to this curriculum report, T. S. Keenan identified four such areas [9]:

> 1. Organization and interaction of equipment constituting an information processing system. The system can include both machinery and people, and its organization will be influenced by the environment in which it is embedded.
> 2. Development of software systems with which to control and communicate with equipment. ...
> 3. Derivation and study of procedures and basic theories for the specification of processes. ...
> 4. Application of systems, software, procedures and theories of computer science to other disciplines.

This report was pivotal in several ways. Besides reaffirming the idea of computer science being a distinct area of study, it gave primary emphasis to the implementation of a computer science curriculum as a distinct mathematical entity, with its own majors. (Similar combinations of subjects were being suggested at about the same time by the Mathematical Association of America's Committee on Undergraduate Programs in Mathematics (CUPM) and by the National Academy of Engineering's Committee on Computer Science in Electrical Engineering (COSINE) as areas of concentration for majors in mathematics and electrical engineering, respectively.) Moreover, while the importance of numerical applications was emphasized (Table 1), there was a tentative but nonetheless explicit effort to depart from the self-limiting notions that computers are numerical instruments and computer scientists are

TABLE 1

Recommended Curriculum for Computer Science Majors.

Courses / Recommendations	Computer Science				Supporting
	Basic Courses	Theory Courses	Numerical Algorithms	Computer Models and Applications	
Required	1. Introduction to Algorithmic Processes 2. Computer Organization and Programming 4. Information Structures	5. Algorithmic Languages and Compilers	3. Numerical Calculus (or Course 7)		Beginning Analysis (12 cr.) Linear Algebra (3)
Highly Recommended Electives	6. Logic Design and Switching Theory 9. Computer and Programming Systems		7. Numerical Analysis I 8. Numerical Analysis II		Algebraic Structures Statistical Methods Differential Equations Advanced Calculus Physics (6 cr.)
Other Electives	10. Combinatorics and Graph Theory	13. Constructive Logic 14. Introduction to Automata Theory 15. Formal Languages		11. Systems Simulations 12. Mathematical Optimization Techniques 16. Heuristic Programming	Analog Computers Electronics Probability and Statistics Theory Linguistics Logic Philosophy and Philosophy of Science

Reprinted by permission from "An Undergraduate Program in Computer Science—Preliminary Recommendations," *Communications of the Association for Computing Machinery*, 8 (1965), 543–552, copyright 1965 by the Association for Computing Machinery.

"superprogrammers" adept at implementing numerical algorithms. The effort was tentative because its curricular uncertainties still were strongly evident: The primacy of algorithmic processes and languages was clearly established (both are required). At the same time, there was an apparent hesitancy in giving up familiar comforts, and so calculus and linear algebra were required. (The numerical courses are shown as computer science courses to emphasize their orientation toward computer usage.)

While the recommendations were not universally adopted, they exerted great influence as a catalyst, prompting accelerated activity in curriculum development throughout the country. A rough index of this growth is obtained by comparing the number of United States undergraduate degree programs in computer science in 1964 (about a dozen) with those of 1968 (close to one hundred). A similar increase is seen at the masters level, and about a fourfold increase (from about 10 to about 40) at the doctoral level. Less evident but still present was the additional growth that occurred within parent departments.

While the ACM's curricular recommendations had some unifying effects, growth in computer science education still continued to be turbulent, pulled in many directions by institutional differences and diverse perspectives. Even where there was agreement that computer science should stand by itself, there was controversy over its placement. If the computer laboratory's key builders were mathematicians (as was true in most cases), the emerging computer science department took shape as a mathematical entity, housed in the school of liberal arts. On the other hand, predominance of engineering usage prompted the establishment of an engineering department, with the curriculum's contents tending to be more pragmatic. A third approach sought to emphasize the interdisciplinary (or pandisciplinary) impact of computing by constructing a curriculum in which basic courses were to be supplemented by a somewhat arbitrary collection of offerings dealing with computers in ______ .

Another, more specific aspect of this turbulence is seen by the fact that there was virtually no consensus on the structure of an introductory course. Sufficient insight already had developed so that the more progressive institutions agreed on what such a course should not be—a cookbook course in high-level language coding—

and on the fact that it should place major emphasis on the study of algorithms and algorithmic processes. Beyond that, opinions diverged on how these concepts should be imparted, with approaches ranging from pencil-and-paper conceptual models through specially designed pedagogical programming languages to the use of existing languages. Moreover, there was wide disagreement on computer science's service role. At one end of the spectrum there was advocacy for a universal introductory course (like General Chemistry I); at the other end, some favored fragmentation (more like statistics) where each department would either provide its own introductory course or send its students to the computer science offering (intended for its own majors but open to others).

At a more fundamental level, many universities, while convinced of computer science's separate identity, felt that an independent program was premature. For them, computer science was a graduate specialty to be preceded by undergraduate concentration in some established area (not necessarily mathematics or science).

The pressures and experiences generated by this explosive growth helped accelerate the refinement of ACM's preliminary undergraduate curriculum so that a fully developed version appeared less than three years later [10]. Even in that brief interval some important conceptual processes matured and, because of this, Curriculum '68 (as it became known) stands as an important landmark for computer science education, perhaps in the same sense that Von Neumann's IAS machine serves as a milestone for conceptual computer design. Philosophically, Curriculum '68 marked the end of debate regarding the separate existence of computer science. Moreover, it clearly placed such occupational areas as computer operations, coding, and data preparation outside the realm of computer science.

The perception of computer science itself underwent important organizational shifts: In a very fundamental reorientation Curriculum '68 identified the representation, structure, and transformation of information as a major focus, conceptually dissociated from specific computer systems or applications.

Consistent with this outlook, hardware and software systems, perceived as separate areas in the earlier recommendations, were redefined in a single framework, i.e., systems capable of transforming information. This also reflected a movement in practice toward the unification of hardware and software design necessitated by the increased capability of new equipment. Effective exploitation of such hardware now began to require integrated design of a computer system's hardware and software instead of superimposing the latter on the former.

Another shift developed with increased recognition of common methodological threads running through computer usage. As a result, a third major computer science focus was articulated, centered around the identification and development of methodologies derived from applications with common processing characteristics, irrespective of the intrinsic relations among them. Thus, for example, the area of computer graphics centralizes techniques distilled from (and useful for) visual display contexts as diverse as medicine, geography, stress analysis, and textile design.

The conceptual division of computer science into these three areas was supported in Curriculum '68 by a fourth category encompassing a wide collection of disciplines involved with computers. (The contents of a given collection, of course, would be dictated by conditions at the particular institution.)

These perspectives were molded into the fully developed core curriculum summarized in Table 2. Computer science offerings are grouped into basic (B), intermediate (I), and advanced (A) levels, commensurate with academic background and maturity. The role of the first two areas (information structures and information processing systems) is emphasized by requiring all students to take the courses addressing those areas. A more flexible approach is taken to the third area (methodologies), with the student selecting those courses most congruent with his interests and objectives.

As implied in Table 2, Curriculum '68 solidified an earlier contention that computer science is primarily a mathematical endeavor and its practitioners can be expected to engage in work requiring predominantly mathematical processes. This was instrumental in producing a clear and consistent picture of computer science's position in a school of arts and sciences, either as a separate department or as a distinct but integral part of a broader mathematical environment. As a result, it provided a very useful source of inspiration, serving as a guide for the formation of numerous new departments as well as for the reorientation of existing ones.

TABLE 2

ACM Curriculum '68 Core Courses.

	Computer Science Courses	Mathematics Courses‡
Basic Courses	B1. Introduction to Computing* B2. Computers and Programming* B3. Introduction to Discrete Structures* B4. Numerical Calculus*	M1. Introductory Calculus* M2. Mathematical Analysis I* M3. Linear Algebra* M4. Mathematical Analysis II*
Intermediate Courses	I1. Data Structures* I2. Programming Languages* I3. Computer Organization* I4. Systems Programming* I5. Compiler Construction† I6. Switching Theory† I7. Sequential Machines† I8. Numerical Analysis I† I9. Numerical Analysis II†	M2P. Probability* M5. Advanced Multivariable Calculus† M6. Algebraic Structures† M7. Probability and Statistics†
Advanced Courses	A1. Formal Languages and Syntactical Analysis A2. Advanced Computer Organization A3. Analog/Hybrid Computing A4. System Simulation A5. Information Organization and Retrieval A6. Computer Graphics A7. Theory of Computability A8. Large Scale Information Processing Systems A9. Artificial Intelligence and Heuristic Programming	

* Required.

† At least two from each of the mathematics and computer science groups.

‡ Based on CUPM recommendations.

Reprinted by permission from "Curriculum '68, Recommendations for Academic Programs in Computer Science, A Report of the ACM Curriculum Committee in Computer Science," *Communications of the Association for Computing Machinery*, 20 (1977), 13–21, copyright 1968 by the Association for Computing Machinery.

Interestingly, Curriculum '68's influence also had a dichotomizing aspect: Its basically mathematical orientation sharpened its contrast with more pragmatic alternatives. Most computer science educators agreed that the proposed core courses included issues crucial to computer science. However, the curriculum brought to the surface a strong division over the way in which these issues should be viewed. In defining the contents of the courses, Curriculum '68 established clearly its alignment with more traditional mathematical studies, giving primary emphasis to a search for beauty and elegance. Pedagogically, this implied a set of academic objectives concerned chiefly with preparation for graduate study leading to a career in research. Consequently, those colleges and universities holding with this perception of computer science saw Curriculum '68 as a reinforcement and endorsement of their orientation and sought to implement it commensurate with their resources.

On the other hand, many educators felt the curriculum to be at odds with their perception of reality. They argued that the uses of computer science and the observed roles of computer scientists militate for an educational approach much closer to that used in professional disciplines. After all, the ultimate outcome of most computer science endeavors is a tangible product (an efficient language processor, chemical process controller, graphic display vehicle, sales analysis system, medical diagnostic aid, or other information processing system) whose primary use is likely to be outside of computer science. The computer science that underlies such a product will be invisible to its users or to its operation. How that science was applied, i.e., the way in which the product was engineered, also will be beyond the user's perception but the effect will be more direct, manifested in terms of the product's cost, performance and reliability. In this light, computer science education should have a strong professional flavor (it was argued), with design principles, general approaches to problem solving, and experiments with current methodologies receiving considerable attention. This would be consistent with the expectation of professional employment starting at the baccalaureate level. Another, related objection pointed out Curriculum '68's neglect of business and commercial data processing, a set of general areas motivating the bulk of the hardware and software industries.

Thus the controversy was not merely a conflict between "theory" and "practice." Rather, the dispute pivoted around the definition of "proper" theoretical material and how closely that material should be tied to actual problems experienced in the field. Strict adherents to Curriculum '68 advocated continued use of material (such as formal language and sequential machine theories) pursued for its own ends in relative isolation from computing contexts. In addition to their innate cultural value, such established and respectable pursuits would continue to lend credibility to the idea of computer science. In this view, computer applications should be picked up elsewhere. (A considerable number of educators favored a curricular model in which computer science would be taken as a joint major with some other discipline; others felt practical knowledge is best acquired on the job either after graduation or via a work-study arrangement.) Moreover, the "relevant" theory, engendered by problems encountered in practice, would be an unsuitable replacement because it still lacked maturity and coherence. Opponents of this view emphasized the importance of establishing a continuing interaction between theoreticians and practitioners, contending that only in this way would it be possible to realize the unfulfilled promise of a continuum from theory to applications.

As a result, computer science growth continued with no decrease in turbulence. Even when basic direction was not an issue, there were problems with implementation. Numerous attempts to install a program based on Curriculum '68 were impeded by its size or by the difficulty in staffing it with qualified faculty. Others found that employers, unable to exploit the background acquired in such a program, would not hire its graduates. On the other hand, efforts to implement a more practically oriented curriculum often went awry because of territorial disputes between computer science and some other area (electrical engineering, mathematics, business, etc.). In some institutions this forced abandonment of the idea of a separate computer science department with the consequent distribution of areas among existing units. Others, feeling that no one program could handle the spectrum of concerns effectively, set up complete programs in more than one department, each with its own orientation and its own majors. (The resources required for such multiple coverage ruled it out for all but a relatively small number of instances.)

In response to this turmoil, alternative curricula began to appear, each intended to answer some class of objections raised by Curriculum '68. A major effort in this regard was a management-oriented curriculum in information systems stressing the information structures side of computer science, with additional emphasis on systems analysis, project management, human communication, and organizational concepts. Curricula also appeared in software engineering, biomedical computer science, information science, applied mathematics (with emphasis on mathematics of computation), computing center management, and computer engineering. The latter term still causes extraordinary confusion in that it evokes a mental image of involvement with computer hardware that is arbitrarily related to the degree of actual emphasis in a given program.

This situation tempts the conclusion that Curriculum '68, despite its solidification of the mathematical viewpoint, aggravated an already existing state of chaos in computer science education. However, there are overriding effects which secure the curriculum's place as a major force in computer science's formative period. First of all, it provided a definite focus for discussion and response, thereby initiating the demise of the ad hoc approach to curriculum development in this area. Thus, while many (probably most) computer science departments (or institutions contemplating such departments) objected to something in the curriculum, it became a reference against which extensions, contractions, replacements, rearrangements, and other "improvements" were formulated and proposed. Moreover, almost everyone interested in computer science education found something in the curriculum not to object to. As a result, various aspects of the curriculum were emulated in many institutions having fundamentally differing viewpoints. The point is that, despite the diversity of vantage points, there was considerably more consistency with regard to the areas of major concern to computer science.

In retrospect, Curriculum '68 was an effective catalyst for intensifying this debate and nudging it in two fairly definite directions. The ACM, harboring no illusions about the permanency of Curriculum '68, remained a central participant in American curricular activity. Through its Curriculum Committee and Special Interest Group on Computer Science Education, it provided a continuing forum for exchange of ideas dealing with the full range of curricular concerns. COSINE and CUPM also remained active, continuing to examine the role of computer science within electrical engineering and mathematics departments, respectively.

The decade since Curriculum '68's announcement has seen the accumulation of a tremendous amount and diversity of educational experience. This, coupled with compelling advances in technology, new and increasingly pertinent theoretical findings, and feedback from a rapidly widening base of employers, has exerted continuing pressure on curriculum designers and developers. A sizeable literature built up on a wide spectrum of topics, including form and content of individual courses, laboratory support for computer science, core sequences, service responsibilities, and entire curricula [11]. One also began to see articles of a more introspective nature, dealing with the direction of academic computer science research, occupational outlets for doctoral graduates, and other more contemplative aspects of computer science. Much of this writing was specialized, concentrating on specific matters in a carefully proscribed context (e.g., implementation of a particular piece of software for classroom use, selection of a laboratory computer, logistics for incorporating outside problems in a class on applications).

This hectic activity was considerably less haphazard than its written products might imply. Disputes and arguments notwithstanding, computer science in the early 1970's was an eminently viable area. A crude but nonetheless interesting indicator of its vigor was the fact that many institutions could claim recurrent employer acceptance of their computer science graduates at all levels. Moreover, scrutiny of computer science programs (especially at the undergraduate and masters levels) outside the perspective of local course differences, choice of teaching language, etc., reveals a coalescent effect that makes it possible to characterize many of the computer science programs that have diverged from the Curriculum '68 model:

1. Computer Science has a strong professional orientation, drawing much of its motivation from practical problems and providing a population of workers uniquely equipped to address these problems.
2. Computer Science has an indispensable experimental aspect. That is, the computer's role extends well beyond its basic use as the vehicle on which application algorithms (expressed as

programs) are implemented. In a very real sense, it is a crucial laboratory for experiments whose purpose, distinct from any application, is to enhance the understanding of information-processing phenomena.

3. Given the two premises stated above, there is no one curriculum that appears to be substantially more effective than others in providing the "proper" growth environment for professional computer scientists. Thus the choice between incorporation of a professionally directed computer science curriculum within another department or establishment of a new administrative unit would appear to be dictated largely by university politics.

The ACM, in dealing with curricular evolution, has become increasingly sensitive to the accelerating growth of professionally focused computer science programs. Accordingly, the organization's second major curricular framework [12] reflects a substantial shift in that direction. The report identifies a combination of knowledge and skills considered to be essential for all computer scientists regardless of the exact curriculum in which these are acquired:

1. The ability to produce correct (operable), clear, well-documented programs.
2. The ability to assess the structural quality and computational efficiency of the program.
3. Background in the applicability of computer techniques to the solution of certain problems.
4. Background in hardware system architecture and component behavior in preparation for configuration analysis and hardware selection.

Superimposed on these attributes is the general requirement that all computer science majors coming through an adequate core curriculum should be sufficiently well grounded in algorithmic techniques, programming languages, hardware and software systems organization, and mathematical foundations to pursue advanced studies in computer science and/or application areas of interest.

Fulfillment of these general objectives is embodied in a series of eight computer science courses required of all majors. As shown in Table 3, these courses assume support from six mathematics

TABLE 3 (continued on p. 47)

Update of the ACM Recommended Curriculum.

	Elementary	Intermediate	Advanced
Computer Science Courses	CS 1 Programming algorithm development, computer organization* CS 2 Structured programming, programming techniques, program testing methods, data structures* CS 3 Assembly language programming, computer structure, assembler construction* CS 4 Basic logic design, computer architecture, data representation, computer arithmetic*	CS 6 Basic structure of operating systems, inter-relation between hardware and software architecture* CS 7 Analysis and design of algorithms and data structures, selection of information processing methods for data base management* CS 8 Organization and formal description of programming languages, analysis of language processing components*	CS 11 Advanced systems programming† CS 12 Minicomputer laboratory† CS 13 Data base management systems design† CS 14 Advanced algorithm analysis† CS 15 Programming language theory† CS 16 Compiler laboratory† CS 17 Automata and computability†

* Required. † Two courses from this category are required. ‡ One of MA2A, MA3 required.

TABLE 3 (continued from p. 46)

	Elementary	Intermediate	Advanced
Computer Science Courses (continued)	CS 5 File structure and organization, data access methods* Courses in specific programming languages‡	CS 9 Design and implementation of operating systems† CS 10 Societal impact of computers†	CS 18 Numerical Analysis† CS 19 Numerical linear algebra† Topics in computer science†
Supporting Math Courses (from CUPM)	MA 1 Introductory calculus* MA 2 Mathematical Analysis I* MA 2A Probability and statistics‡ MA 3 Linear algebra‡	MA 4 Discrete Structures* MA 5 Mathematical analysis II	

* Required. † Two courses from this category are required. ‡ One of MA2A, MA3 required.

Reprinted by permission from R. H. Austin et al., "Curriculum Recommendations for the Undergraduate Program in Computer Science, A Working Report of the ACM Committee on Curriculum in Computer Science," *SIGCSE Bulletin*, 9, no. 2 (1977), 1–16, copyright 1977 by the Association for Computing Machinery.

courses (four required) to provide mathematical maturity and acculturation.

Once the core is assured, the new curriculum expects a flexible approach to the remainder of the program, with emphasis dictated by local preferences. As indicated by the breadth of the optional computer science courses (CS 10 through CS 19 and the "topics" courses) the core may be complemented by conceptual work in a variety of directions. Thus Curriculum '77 (as this revision is called) has moved toward a more balanced program in which (predoctoral) professional employment is an explicitly expected (and perhaps the predominant) possibility. Beside the practical orientation unavoidably obtained from the beginning courses, many of the proposed higher level courses are split between lecture and laboratory sessions, thereby reflecting increased recognition of the laboratory's importance.

CURRENT STATUS AND TRENDS

Because of the widely different contexts, it is not particularly helpful to make a detailed comparison between Curriculum '77 and its predecessor. While Curriculum '68 served as a very useful point of departure and helped crystallize two basic alternatives for computer science education to follow, the revised set of recommendations reflects the reality of roughly 65 American doctoral programs in computer science and at least twice that number of undergraduate and/or masters programs, with a substantial fraction of these being professionally oriented.

The overall stabilization of undergraduate computer science in an engineering context is indicated by the recent (1977) inclusion of such departments as eligible candidates for accreditation by the Engineering Council for Professional Development. Consequently, it will be more interesting to consider the major issues that remain open as computer science education completes its second decade.

It still is premature to ascribe to computer science a coherent set of principles that unify its major aspects. The beginnings of such structures are emerging as behavioral data obtained from computing systems and are being accommodated by more and more comprehensive models which are catalyzing the formulation of more effective design methodologies. Improvements in resulting systems, assessed in engineering terms (e.g., shorter implementation times, lower software failure rates), have accelerated the transfer of this new knowledge to workaday contexts. The overall result has been a noticeable increase in computer science research engendered by problems encountered in practice, and an accompanying convergence of theory and applications. Of course the effects of this convergence will vary widely among academic institutions.

There has been no slowdown in the flow of problems. Success with a given application, coupled with growing insight into further improvement, usually encourages a more ambitious undertaking. In many instances the concomitant increase in complexity stresses current models beyond their effectiveness, thereby necessitating further conceptual work. For example, there are numerous applications in which the advantages of complex configurations involving multiple computers are easily perceived. Moreover, the construction of such configurations is well within current technological capabilities. However, the behavior of information processes on many of these complex constellations is insufficiently understood to allow their systematic exploitation.

This type of situation has a more fundamental aspect, stemming from the fact that the hardware revolution is far from over. The current phase, centered around microelectronics, is producing technologies that can place the processing capability of thousands of ENIACs on a single silicon chip. Moreover, the cost would make it feasible to configure systems in which such a device is but a single component replicated an arbitrary number of times. The excitement implied by this "silicon miracle" is vitiated by the thought that meaningful use of such computational power must be built on an understanding of how such systems go together, the structure of information consistent with such systems, and the behavior of algorithms operating in such environments. Consequently, the pressure continues to mount for computer scientists to develop ways for describing and examining these complexes. It is highly unlikely that the ad hoc usage of sequential machines, initially built up as folklore, can be repeated for these highly parallel systems.

The consequences of this cultural lag already have raised educational issues sharpened by Curriculum '77. Perhaps the most conspicuous of these is the one stemming from the rapid blurring of the boundary between hardware and software. Accordingly, the use of one vehicle or another to implement a particular algorithm in a particular context no longer is a clear-cut matter. Experience with systems involving such decisions is producing evidence that major responsibility for these choices may fall to computer scientists. The "proper" place for this responsibility will be the subject of continuing curricular debate, with further dichotomization the likely result. Willingness to assume this responsibility implies an extended commitment to hardware within a computer science program, thereby requiring substantial laboratory support independent of the institution's central facilities and unaffected by the operating restrictions that such facilities must impose. Consequently, the inclusion of active hardware pursuit at the functional level will become an important attribute to help characterize a program's position within the spectrum of computer science curricula. But the breadth of that spectrum is not likely to narrow. This ongoing turmoil, fueled by a diversity of viewpoints, will continue to enrich the discipline and could well lead to convergence on not one but several viable identities.

REFERENCES

1. B. Randell, ed., *The Origins of Digital Computers*, 2nd ed., Springer-Verlag, New York, 1975.
2. B. B. Bowden, ed., *Faster Than Thought*, Pitman, London, 1953.
3. H. H. Goldstine, *The Computer from Pascal to Von Neumann*, Princeton University Press, Princeton, N.J., 1972.
4. C. Eames and R. Eames, *A Computer Perspective*, Harvard University Press, Cambridge, Mass., 1973.
5. J. Von Neumann, *First Draft of a Report on EDVAC*, Moore School of Electrical Engineering, University of Pennsylvania, Philadelphia, 1945.
6. P. Hammer, ed., *The Computing Laboratory in the University*, University of Wisconsin Press, Madison, 1957.
7. E. W. Dijkstra, *A Discipline of Programming*, Prentice-Hall, Englewood Cliffs, N.J., 1976.
8. "An undergraduate program in computer science—Preliminary recommendations," *Communications of the Association for Computing Machinery*, **8** (1965), 543–552.
9. T. A. Keenan, "Computers and education," *Communications of the Association for Computing Machinery*, **7** (1964), 205–209.

10. "Curriculum 68, Recommendations for academic programs in computer science: A report of the ACM Curriculum Committee in Computer Science," *Communications of the Association for Computing Machinery*, **11** (1968), 151–197.

11. R. H. Austing, B. H. Barnes, and G. L. Engel, "A survey of the post-Curriculum 68 literature in computer science education: A report of the ACM Curriculum Committee on Computer Science," *Communications of the Association for Computing Machinery*, **20** (1977), 13–21.

12. R. H. Austing, et al., "Curriculum recommendations for the undergraduate program in computer science: A report of the ACM Committee on Curriculum in Computer Science," *SIGCSE Bulletin*, **9**, no. 2 (1977), 1–16.

PROGRAMMING LANGUAGES AND SYSTEMS

William E. Ball

1. INTRODUCTION

Whenever a person needs to use a digital computer, it is necessary to communicate to the machine the step-by-step procedure to follow. This act of communication requires some means of expressing the desired computational steps in a form that is understandable to both the person specifying the steps and the computer that is to execute them. In this paper we shall look at a range of possibilities, from direct specification in a language that the machine is designed to understand, to attempts for getting the machine to understand a language as close to natural language as possible. Historically, the trend has been from simple machine-oriented methods to more and more sophisticated programming systems that can better support the needs of a general-user community. In order to put some of these current systems into proper perspective, we shall first review a few of the high points in the development of languages for man-machine communication.

1.1. Algorithms and Programs. The concept of an algorithm is basic to computer science. It represents the fundamental information that we must communicate to any computing device for it to function adequately. An algorithm is defined [1] by the following five properties:

1. *Finiteness.* For a computational procedure to be considered an algorithm, we must guarantee that it will terminate after a finite number of operations.
2. *Definiteness.* The step-by-step operations that specify how our computation is to be made must be described in terms of actions that are rigorously and unambiguously defined.
3. *Input.* Values that start the computation must be specified clearly.
4. *Output.* The normal function of any computational procedure is to operate on the input values to produce some specified output.
5. *Effectiveness.* The basic operations that comprise the computational sequence representing an algorithm must be such that they can be performed in an effective manner. This concept of effectiveness is somewhat vague, but in principle it implies that a person using pencil and paper actually can perform the operations specified. That is, an effective and definite operation might be "multiplication." An operation that is definite, but not effective, would be "solve a complex nonlinear three-dimensional partial-differential equation in five minutes."

It is the purpose of programming languages to let us present our computational requirements to a computing device in such a manner that the algorithmic aspects are clearly specified. That is, a programming language must enable the user to describe the sequence of definite operations which will guide the computing device from the input values through the specified steps to the production of the final output values.

Two related questions are outside the scope of this paper: What is the finite number of steps required for the termination of the algorithm? Is the program output correct based on the original function's specifications? These items are discussed in [2].

1.2. Program Representation. Perhaps the ideal programming language would be English, or some other natural language: The user simply could write the problem specification as if conferring with a colleague, submit it to the computer, and have the desired computation performed. Unfortunately, this brings us immediately to the question of just what an effective operation is. Normally, the transformation from a vague problem specification in natural language to a precisely stated step-by-step computational procedure (i.e., an algorithm) is a very difficult one, involving a significant amount of mathematical maturity, experience, and training. Although there has been considerable research in this area, our present systems are still not capable of "understanding" a natural language dialogue to an extent sufficient to automate this transformation. Some aspects of this work are discussed in Dr. Slagle's article. Weinberg [3] considers why natural languages are not programming languages.

At the other end of the language spectrum, the early days saw an operational sequence specified by actually plugging wires into appropriate locations in the computing device. To carry the information from, say, an arithmetic unit to a printer required physically connecting the two units. Clearly, the flexibility and power of a hard-wired or plug-wired device leaves a lot to be desired. The concept of the stored program computer, in which a general memory served to store both the data to be processed and the instructions that described the operations to be performed, was the key that allowed the versatility needed for the development of our modern computational methods. However, the programmers of the early devices ran into a serious problem: They had to actually set the exact values of the stored program instructions into the memory in a crude, laborious, manual fashion. Significant conceptual breakthroughs occurred when it was realized that once a person specified such a program, a major savings of effort could be achieved if another person could access this prior work easily and directly. Thus, early in the 1950's, systems based on libraries of reusable programs came into existence. These made it convenient for one person to use the programs created by someone else, either as independent processing procedures or as components in larger programs. Rosen [4] and Sammet [5], [6] have presented extensive and detailed discussions of some of these early developments, along with the factors motivating them.

In order to clarify these language concepts and to illustrate further the relation between program language specification and algorithmic representation, we shall look at a number of particular approaches that have been developed over the past few decades.

2. ASSEMBLY LANGUAGES

2.1. The Assembly Process. If we define the operations in our algorithmic specification to be the instructions executable directly by a computing device, then there is no question but that our operations will be definite and effective. An operation such as "load a value from a memory location into a register" is a clearly specified and effective operation. The actual machine instruction usually will be a pattern of zeros and ones (i.e., a binary number interpreted as a sequence of bits) which, when properly translated by the control mechanism in the computer, first will direct the hardware to select a particular memory location. The information contained in that address would then be transferred into an accumulator, a register in which all arithmetic and logical operations take place. Further operations may then be performed on the datum moved into the accumulator, such as "add on a value from another memory location."

It was soon found that the (conceptually) simple task of assigning specific memory locations for various items of data is a task with a high potential for clerical mistakes. However, it is reasonably straightforward to program a computer to perform this job: First, make all of the memory references symbolic; then tabulate all of the symbolic references; finally, assign sequential memory addresses to the list of symbolic addresses. As a consequence, one of the very earliest software tools (i.e., a program developed to help people work with the computer) was the assembler. This program intercedes on behalf of the programmer, who uses symbolic addresses and mnemonic instructions and takes over most of the routine chores of preparing a detailed set of machine instructions for computer execution on a particular type of equipment. Steinhart and Pollack [7] present detailed directions for using one particular assembler program, but the principles of using an assembly language are easily learned and transferred to other machines and systems.

It is far easier for a programmer to remember and to write

"LDA" to request the operation of "LoaD Accumulator" than it is to remember the series of bits that must be stored as a machine language instruction. In a similar fashion, it is far easier to write the symbolic reference "X" than it is to remember where the item of information called X is really stored and to always use that assigned numerical location. Thus an assembler is a program that can accept as input a line such as

LDA X

and generate as output the corresponding bit pattern representing the actual machine operation that must be executed, including the assignment of an appropriate location to store "X".

Figure 1 illustrates the assembly process for one specific machine, the Texas Instruments Model 980B minicomputer, but the

As part of a larger problem, compute the sum of A + B + C and store the result in D.

Input Assembly Language			Output Machine Language (numbers are base 16)	
<label>	<op code>	<address>	<location>	<contents>
	⋮		⋮	
	LDA	A	0100	0062
	ADD	B	0101	2060
	ADD	C	0102	205E
	STA	D	0103	8060
	⋮		⋮	
C	DATA	7	0161	0007
B	DATA	9	0162	0009
A	DATA	11	0163	000B
D	BSS	1	0164	0000
	⋮		⋮	

FIG. 1. The assembly process.

principle is the same from microprocessors to extremely large machines. The TI980B uses a memory organized into 16-bit addressable units (words). One word may contain an instruction of the form:

0 4	5 7	8 15
⟨op code⟩	IXB	⟨displacement⟩

where: ⟨op code⟩ = 5-bit code indicating the operation to be performed.

IXB = 3 bits to describe how the machine will interpret the address specification.

⟨displacement⟩ = 8-bit field whose exact meaning depends on the setting of the IXB bits but whose basic purpose is to indicate the address of the datum involved in the operation. In the example of Figure 1 (IXB = 0) the address is given as a relative displacement from a reference address.

Memory words also may contain character codes, logical switches, integer numbers, floating point numbers, or anything that we wish the bit patterns to represent. In Figure 1, integers were assigned and stored for the values of A, B, and C. Since one hexadecimal digit (base 16) represents any one of 16 possible values, it corresponds exactly to 4 binary bits. Thus, each of these 16-bit binary numbers is expressed as 4 hexadecimal digits simply to shorten the number of characters required to write them.

The assembler statements consist of three fields:

⟨label⟩ ⟨op code⟩ ⟨address⟩

where the fields are each terminated by one or more blanks. The operation codes used in the example consist of the instructions:

LDA X = Load the accumulator with the contents of memory location X.

STA X = Store the contents of the accumulator in memory location X.

ADD X = Add the contents of memory location X to the contents of the accumulator, leaving the sum in the accumulator.

The TI980B assembler actually uses a total of 99 such instructions. There also are assembler directives (i.e., instructions for the assembler program itself to direct the code generation process). For instance,

⟨label⟩ DATA ⟨value⟩ directs the assembler to assign the current location as the value of the symbol in the label field and then to store the given value in the current location.

⟨label⟩ BSS ⟨count⟩ assigns the current location as the value of the symbol in the label field and then advances the current location by the number contained in the count field in order to reserve that many locations for data to be stored later.

The TI980B assembler has 21 such directives.

One computer software tradition was established soon after the creation of the first programs: Once a program is running, somebody will think of an improvement or an enhancement that must be made to obtain a new and better program. Thus, the basic assembler had many features added to it to make it more convenient for the human to interface with the computer hardware. Although the assembler already had built into it every possible machine instruction, assembler directives were added to aid it in allocating storage for data, printing, and spacing of output into more easily read format, conditional features to allow code variation, and various options for saving the generated machine code for later reuse. All of these enhancements were directed toward making the human/machine interface a smoother, more easily navigated boundary.

2.2. Macro Definition. An interesting early observation made by programmers was that they were continually writing the same program fragments over and over again. As long as there is a block of code that is to be duplicated, with perhaps only a few changes to be made from one copy to another, it again seems reasonable to let the computer do this essentially mechanical work. As a consequence, the concept of a macro definition was implemented in most assembly language processors. This procedure allows a code string to be defined once in terms of a set of parameters and an assigned name. Parameter values then would be specified later in terms of variables to generate the actual desired code. Wegner [19] describes the details of how this may be accomplished.

Figure 2 illustrates this process and demonstrates some of the power and flexibility that the macro capability provides to the programmer. In the example the name of the macro is SUM, with parameters A, B, C, and D. All of these parameters enter into the address fields of the instructions contained within the macro body, but we could just as well have written an operation code or even an entire instruction as a parameter. The macro body may be considered as a template in which the formal parameters are replaced by the actual parameter values at the time of the call of the macro.

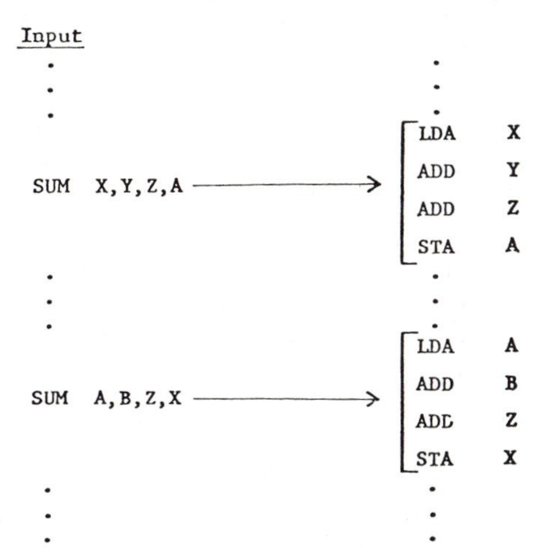

FIG. 2. Macro definition.

The correspondence between formal and actual parameters is determined strictly by the ordinal position of the parameter in the parameter list.

Although the simple substitution discussed above may be very convenient, most of the real power of the macro concept arises when other features are added. For example, a macro body also may contain a macro call, producing nested macro calls. Conditional directives may be added, in which case a given block of code may be generated only if certain conditions are satisfied at the time the macro is called. Finally, the macro definitions themselves may be nested, allowing parameters to be used to change dynamically the macro definitions.

2.3. Summary. By implementing algorithms in assembly language, the programmer has complete access to all of the capabilities of the hardware. As a consequence, a great many of the operating system programs (i.e., programs dealing with the details of memory management, input/output, control of task execution, etc.) and also the utility programs (i.e., copying and reorganizing collections of data, sorting programs, etc.) are frequently still written in assembly language. This also allows the programmer to obtain the utmost computer efficiency for execution. The process is one in which the programmer specifies his algorithm in terms of the allowed symbolic language statements, passes these statements as input into the assembler program, and then collects the output as a series of machine instructions for execution. The output code may be stored on some external storage device such as disc or tape. At a later time, when it is desired to execute the program thus assembled, another program (the loader) will be instructed to go to the external device, pick up the appropriate assembler output, load these instructions into the computer memory at specified locations, and then transfer control to those instructions for final execution. The entire process consists of the four sequential stages:

1. program preparation
2. assembly
3. loading
4. execution

3. PROCEDURE-ORIENTED HIGH-LEVEL LANGUAGES

A line of assembly language code produces essentially one machine language instruction. A line of code from a "high-level language" may produce an arbitrarily large number of machine language instructions. Moving from an assembly language program to add two numbers and to store the result:

```
LDA X
ADD Y
STO Z,
```

to a higher-level language statement to perform the equivalent operation,

$$Z = X + Y,$$

seems like a relatively small step. In practice, of course, we want to go in the reverse direction, from the higher level to the lower. However, higher-level languages accept statements that may contain a complex structure that must be analyzed before the output (essentially a set of assembly language statements) may be generated. Use of a high-level language may achieve a large improvement in the ease of communication from man to machine, but only at the expense of requiring the computer to perform the extra work of structure analysis. However, this approach still requires the effort on the part of the programmer to understand the intellectual concepts involved for the program preparation, since the necessary algorithmic steps still must be described in detail. Essentially only machine-dependent mechanical details are saved by the use of a high-level language.

There is a basic difference between the formal artificial languages of the computer world (PL/1, FORTRAN, PASCAL, COBOL, etc; see [8] for a more detailed description of many of these languages) and a natural language such as English. We know how to do a complete structural analysis of a formal language (Section 4), but we do not know how to do the equivalent analysis of a natural language. The program that performs this language analysis and generates the corresponding machine language code is called the *compiler*. Higher-level languages continue to evolve, along with our ability to create compilers, so that source languages are becoming more convenient for human use. However, we are still far from being able to write compilers that accept a completely "natural" language as input. Variations of FORTRAN will be with us for a long time to come.

3.1. Data Elements and Variables. We shall use the expression "elementary datum" to describe a basic unit of information that can be accessed and modified as a single value within a given programming language. Normally, elementary data items will consist of such things as numbers, either integer or floating point, character strings, Boolean values, bit strings, or references to actual machine addresses (called *pointers*). Since these items represent the fundamental units of information that must form the basis for any algorithmic development, it is essential that a high-level language provide a mechanism by which the programmer may store, retrieve, and manipulate these items. The standard mechanism is to provide symbols called *variables* to name the datum. Additional symbols, the *operators*, indicate what processing is to be done.

The normal basic variable in a standard language such as FORTRAN or PL/1 essentially names a given location in the computer memory. A reference to this variable, then, is a reference to the value stored at the indicated location. Thus, a statement of the form

$$X = Y$$

may be interpreted as telling the computing system to go to the location named by Y, reproduce the value of the item stored there, and move it into the location named by X. This seemingly simple operation may invoke a number of obvious and also subtle problems. For example, if the datum named by Y happens to be an integer numeric value, and that named by X is supposed to be a character string value, then what does the assignment of values really mean? Clearly, in order to maintain a consistency of information described by the symbols, one of two alternative disciplines should be enforced:

1. Data elements, on being stored into named locations, must take on exactly the characteristics implied by the variable name with its attributes.

2. Data elements, on being stored into named locations, must also carry descriptive information giving their exact characteristics.

As a consequence, all languages must have specific rules as to what elementary data items they allow, what elementary data items may be automatically translated from one (internal) form to another (and when), and how the variables take on (i.e., how the machine "knows") the appropriate attributes of the assigned data items.

Thus we see that a fundamental decision must be made in designing a programming language: Are the variables to have fixed attributes defined before the compiler processes the program (as is the case in languages such as PL/1, FORTRAN, and COBOL) or may the attributes vary, as defined by the data stored in them at the time the program executes (as in APL, SNOBOL, LISP)? As a gross generalization, we can state that the former tends to produce more efficient code, since the compiler knows more about the data types to be processed. The latter tends to produce more flexible, easily used systems, since the programmer does not need to be as concerned with the details of the data while still writing the program.

To illustrate this dichotomy further, the PL/1 language requires a complete specification at compile time of the type of information that will be associated with a particular variable name (the type information may be provided by the programmer or by the language default values if not explicitly specified). That is, during the compilation of the program, the PL/1 compiler knows that a reference to the symbol X stands for exactly one type of datum, such as numeric value with the attributes' floating point, base 10, 12 significant digits, etc. This set of attributes will be unique and unchanging throughout the execution of the program. All generated code that refers to this variable will be specific, thereby minimizing the amount of data-checking that must be performed. In contrast, SNOBOL, primarily a string manipulation language, does not have any preassigned attributes associated with a variable name. Consequently, a very dynamic type of environment is created in which the attributes associated with a particular name will be defined (bound) only when an actual value is stored. We may write an assignment which stores an integer number as the value of the variable X at one point in a program, and then a later assignment may store a character string for that same variable. To handle this situation, the executing code must at all times check the type of information stored for the particular variable to ensure that consistent operations are being performed. If the variable X contains a character string and X appears in an expression requiring numerical addition (X + 1), appropriate instructions will be required to convert the character string form into a numeric form for the addition operation. If X was then reassigned a numeric value, the code required for the evaluation of the same source expression (X + 1) would be quite different.

It is an unfortunate situation, but a very common practice in many current programming languages, to use a single symbol to stand for two entirely different operations. The exact meaning may be inferred only from the context in which the symbol appears. One example of the problems that this practice creates comes from the use of the equal sign, " = ", to mean both an assignment of value operation and a relational test for equality. Thus we have that strange-looking assignment statement that has confused generations of fledgling programmers,

$$X = X + 1,$$

where the " = " really means "assignment of value." In PL/1, this dual meaning produces an even odder looking statement:

$$A = B = C.$$

The first equal sign is an assignment operator and the second one is a relational operator. Thus A is assigned the value of "true" or "false" depending upon whether or not B has the same value as C.

3.2. Data Structures and Storage Structures. Normally, the logical constructs involved in solving a problem or implementing an algorithm require more than just the elementary data items. We use the expression "data structure" to describe that particular logical entity that represents the information that we actually wish to manipulate in the algorithmic process. For example, we may wish to deal with the concept of a set or a graph as the basic structure in our process. Two particularly common types of data structures occurring in scientific and mathematical computations are vectors and matrices. These arrays normally are composed of elementary data items as previously described. In the algorithmic sense, if our data structure consists of sets, then the operations we might wish to perform probably would include such things as "is a particular element a member of a particular set," "find the union of two sets," "find the intersection of two sets," etc. These are well-defined operations on well-defined data structures. However, when it is time actually to implement an algorithm on a real computer we are faced with the question of how to store a "set" in a computer memory that has only a linear addressing capability. Also, we must determine how set operations are actually executed, given a particular method of storing the set information. The implementation of an algorithm always requires an answer to the question of what storage structure should be used to represent a particular abstract data structure. One of the most common stumbling blocks in algorithmic implementation stems from failure to differentiate between the data structure, which is conceptual, and the storage structure, which is an actual realization.

Figure 3 illustrates the comparison of data structures and storage structures for a vector (a one-dimensional array) and a matrix (a two-dimensional array). The notation is to indicate that the first element of the vector, V(1), may be stored in the addressable storage location "L". If an element occupies "s" addressable locations, then the second vector element, V(2), will be stored at location "L + s". For example, the IBM System/370 computer attaches a separate numerical location to each byte of storage, and a double precision floating point number occupies 8 bytes. Hence if V(1) is stored at location 1000, then V(2) must be stored at location 1008, V(3) at 1016, etc.

In the full linked list form the notation means that one cell contains:

1. The datum
2. The address of the next cell.

The last cell contains a special value in lieu of the next cell address to indicate the termination of the list. The matrix form also may be stored in a solid column major order, or in one of a number of possible variations of the linked list forms. Extensions to higher dimensional arrays may easily be accommodated.

Given a storage structure implementation of a data structure, it

Vector Data Structure: $(v_1, v_2, \ldots, v_n)$

Vector Solid Storage Structure:

location	content
L	v_1
L+s	v_2
.	.
.	.
.	.
L+s*(n-1)	v_n

Vector Full Linked List Storage Structure (v_i in the ith cell):

$\rightarrow$ [v_1 | •] $\rightarrow$ [v_2 | •] $\rightarrow \cdots \rightarrow$ [v_n | /]

Vector Sparse Linked List Storage Structure (i, v_i in the jth cell):

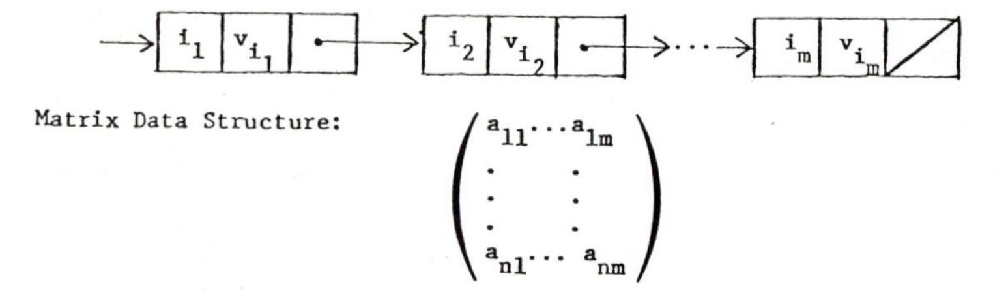

Matrix Row Major Solid Storage Structure:

location	content
L	a_{11}
L+s	a_{12}
.	.
.	.
.	.
L+s*m	a_{21}
.	.
.	.
.	.
L+s*(n*m-1)	a_{nm}

FIG. 3. Data structures and storage structures.

then is necessary to provide the *mapping function* that lets the system access a particular element of the storage structure whenever the corresponding element of the data structure is referenced in the algorithm. For example, given the data structure for a matrix, how does the program actually find the location of a(i, j) when that datum is required? The matrix example is an interesting one in the sense that, for a solid representation, it is possible to compute a specific location for the desired element directly from the indices in the array reference. The mapping function in this case would be (for row-major storage order):

$$L[a(i, j)] = L[a(0, 0)] + s * (d(2) * i + j)$$

where: L[·] = "the location of · "

s = size, in addressable storage units, of one datum item (normally fixed by the computer architecture).

d(2) = number of elements in the second dimension (the rows) of the matrix.

Thus it is sufficient to know the values of L[a(0, 0)], d(2), and s to store or retrieve all of the elements of any two-dimensional array, based on this storage structure.

In some storage structures, such as in the vector representation using a full-linked list, it is not possible to get the fifth element in the list without actually stepping through the list by finding the location of the first, second, third, and fourth elements. Only the fourth element contains the required location of the fifth element. In the sparse form, the fifth element will appear only if it is needed (say for nonzero values being stored). Consequently, the subscript values must be checked during a linear scan of the list in order to determine whether the desired element has been found or if it is missing.

Element retrieval questions, however, are not the only important aspect of the storage structures used to implement a particular data structure concept. Consider the case in which it is desired to insert a new element between the fourth and fifth elements in an existing vector. Given a packed representation of the vector, the only way such an element could be inserted is by physically moving all of the following elements down one location in the storage structure. In a linked list, however, such an insertion can be made simply by creating a new element and changing one address pointer. It is considerations such as these that indicate that the study of data structures and storage structures has a tremendous impact on the efficiency with which computations are designed. Pursuing this topic further is beyond the scope of this paper: Here we shall limit ourselves to an examination of the language techniques by which a desired structure may be implemented.

3.3. Control Structures. When an abstract algorithm is implemented, a specification must be introduced in the implementation language to indicate precisely the sequence of operations to be performed. We shall assume that the sequence in which statements are written will be the sequence in which they would normally be expected to be executed. However, variations from this normal sequence must be allowed so that cyclic operations, conditional execution of statements, and other such decision-related events can be accommodated.

In some of the early higher-level languages, such as FORTRAN, it was thought sufficient simply to carry over the assembly language control structure, embedding it in the higher language. Thus, we find such statements as

```
GO TO 10
```

for the unconditional transfer of control, and the arithmetic IF,

```
IF (X) 10,20,30,
```

that tests a numeric value for negative (GO TO 10), zero (GO TO 20), or positive (GO TO 30), to produce a conditional transfer of control. These statements essentially represent a direct carry-over from the machine language of the computer on which FORTRAN originally was designed and implemented. Unfortunately, statements of this type induce a tendency to write programs involving a convoluted logical flow of control, with transfers back and forth between small segments of coding being quite common. It has been found in practice that programs having this fragmented characteristic are very difficult to debug and maintain. Specifically, it is very difficult for someone (even the original programmer!) just to trace through the logic of a program in order to establish exactly what the code will do in a given situation. We shall return to this point shortly.

The sequence of operations executed by a computer code resulting from compiling a program depends upon a number of separate and distinct factors:

1. The actual sequence of statements as written by the programmer.
2. Within a statement, the order of evaluation of the operators as defined in the language itself.
3. Techniques used within the compiler to produce the final (optimized) output code.

The first item is clearly under the control of the programmer, but the other two have effects that are sometimes surprising.

The result of a sequence of operations from a single statement can depend heavily on the properties of the computing device on which the statement code is executed. One of the significant problems is that computer arithmetic is based on finite size numbers and hence differs from the normal mathematical definitions. For example, floating point addition is not associative as we would assume normal arithmetic addition would be. That is, consider the sum A + B + C, where the values of B and C are slightly smaller than the smallest number that would affect the value of A. If we perform the addition in the order (A + B), the resulting value will be the same as A and hence adding on C will also produce the value of A. However, if we perform the addition from right to left, the value of (B + C) may be of sufficient size to change the value of A. Thus we have, for the IBM 370 computer, the results:

$$\begin{array}{ll} \text{if} & A = 1.0;\ B,C = 0.9E - 6; \\ \text{then} & ((A + B) + C) = 1.000000 \\ \text{and} & (A + (B + C)) = 1.000001. \end{array}$$

As a consequence, the code generation algorithms used by the compiler may affect the computed results even though in a mathematical sense the order of execution should not matter.

Another significant cause of problems, still within a single statement, is the use of functions that involve side effects. Consider the simple example of a function, called HALF(X), that returns the value (X + 1)/2, but also changes the value of the argument X to X + 1. In PL/1 this function would be written as:

```
PROCEDURE HALF (X) RETURNS (FLOAT);
          X = X + 1.0;
          RETURN(X/2.0);
END HALF.
```

The expression X * HALF(X) now becomes ambiguous since its value depends upon the order in which the operations are performed. That is, first finding the value of HALF(X) and then multiplying by the value of x will not produce the same result as taking the value of X and then multiplying it by the value of HALF(X). The sequence:

```
X = 3;
Y = X * HALF(X);
```

will produce 8 for the former case and 6 for the latter.

Within the past decade, a major question in computer science has been to decide what types of control structure language features should be included in new language designs. The objective of these structures would be to:

1. Maximize the probability of writing a correct program.
2. Maximize the readability of the code.
3. Make it easier for a compiler to analyze the flow of control to help optimize the machine code that is generated.

The results of numerous studies clearly indicate the program logic can be more clearly written and explained if the possible language constructs are restricted to a few fundamental control structures. These control structures all have one very significant feature in common: There is only one entry point into the structure and only one exit point from the structure. This is in contrast to the unrestricted use of the traditional GO TO method of control transfer. McGowan and Kelly [9] discuss these motivations extensively.

The basic allowable control structures may be selected from a number of potential types, but the following set is sufficient for most algorithmic design or programming tasks:

1. Normal sequential sequencing
2. IF – THEN – ELSE
3. DO – WHILE
4. REPEAT – UNTIL

Figure 4 illustrates how these structures affect the flow of control in a normal program.

Not only do these structured constructs enhance clarity of expression of an algorithm, they also aid in designing a correct program by using a top-down design approach [9]. If we start by

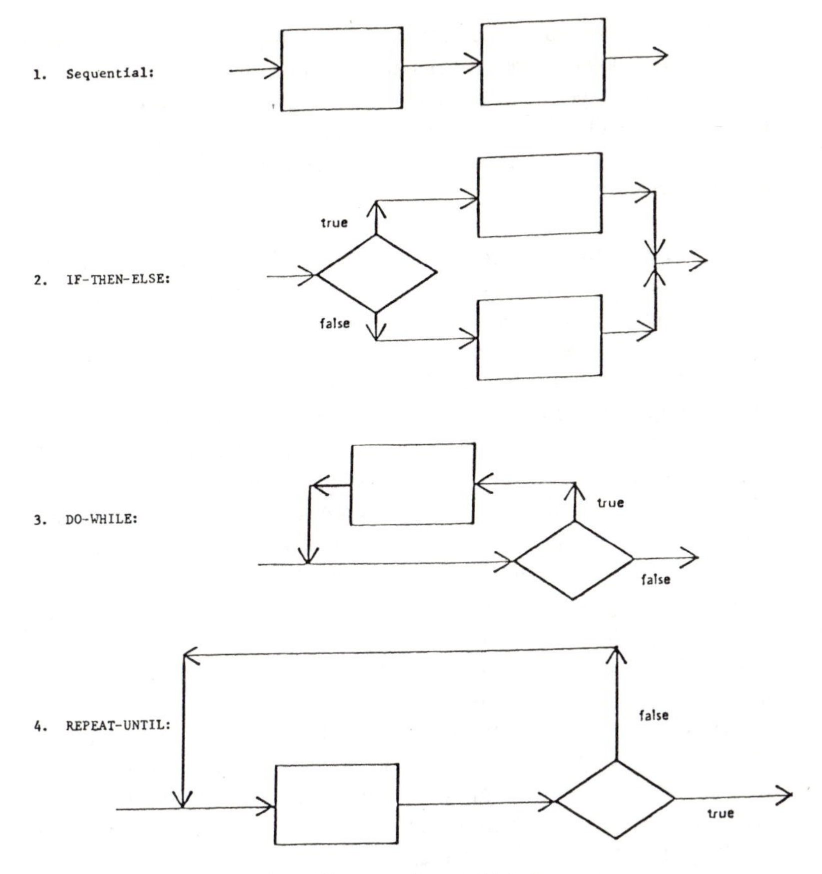

FIG. 4. Structured control blocks.

expressing our algorithm in terms of very large scale operations, thereby limiting its description to a few steps and simplified control structures, we may essentially judge the correctness of the program that we have written. A series of stepwise refinements in which, at every level, a complex operation is restated in terms of additonal detail, will still maintain the original correctness of the program if each refinement is itself correct. This stepwise refinement may be continued in ever-expanding detail until all the operations have been specified at a level corresponding to that of the programming language we are using. One important advantage of using a high-level language is that this refinement may stop much sooner than it would have if we were compelled to refine all the way down to machine language instructions. Thus, it usually follows that the higher the level of the language available to us (the richer the language and data constructs), the easier it is to implement algorithms correctly.

3.4. An Example Program. In order to clarify some of the previous remarks, let us look at one sample problem. We shall illustrate the development of an algorithm, the conversion of the algorithm into a program, and the corresponding control and data structures needed for the implementation. Sorting by insertion is the problem chosen to illustrate these principles. In Figure 5, column 1 represents a gross statement of the problem. Column 2 represents an expansion of column 1 that includes some of the details of the algorithm being implemented, but still not in sufficient detail to code directly. Column 3, a further refinement of column 2, is much closer to the level of detail required for coding. Column 4 represents the final conversion of the flowchart into the PL/1 programming language.

At each stage in the stepwise refinement of the flowchart, one operation was expanded into a more detailed description of the process using only the structured blocks listed in Figure 4. Thus the simple one-in/one-out control logic is always maintained during the development of the program. After the algorithm has been detailed sufficiently by this process, it may be converted into any suitable programming language. A language with operators and data structures closely resembling what appears in the real problem will clearly make this final coding task easier than if a totally unrelated language was used.

4. HIGH-LEVEL LANGUAGE TRANSLATION

Detailed specifications for two separate aspects of a programming language must be presented in order to describe it completely. The SYNTAX (i.e., what symbols may appear in what order) must be unambiguously defined in order to establish what a legal sentence is. Furthermore, once we have identified a legal string, we must also specify its *SEMANTICS* (i.e., the meaning of the par-

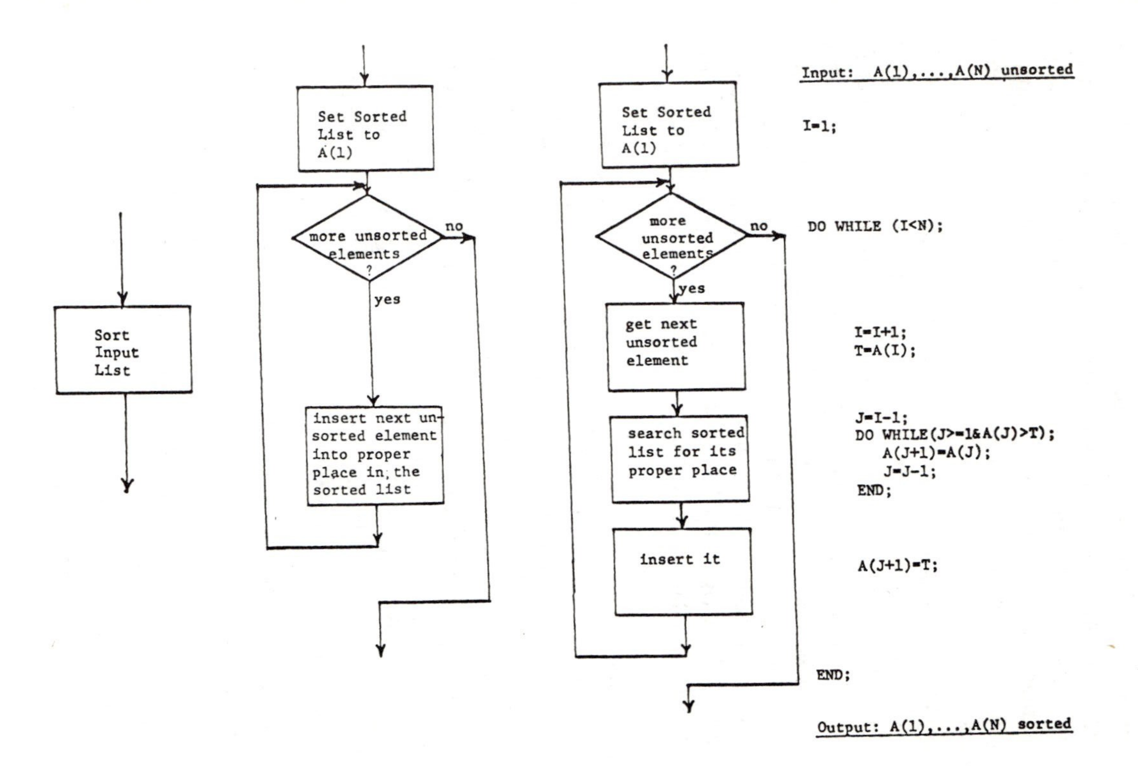

FIG. 5. Sort program development.

ticular phrases that are introduced into the language). These two sets of specifications then must be encoded into a program (the "compiler") which will actually translate the source language statements provided by the programmer (described by the syntax) into a set of machine language instructions (described by the semantics). Aho and Ullman [10] present an excellent treatment of this entire process.

As our knowledge of language constructs has improved over the past few years, we have been able to develop software tools that incorporate algorithms for the transformation of these two specification sets directly into a working compiler. This type of software development tool makes it possible for us to create and test new languages and language features at a minimum cost. The creation of a new compiler today is not the vast multiyear, multiperson project of the 1950's and 60's, but, instead, a project in which the developers need concentrate only on the principles of interest.

4.1. Syntax Specification. How do we know the statement $X = A + B$ is a valid FORTRAN assignment statement? This question could be answered by listing every possible valid assignment statement and then searching the list every time we wanted to see if a given statement was included. However, the size of an exhaustive list makes this approach impractical.* An equivalent check may be done by a constructive approach, sufficient for the types of languages that we are considering, by defining the languages as generated from a *context free grammar*. A context free grammar (CFG) consists of four parts:

1. *Terminals*—the basic symbols which compose strings in the language (i.e., A,B,0,1, +, −, . . .).
2. *Nonterminals*—special symbols that denote sets of strings (i.e., ⟨variable⟩, ⟨assignment statement⟩. The "⟨" and "⟩" are conventional brackets for these symbols.). Also referred to as phrase names.

---

* A typical FORTRAN variable name consists of a letter followed by up to five additional letters or digits. There are 1,617,000,000 legal variable names that may begin an assignment statement!

3. *Start Symbol*—one nonterminal symbol that denotes the set of strings consisting of all of the valid statements in which we are interested.
4. *Production Rules*—these rules define the ways in which symbols and strings may be legally combined to produce new strings. The name "context free" implies that the production rules are restricted to be of the form:

   "nonterminal symbol" → "string of symbols"

   where this rule implies that, for any occurrence of the symbol on the left, it may be replaced by the string on the right.

To illustrate how such a CFG allows us to recognize valid statements in a language, consider a grammar defining a simple arithmetic assignment statement language:

1. Terminal symbol set $= \{A, B, C, D, \ldots, X, Y, Z\}$
2. Nonterminal symbol set $= \{\langle V \rangle, \langle E \rangle, \langle A \rangle\}$
3. Start Symbol $= \langle A \rangle$
4. Production Rules =

P1: (1) $\langle A \rangle \rightarrow \langle V \rangle = \langle E \rangle$

(2) $\langle E \rangle \rightarrow \langle E \rangle + \langle E \rangle \mid \langle V \rangle$

(3) $\langle V \rangle \rightarrow A \mid B \mid C \mid \ldots \mid X \mid Y \mid Z$

An additional notational convention has been introduced to shorten the writing of these rules. When two rules have the same left-hand side symbol, the rules are merged using the symbol "|" to represent the alternate right-hand side strings from the original rules. Thus, the nonterminal symbol $\langle V \rangle$ denotes the set of strings consisting of upper-case single letters. The symbol may also be descriptive of the phrases it names, such as $\langle$Variables$\rangle$ or $\langle V \rangle$.

If we begin with the start symbol, we may use the production rules to substitute strings for nonterminal symbols to produce strings called *sentential forms:*

$\langle A \rangle$	$\Rightarrow$	$\langle V \rangle = \langle E \rangle$	[sentential form, rule P1.1]
	$\Rightarrow$	$X = \langle E \rangle$	[sentential form, rule P1.3x]
	$\Rightarrow$	$X = \langle E \rangle + \langle E \rangle$	[sentential form, rule P1.2a]
	$\Rightarrow$	$X = \langle V \rangle + \langle E \rangle$	[sentential form, rule P1.2b]
	$\Rightarrow$	$X = A + \langle E \rangle$	[sentential form, rule P1.3a]
	$\Rightarrow$	$X = A + \langle V \rangle$	[sentential form, rule P1.2b]
	$\Rightarrow$	$X = A + B$	[sentential form, rule P1.3b].

When only terminal symbols remain in the sentential form, no production rule can apply and hence these substitutions must stop. The chain of zero or more substitutions is called a *derivation* of the sentential form and denoted thus:

$$\langle A \rangle \Rightarrow * \; X = A + B.$$

A sentential form consisting of only terminal symbols is called a sentence. The set of all sentences derivable using the CFG is called the language generated by the grammar.

Since we have worked through a derivation showing how the string X = A + B may be developed from the start symbol of our sample grammar, we see that this string is in the set of strings denoted by ⟨A⟩, and hence it is a sentence in the language generated by this grammar. That is, we conclude that this string is a valid assignment statement! Note, however, that we still have not said anything about what this sentence *means*.

A *parse tree* is an alternative method of showing the derivation that establishes a particular sentence in a language. This tree is constructed by using a nonterminal symbol as a tree node, with the right-hand-side string symbols as dependent nodes. At any stage in a derivation, the tree leaves make up the sentential form. Corresponding to the previous derivation we have the following parse tree:

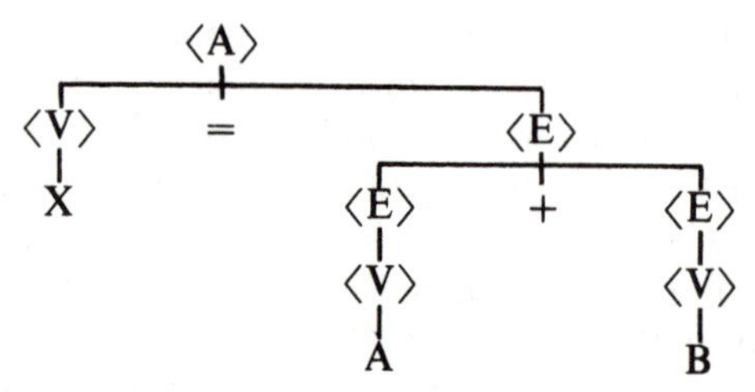

Note that the derivation

$$\begin{aligned} \langle A \rangle \;&\Rightarrow\; \langle V \rangle = \langle E \rangle &&\Rightarrow\; \langle V \rangle = \langle E \rangle + \langle E \rangle \\ &\Rightarrow\; \langle V \rangle = \langle E \rangle + \langle V \rangle &&\Rightarrow\; \langle V \rangle = \langle E \rangle + B \\ &\Rightarrow\; \langle V \rangle = \langle V \rangle + B &&\Rightarrow\; \langle V \rangle = A + B \\ &\Rightarrow\; X = A + B \end{aligned}$$

applies the same rules as before, but in a different order. However, the parse trees for these alternative derivations are exactly the same; hence the recognized phrases in the sentence also are exactly the same.

If we wish to add multiplication to the set of legal operations in our language, we might try the following modified production rule set:

P2: (1) $\langle A \rangle \rightarrow \langle V \rangle = \langle E \rangle$
(2) $\langle E \rangle \rightarrow \langle E \rangle + \langle E \rangle \mid \langle E \rangle * \langle E \rangle \mid \langle V \rangle$
(3) $\langle V \rangle \rightarrow A \mid B \mid C \mid \ldots \mid Y \mid Z$

The following two derivations, equally valid based on the above rule set, produce exactly the same final terminal string:

(1)
$$
\begin{aligned}
\langle A \rangle &\Rightarrow \langle V \rangle = \langle E \rangle \\
&\Rightarrow \langle V \rangle = \langle E \rangle + \langle E \rangle \\
&\Rightarrow \langle V \rangle = \langle E \rangle + \langle E \rangle * \langle E \rangle \\
&\Rightarrow * \quad X = A + B * C
\end{aligned}
$$
(i.e., add A to the product of B and C)

(2)
$$
\begin{aligned}
\langle A \rangle &\Rightarrow \langle V \rangle = \langle E \rangle \\
&\Rightarrow \langle V \rangle = \langle E \rangle * \langle E \rangle \\
&\Rightarrow \langle V \rangle = \langle E \rangle + \langle E \rangle * \langle E \rangle \\
&\Rightarrow * \quad X = A + B * C
\end{aligned}
$$
(i.e., multiply the sum of A and B by C)

but with the two different parse trees:

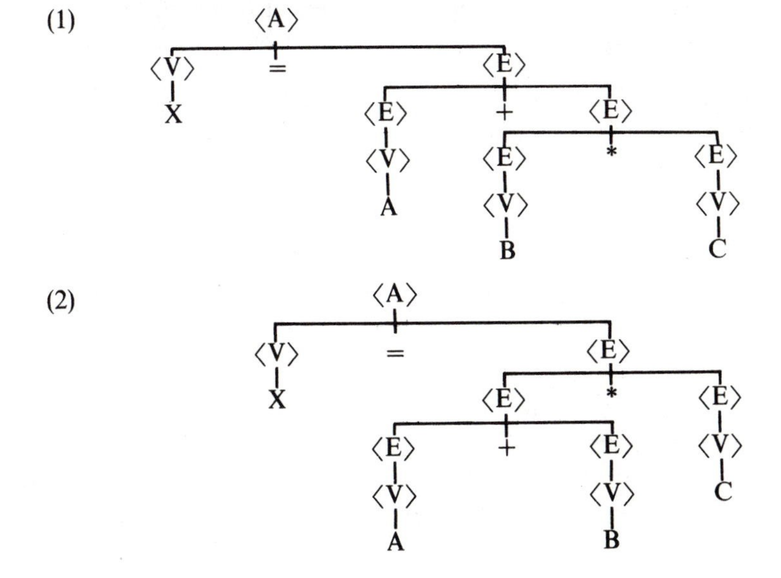

In general, if the same terminal string can be produced by two different derivations that correspond to two different parse trees, then it is not clear what symbols must be associated with each nonterminal string set. This situation is termed ambiguous; and the grammar that allows it, an *ambiguous grammar*. The definition is intuitively satisfying since we saw that in the last example, using P2, it was not clear which of the possible parses was intended:

$$(1) \qquad X = (A + (B * C))$$

or

$$(2) \qquad X = ((A + B) * C)$$

Thus, the question of ambiguity centers around the uniqueness of the parse trees, not the number of possible derivatives.

As a general principle, we want our languages to be unambiguous so that when we write a particular statement in the language we know precisely the interpretation that the compiler will place upon that statement. We saw an example of a context-free grammar that contained both addition and multiplication operators in which the grammar was ambiguous. Can we resolve this ambiguity problem and still retain all of the desired features of the CFG approach? The solution requires that we introduce the concept of *operator precedence*; in this case multiplication is an operation that should take precedence over addition. Consider the following revised set of production rules which include both operator precedence and the ability to use parentheses to force a desired order to a calculation:

$$\begin{array}{lll} \text{P3:} & (1) & \langle A\rangle \rightarrow \langle V\rangle = \langle E\rangle \\ & (2) & \langle E\rangle \rightarrow \langle E\rangle + \langle T\rangle \mid \langle T\rangle \\ & (3) & \langle T\rangle \rightarrow \langle T\rangle * \langle F\rangle \mid \langle F\rangle \\ & (4) & \langle F\rangle \rightarrow (\langle E\rangle) \mid \langle V\rangle \\ & (5) & \langle V\rangle \rightarrow A \mid B \mid C \mid \ldots \mid X \mid Z \end{array}$$

The following parse tree illustrates that, in this grammar, multiplication takes precedence over addition. The phrases involving the multiplication operator are forced by production rules P3.2a and P3.3a to be expressed as subtrees under the addition operator. That is, multiplication must be performed before any addition may be undertaken.

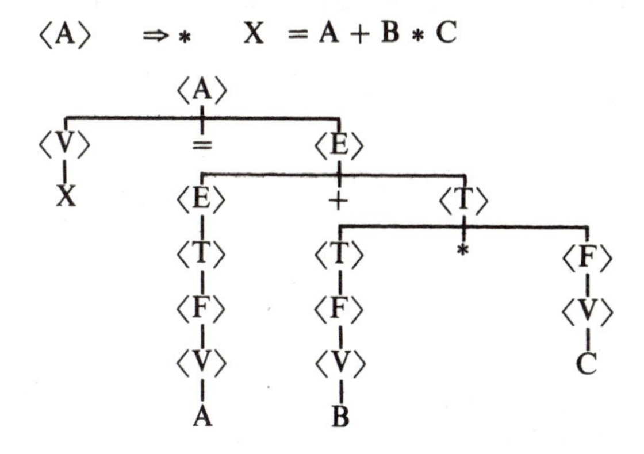

4.2. Language Recognition. We have seen how to generate sentences; but the real problem we face is that the input to the system is a given terminal string, and we must find out if there is a parse tree that will connect these terminal symbols to the start symbol of the grammar. One of the most direct methods of establishing the parse tree, if it exists, is to use the *recursive descent* algorithm, a top-down method. That is, it begins with the start symbol of the grammar as the root of the parse tree and tries to build the tree down to connect with the symbols in the source string as leaves in the tree. It accomplishes this by repeatedly asking the question, Can an instance of the desired nonterminal string class be found in the input string?

Unfortunately, direct application of this top-down method to the sample grammar produces an infinite recursion for the rule $\langle E\rangle \rightarrow \langle E\rangle + \langle T\rangle$. In this case the recursive descent algorithm would look for an instance of the string class $\langle E\rangle$ at the current position in the input string by first trying to find an instance of $\langle E\rangle$ in the input string. This is a disastrous situation for a computer program! The problem arises because the rule in question is *left recursive*, a situation in which a nonterminal is defined by a symbol string that begins with the same symbol being defined.

There are other parsing algorithms that can accept left recursive production rules directly. However, let us modify our sample grammar so that it still generates the same language and is also in a form suitable for recursive descent parsing. The left recursion in a production rule of the form $\langle \mathrm{A} \rangle \rightarrow \langle \mathrm{A} \rangle \mathrm{x} \mid \mathrm{y}$ may be eliminated by converting the single rule into a pair of rules via the addition of a new nonterminal symbol;

$$\begin{aligned} \langle \mathrm{A} \rangle &\rightarrow \mathrm{y} \langle \mathrm{A}' \rangle \\ \langle \mathrm{A}' \rangle &\rightarrow \mathrm{x} \langle \mathrm{A}' \rangle \mid \mathrm{null}, \end{aligned}$$

where "x" and "y" are arbitrary strings and "null" represents the empty string. A convenient iteration form that is equivalent to this pair is

$$\langle \mathrm{A} \rangle \rightarrow y \{\mathrm{x}\}$$

where the braces { } indicate that the included symbols may be repeated zero or more times. All three formulations produce exactly the same language, the set of strings {y, yx, yxx, ... }.

P4 is a version of the sample grammar, as modified by removing left recursion, that is suitable for a recursive descent recognizer:

$$\begin{array}{lll} \mathrm{P4:} & (1) & \langle \mathrm{A} \rangle \rightarrow \langle \mathrm{V} \rangle = \langle \mathrm{E} \rangle \\ & (2) & \langle \mathrm{E} \rangle \rightarrow \langle \mathrm{T} \rangle \{ + \langle \mathrm{T} \rangle \} \\ & (3) & \langle \mathrm{T} \rangle \rightarrow \langle \mathrm{F} \rangle \{ * \langle \mathrm{F} \rangle \} \\ & (4) & \langle \mathrm{F} \rangle \rightarrow (\langle \mathrm{E} \rangle) \mid \langle \mathrm{V} \rangle \\ & (5) & \langle \mathrm{V} \rangle \rightarrow \mathrm{A} \mid \mathrm{B} \mid \mathrm{C} \mid \ldots \mid \mathrm{Y} \mid \mathrm{Z} \end{array}$$

Figure 6 contains a flowchart that implements the recursive descent recognizer for this version of the sample grammar. A trace of the operation of this program is presented in Figure 7, after code generation capabilities have been added to the recognizer. The diagram, however, does not contain all of the necessary detail: V is the procedure that checks to see if the current input symbol is a letter; NEXT is a procedure to move on to the next input symbol and to store that symbol in the variable IS; ERROR is a procedure that issues a message when an error has been found in the input string and also terminates execution, since the parse tree cannot be constructed.

4.3. Semantic Specification. The construction of the parse tree for a statement is sufficient to show the validity of both the statement and its organization into phrases. However, parsing does not produce executable instructions—we must provide a "meaning" for

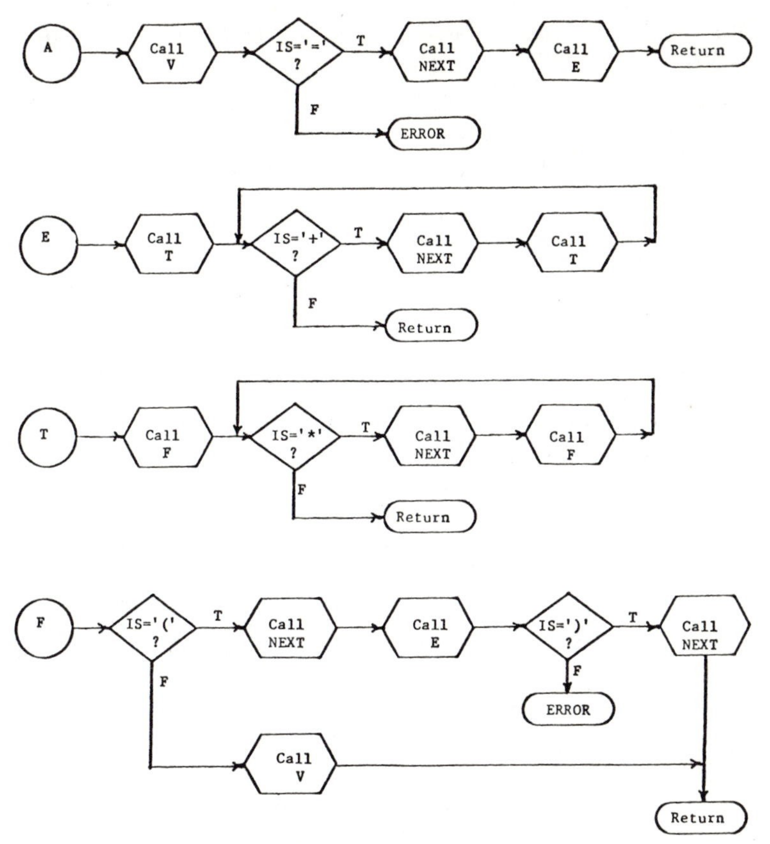

FIG. 6. Recursive descent recognizer.

each of the phrases to be found in the language. For example, is the plus sign a unary operator or a binary operator? If it is binary, does it refer to integer addition, floating point addition, matrix addition (integer or real), or perhaps the logical OR operation? Such a meaning must be supplied for every symbol used in the language so that the compiler may generate the appropriate code for the local context of the symbol. Notationally, we could provide an escape character which would allow us to intersperse semantic commands within the production rules. Then, as a production rule is applied during the parsing of the input string, the semantic commands would indicate the semantic routine that should be applied

```
Input String   Program Execution              Parse String                Code
X=A+B          Call A
                 Call V
=A+B             return from V                <V>(X)
                 IS='='? yes
A+B              Call E                       <V>(X)=
                   Call T
                     Call F
                       IS='('? no
                       Call V
+B                     return from V          <V>(X)=<V>(A)
                     return from F            <V>(X)=<F>(A)
                     IS='*'? no
                   return from T              <V>(X)=<T>(A)
                   IS='+'? yes
B                  Call T                     <V>(X)=<T>(A)+
                     Call F
                       IS='('? no
                       Call V
null                   return from V          <V>(X)=<T>(A)+<V>(B)
                     return from F            <V>(X)=<T>(A)+<F>(B)
                   return from T              <V>(X)=<T>(A)+<T>(B)
                   IS='+'? no                 <V>(X)=<T>(t1)              LDA A
                                                                          ADD B
                                                                          STA t1
                 return from E                <V>(X)=<E>(t1)
               return from A                  <A>                         LDA t1
                                                                          STA X
```

FIG. 7. Recursive descent trace and code production.

at that point in the code generation process. This is the technique that will be used in a later example. An extension of this concept offers even more flexibility: Provide the capability of introducing the semantic routine code directly into the production rules by allowing a complete block of code to be added to each rule. This code block then provides for whatever code generation actions are necessary. This is the approach used in the YACC [11] system, for example, to implement a number of highly successful compilers.

The following example illustrates how a simple set of machine instructions may be produced for the sample arithmetic assignment language. The development is based on a variation of the methods described by Graham [13].

First, let us assume that the target computer code, similar to that discussed in Section 2.2, consists of the instructions listed there plus:

MPY X = Multiply the contents of the accumulator by the contents of memory location X and leave the product in the accumulator.

We also need to extend the production rules in two ways:

1. Add the escape character "." to flag semantic commands. These will have the form ".n" to indicate the execution of the *n*th semantic routine.
2. Associate each phrase name in the parse tree with a symbolic address that will eventually correspond to the location in memory that contains the value of the phrase. Thus $\langle V\rangle$ may be associated with X, and $\langle E\rangle$ associated with T3, the third generated temporary symbolic address. The generated addresses will be needed to hold values during the evaluation of an expression.

The final grammar for our example language, including the semantic calls:

P5: (1) $\langle A\rangle \rightarrow \langle V\rangle = \langle E\rangle .1$
(2) $\langle E\rangle \rightarrow \langle T\rangle \{+\langle T\rangle .2\}$
(3) $\langle T\rangle \rightarrow \langle F\rangle \{ * \langle F\rangle .3\}$
(4) $\langle F\rangle \rightarrow (\langle E\rangle) \mid \langle V\rangle$
(5) $\langle V\rangle \rightarrow A \mid B \mid C \mid \ldots \mid Z$

where the indicated semantic routines must produce the actions:

.1: Generate the code to store the value of the right hand expression in the left side variable.
.2: Generate the code to add the two terms together and to store the result in a new temporary location.
.3: Generate the code to multiply the two factors together and to store the result in a new temporary location.

In all cases the last symbolic address used or generated on the right side of the production rule should be associated with the phrase name on the left side. Figure 7 demonstrates this process by illustrating the step-by-step procedure for the development of a parse tree and its associated generated code.

4.4. Code Optimization. The code generation example of the previous section suffers from two major deficiencies in terms of the required performance of real compilers. First, it assumes a computational machine that is far simpler than most current computers. The semantic routines must be augmented significantly to account for items such as multiple accumulators and different modes of addressing. Second, the generated code contains a number of instructions that are actually unnecessary. The redundant instructions were created because of the treatment of each phrase as an independent entity irrespective of the context in which it appears. This latter situation may be improved by applying the techniques of code optimization to the code as produced by the semantic routines. We find, however, that when we look at the subject of code optimization, it immediately splits into two distinctly different areas:

1. Local Code Optimization—improving the code efficiency by working within the immediate local context of the particular phrase being parsed, essentially within the confines of a single statement.
2. Global Code Optimization—improving the code efficiency by looking at the interaction between statements. For example, by not repeating the evaluation of a duplicated expression or by factoring loop-independent expressions outside the loops in which they appear.

The code as generated for the example in Figure 7 was:

```
LDA  A
ADD  B
STA  T1
LDA  T1
STA  X
```

The value of the sum is available in the accumulator when the time comes to store that value in X. Consequently, the "LDA T1" instruction is clearly unnecessary. Is the "STA T1" instruction also not needed? If the original source language program never used the value again, then there would be no advantage to saving it. The reason for storing any value is to save it for later use, thus making it unnecessary to reevaluate the same expression. Consider how the following program fragment could be handled by an optimizing compiler:

```
X = A + B
Y = (A + B) * (A + B)
```

Now the temporary saving of the sum could easily produce the code

```
LDA  X
ADD  B
STA  T1
STA  X
MPY  T1
STA  Y
```

with a correspondingly significant improvement in the object code efficiency.

The theory of formal language recognition has been studied for some time, and it is now possible to produce software tools that provide most of the detail work necessary to develop compilers for new languages. The problem of code generation is not nearly as clean and well defined! It is highly dependent on the machine and language, and hence has been solved by using many heuristic and ad hoc methods. Graham [13] describes many of the local code optimization techniques used in the early compilers. Lowry and Medlock [14] describe the global code optimization methods used in the IBM FORTRAN H compiler, a program that produces a very efficient object code. Aho and Ullman [10] describe more recent developments, including some of the efforts being made to systematize the field.

5. PROBLEM-ORIENTED HIGH-LEVEL LANGUAGES

While general-purpose high-level languages were being developed, another parallel stream of language development was taking place. Computer users with specific requirements tended to create languages that were more oriented toward their particular problems. Some of the characteristics of these problem-oriented languages are sufficiently different from the general-purpose procedure-oriented languages that we shall discuss a representative sample of them in a number of separate application specific categories.

5.1. Report Generators. RPG (Report Program Generator) [15] is a specific language that is widely used on small computer systems. We shall use it as an example of the general class of languages devoted to writing reports. The basic assumption in a report generator system is that a series of records will be read sequentially from one or more input files, some information will be extracted from a selected set of these records, and then one or more output files, along with a final printed summary, will be produced. As a result, the overall logic of the program is built into the system and the programmer does not need to specify the exact sequence of steps; rather, he indicates what operations are to be done. In fact, one of the main concerns in the design of procedure-oriented languages (the definition of control structures for flow of control specification) is generally of no concern in a report generation language. This type of information is completely defined within the language compiler.

To simplify the source program preparation further, a fixed format of some type usually is specified for the source language. This may be seen most easily by studying a few of the common RPG statement types:

1. File Description Specification—Declares the names and physical descriptions of all files that are used in the program.
2. Input Specification—Declares the names and attributes of all items of information that are read from the input files.
3. Calculation Specification—Describes the calculations to be applied to the input data.
4. Output-Format Specification—Describes all output information to be produced, including formats and file designations.

The information fields on a file description specification include:

column 6: 'F', to indicate statement type.
7-14: file name.
15: indication if input, output, or update file.
24-27: number of characters in each record.
40-46: physical device holding file (tape, disk, etc.).

A specification for a file of input punched cards would be:

```
F INPUT    I    80    READ01
```

This statement describes the File named INPUT as an Input file consisting of 80 character records (one 80-column punched card) on the card reader READ01.

The information fields of an input specification statement include:

column 6: 'I', to indicate statement type.
7-14: file name.
19-20: record identifying indicator.
21-41: record identification codes.
44-51: field location.
53-58: field name.

This statement introduces a fundamental data type in an RPG program: the *indicator*. Each indicator is identified by a two-digit code (01-99) and represents a two-position switch (on/off). The indicators are normally turned off before an input record is read. The record identification code, detailed in columns 21–41, will turn on the appropriate indicator if the data in the record meets the specified condition (such as "there must be a 'D' in position 10 of the input record"). If the conditions are satisfied, the variables listed in the name fields also will be assigned values from the positions specified by the location fields. The lines:

```
I INPUT    14   10 C D    11 30 NAME
I                         40 60 ADDR
```

will turn on indicator 14 if the contents of position 10 is the Character 'D', and also will store positions 11-30 in the variable NAME and positions 40-60 in the variable ADDR. The output specification is similar, but with the added complication of extensive editing and formating capabilities.

The calculation specification includes:

column 6: 'C', to indicate statement type.
7-17: Indicator logical expression.
18-27: Factor 1.
28-32: Operation.
33-42: Factor 2.
43-48: Result field.

The indicator logical expression evaluates to "yes" or "no" based on the settings of the indicators (such as "if indicator 20 is 'on' and indicator 22 is 'off,' then 'yes,' otherwise 'no'"). If the answer is "no," the balance of the statement is ignored. If "yes," the statement is executed. For example,

RATE MULT OVERTM SAVE

will multiply the two factors RATE and OVERTM and store the result in SAVE.

The execution of an RPG program consists of reading an input record, setting the indicators, doing what is desired based on the indicators, and then repeating this cycle until the input data have been exhausted. The simplicity and directness of this approach has been so successful that the language has expanded far beyond what this short introduction can describe.

5.2. Database Query Language. The ability of computer systems to store large amounts of information has affected all of us in many ways. However, just storing information is not sufficient to make it useful—we must also be able to locate it again when we need it. It is the function of a database query language to allow a user that is not a computer specialist to access and manipulate the information stored within a database system. As a representative of this class we shall look briefly at SEQUEL 2 [16]. Wiederhold [17] contains an extensive list of references to additional systems of this type, as well as a detailed discussion of the data structures required to support such a language system.

SEQUEL 2 assumes that the database has a *relational* [18] form. That is, information is stored in files that have a tabular organization:

EMPLOYEE

NAME	DEPTNO	SALARY	JOBTITLE
John Jones	40	15000	Programmer
Shirley Smith	20	16000	Mathematician

The name of this relation is EMPLOYEE, with the column names (or attributes) NAME, DEPTNO, SALARY, and JOBTITLE. This descriptive information about the relation may be summarized as:

EMPLOYEE(NAME,DEPTNO,SALARY,JOBTITLE).

Each line in the table is called a tuple. A tuple represents one instance of the relation, with each tuple assumed to be unique. The tuples are unordered in the relation. Normally a database would contain many such relations in order to express all of the desired information.

The basic operation of the SEQUEL 2 language is a *mapping*, in which a known quantity is transformed into a desired quantity by means of a given relation. The structure of this mapping statement is:

```
SELECT  ⟨list of attributes to be returned⟩
FROM    ⟨name of relation to use⟩
WHERE   ⟨predicate to determine what tuples to select⟩
```

Thus to find the personnel list of department 20 we could use the query:

```
SELECT NAME FROM EMPLOYEE WHERE
                   DEPTNO = '20'.
```

We could obtain a list of all mathematicians making more than $15,000 by using the query:

```
SELECT NAME FROM
  EMPLOYEE WHERE SALARY > '15000'
                 AND JOBTITLE = 'Mathematician'.
```

A relation may be viewed as a set of tuples. Since a mapping returns a desired set of values, a mapping may also be considered as producing a new relation as its output. Thus the simple SEQUEL 2 mapping operation may be easily extended by nesting mappings and by using set operations such as union and intersection. To illustrate these extensions, assume that our database also contains the relation:

DEPT(DEPTNO,LOCATION,MANAGER)

We may find all of the people that work in St. Louis by using the query:

```
SELECT NAME FROM
  EMPLOYEE WHERE DEPTNO =
SELECT DEPTNO FROM DEPT
  WHERE LOCATION           = 'St. Louis'.
```

What if we were interested in the names of all of the department managers that had salaries greater than $20,000? This information could be obtained by finding both the set of all managers and the set of all employees with the desired salary, and then determining the intersection of these two sets:

```
SELECT NAME FROM EMPLOYEE
WHERE SALARY > '20000'
INTERSECT
SELECT MANAGER FROM DEPT.
```

Since the last mapping does not contain a WHERE clause, every tuple from the DEPT relation will be selected, thus producing a list of all of the managers.

There are many additional features in the language ranging from simple query facilities easily learned by nonspecialists to more complex facilities intended for professional programmers. Functions such as AVG, SUM, COUNT, MAX, and MIN are provided for application to the set of values found by the SELECT statement. A relation may be partitioned into separate tuple sets by the GROUP command or sorted by the ORDER command.

Relations may be created, modified, or destroyed. The user who creates a relation is fully and solely authorized to perform actions upon it, but may grant access rights to the relation for other users. These access rights include, among others, the ability to read, insert, delete, and update tuples in the relation. This facility allows the owners of the information total control (hopefully!) over who may have access to it—a critically important feature for any database system.

5.3. Graphic Languages. The ability to sit at a computer terminal that contains a graphics display tube has created a new world of possibilities for the mathematical modeling of curves, surfaces, and figures. LG [20] is a language that allows the definition of geometric objects and elements, computes their parameters, and displays the results. It was designed specifically for use by nonprogrammers, being easy to learn and very close to the natural language used in geometry.

LG is a conversational language, providing for a user dialogue with the graphic terminal by means of a command-answer type of processing. A command is given by the user, and the system responds with two forms: the computed standard parameters of the desired geometric element and a display of the element itself on the screen.

The LG instruction line that establishes what the user wants done consists of three parts separated by delimiters:

⟨command⟩; ⟨name⟩: ⟨definition⟩.

1. ⟨command⟩—Valid commands are either built in by being defined at system generation time or added on-line by the user. Many of the standard types of goemetric items (POINT, LINE, SPHERE, etc.) would normally be built in, as well as a number of standard functions to be performed (LOCUS, COMPUTE, MACRO, etc.). The user may add new commands at any time by using the MACRO capability.
2. ⟨name⟩—The name to be assigned to the newly specified geometric element.
3. ⟨definition⟩—Specifies a list of data elements, separated by delimiters, which defines the geometric element. Each element in the list is either the name of a previously defined geometric element or a keyword set equal to an arithmetic expression.

Some sample instruction lines:

POINT; P1: X = COS(3.5), Y = L, Z = 2 * ALPHA

The point named P1 is defined by giving its X, Y, Z coordinates. X, Y, Z are the keywords indicating which coordinates are set by the respective arithmetic expressions. These expressions may contain scalar elements, FORTRAN operators and functions. P1 is called a *fixed element* because it has been completely specified.

LINE; L: P1, P2

The line L is defined by specifying two previously defined points.

COMPUTE; PI: = 4 * ATAN(1.0)

This command represents essentially a standard assignment statement.

POINT; A: L, X = 3

The point A has been defined as the intersection of the line L and the plane x = 3.

SPHERE; S: R = 1, X = 0, Y = 0, Z = 0:

the definition of a sphere of unit radius centered at the origin. All of the examples above represent fixed elements and hence will be displayed automatically.

Consider the following sequence of commands:

```
PARAMETER ; T : = 0
POINT     ; P1: X = T, Y = 0, Z = 0
POINT     ; P2: X = 0, Y = T, Z = 5
LINE      ; L : P1, P2
LOCUS     ; L : T, MIN = -10, MAX = 10, STEP = 0.5
```

The first line creates the scalar variable T as a parameter and initializes its value. P1, P2, and L are called *variable elements* since they contain a parameter in their definition. Variable elements will be displayed only on specific instructions, such as the last line containing the LOCUS command. The display will be a hyperbolic paraboloid composed of the line L drawn for the values of T equal to $-10, -9.5, -9.0, \ldots, +10$. The use of

TRACE; L

would display L for the current value of any parameters. A parameter value may be changed at any time by the command:

REPLACE; T: = 4.

The current implementation contains approximately 80 commands. There are also many additional ways to define geometric elements, including models such as intersections, transformations, and alternative coordinate systems. The language has been used in studying variations in complex geometric figures, problems in kinematics and mechanics, and for some applications in optics and topography.

6. CONCLUSIONS

Sammet [21] lists a total of 166 programming languages that are in use today, 76 in the categories of numerical scientific, business data processing, list processing, string processing, formula manipulation, and multipurpose. More specialized application areas (such as accounting, circuit design, editing and publishing, simulation, etc.) cover the remaining 90 languages.

There are some interesting and contradictory trends taking place that help to explain the proliferation of languages. First, it should be noted that only a small number of the available languages are in widespread use. There is an important economic motivation for this: We do not want to reprogram a problem every time we want to run it on a different machine. If we use a high-level language that is machine independent, then we may run the program on any computer that has a compiler for that language. Thus, this portability issue argues strongly for a few powerful standardized languages that are implemented on as many different computers as possible. Since programs that are widely distributed are frequently production-type tasks (everybody runs a payroll program regularly!), it is important that the compilers produce efficient object code. This, in turn, means that the compilers themselves are large expensive programs, and hence tend to be restricted in numbers. We now have standard COBOL, standard FORTRAN, and a few other languages either already standardized or being studied in preparation for standardization.

On the other hand, the software tools now exist that make it possible to design and implement a language quickly. The price paid for a fast implementation may be slow execution, as compared to what could be achieved with more effort spent on code optimization for the particular hardware in use. However, consider that in a research environment, say where language constructs are being studied, code efficiency may not be really important. Also, as hardware costs decrease and programmer costs increase, it becomes more important to concentrate on human efficiency (i.e., language and system design) than on hardware efficiency. This is particularly true for programs that are to be executed just a few times, since most of the costs will be concentrated in program preparation and testing.

It looks as if both trends will continue in full force: A few very general languages will be fixed by standardization to achieve a maximum degree of portability and efficiency. There also will be a continuing stream of new languages, both general purpose and problem oriented, as we attempt to make it as convenient and as easy as possible for everyone to use a computer.

REFERENCES

1. D. Knuth, *Fundamental Algorithms*, The Art of Computer Programming, vol. 1, 2nd ed., Addison-Wesley, Reading, Mass., 1973.
2. T. Pratt, "Formal Analysis of Computer Programs," in this study.
3. G. Weinberg, *The Psychology of Computer Programming*, Van Nostrand Reinhold, New York, 1971.
4. S. Rosen, "Programming systems and languages—A historical survey", in S. Rosen, ed., *Programming Systems and Languages*, McGraw-Hill, New York, 1967, pp. 3–22.
5. J. Sammet, *Programming Languages: History and Fundamentals*, Prentice-Hall, Englewood Cliffs, N. J., 1969.
6. ———, "Programming languages: History and future," *Comm. ACM*, **15**, no. 7 (July 1972), 601–610.
7. R. Steinhart and S. Pollack, *Programming the IBM System/360*, Holt Rinehart and Winston, New York, 1970.
8. T. Pratt, *Programming Languages: Design and Implementation*, Prentice-Hall, Englewood Cliffs, N. J. , 1975.
9. C. McGowan and J. Kelly, *Top-Down Structured Programming Techniques*, Petrocelli/Charter, New York, 1975.
10. A. Aho and J. Ullman, *Principles of Compiler Design*, Addison-Wesley, Reading, Mass., 1977.
11. S. Johnson, "YACC—Yet another compiler compiler," *Comp. Sci. Tech. Rep. 32*, Bell Laboratories, Murray Hill, N. J., 1975.
12. L. Heindel and J. Roberto, *LANG-PAK = An Interactive Language Design System*, American Elsevier, New York, 1975.
13. R. Graham, "Bounded context translation," *Proc. AFIPS Spring JCC*, **25**, Spartan, N. Y., 1964, pp. 17–29.
14. E. Lowry and C. Medlock, "Object code optimization," *Comm. ACM*, **12**, 1 (January 1969), 13-22.
15. J. Murray, *Programming in RPG II: IBM System/3*, McGraw-Hill, New York, 1971.
16. D. Chamberlin et al., "SEQUEL 2: A unified approach to data definition, manipulation, and control," *IBM Jour. Res. Dev.*, **20**, no. 6 (November 1976), 560–575.
17. G. Wiederbold, *Data Base Design*, McGraw-Hill, New York, 1977.
18. E. Codd, "A relational model for large shared data banks," *Comm. ACM* **13**, 6 (June 1970), 377–387.
19. P. Wegner, *Programming Languages, Information Structures, and Machine Organization*, McGraw-Hill, New York, 1968.
20. J. Raymond, "LG: A language for analytic geometry," *Comm. ACM*, **19**, no. 4 (April 1976), 182–187.
21. J. Sammet, "Roster of programming languages for 1976-77," *SIGPLAN Notices*, **13**, no. 11 (November 1978), 56–85.

SPECIFYING FORMAL LANGUAGES

Ronald V. Book

1. INTRODUCTION

What is formal language theory? There are several themes that provide partial answers to this question and that together provide a first approximation to the answer.

(1) Mathematical models for natural language provide a formal basis for studying the nature of languages such as English and French. It is important to have such models if linguists are to understand the common features of such languages and their generative structures, and if automatic translation between such languages is to be achieved [7], [19]–[21], [23], [60].

(2) The essential features of the syntax of many programming languages can be faithfully modeled by means of context-free grammars, the same structures that have been used to describe the syntax of natural languages. By using context-free grammars to describe the syntax of a programming language, tools for trans-

---

During the preparation of this paper, the author's research was supported in part by the National Science Foundation under Grants MCS76-05744 and MCS77-11360.

lating between source programs and machine code have been developed and the automatic generation of certain portions of compilers has been facilitated [3], [18], [51], [61], [94], [95].

(3) Abstract automata provide mathematical models for certain aspects of the process of computation; e.g., there are Turing machines, devices which capture the notion of computation as described by logicians, and there are finite-state machines which model the behavior of switching circuits. Other devices, such as pushdown store transducers, are used to model aspects of the compiling process. Formal language theory provides tools for describing precisely the power and the limitations of such models and thus is extremely useful in relating studies in computability theory, automata theory, and computational complexity [2], [28], [37], [41], [58], [67], [81], [82].

(4) Formal language theory as an abstraction of many aspects of real computation is a new branch of mathematics related to logic and to combinatorial algebra [29], [96].

In this paper, language theory is presented from the standpoint of (3) with the understanding that (1) has provided much initial motivation, that (2) has provided motivation and applications, and that (4) has not yet fully emerged. It is the use of formal language theory as a tool for investigating a wide variety of topics within theoretical computer science that has brought forth the rich family of generative and automata-theoretic structures and the many properties of languages and classes of languages that have given structure to the theory. Thus the classical connections between generative structures, automata, and properties of languages and classes of languages are surveyed here with the intent of exemplifying for the nonspecialist the questions asked in the study of formal language theory.

A language is a set of strings of symbols where the symbols are taken from some finite set, the *alphabet*. A language may have structure because its strings are related by some common property, e.g., $\{w \in \{a, b\}^* \mid$ the number of a's occurring in w is equal to the number of b's occurring in w$\}$ or $\{w \in \{a, b\}^* \mid$ the number of a's occurring in w is equal to the number of b's occurring in w and for any prefix y of w, the number of a's in y is greater than or equal to the number of b's in y$\}$. A language may be finite or infinite. In the case that a language is infinite we encounter the notion underlying much of the theory: To discuss an infinite language in terms of effective computation, the language must have some finite representation; and this representation must give information about the language that will allow one to construct effective procedures for generation or acceptance.

There are three methods that have been used extensively for finitely specifying languages.

(i) Define a finite structure that *generates* the strings of the language. Thus one must specify an effective procedure that enumerates the language and nothing outside of the language. The most familiar example of such a system is a phrase-structure grammar and the notion of derivation in such a grammar. Generative structures are similar to formal systems as studied by logicians.

(ii) Define a finite structure that *recognizes* or *accepts* the words of the language. Examples here include finite-state acceptors, Turing machines, pushdown store acceptors, etc. Generally, the device examines an input string and computes the characteristic function of the language, giving an answer "yes" or "no" to the question "Is w in L?"

(iii) Explicitly specify a language by defining it from given languages and certain operations on languages.

The class of *context-free* languages lies at the core of the theory of formal languages, and so it is useful to explain just how context-free languages can be specified. In keeping with (i), a language is context-free if and only if it is generated by a context-free grammar. (These notions are developed in Section 3.) In keeping with (ii), a language is context-free if and only if it is accepted by a nondeterministic pushdown store acceptor. (These notions are developed in Section 4.) In keeping with (iii), a language is context-free if and only if it can be obtained from the Dyck set on two letters by a finite number of applications of the following operations: intersection with regular sets, inverse homomorphism, and homomorphism. (These notions are developed in Section 3.4, particularly Theorem 3.14.)

Just as individual languages are specified by grammars, automata, or algebraic characterizations, classes of languages may be specified by considering all languages generated by a specific class of grammars or all languages accepted by a specific class of automata or all languages defined algebraically from a certain basis class of languages.

In this paper the class of context-free languages is used to exemplify themes that arise in formal language theory. Extensions and restrictions of this class are described and the study of complexity classes is introduced. However, this overview is not an exhaustive survey. The topics mentioned here are central to the study of formal language theory, but the topics omitted are too numerous to list in this brief paper.

The practical role of context-free languages in compiling a program (written in a context-free language) is discussed in the preceding article by Ball.

The primary references given here are included because of their historical interest, because they have played a key role in the development of the subject, because the material included is not described in any secondary source, or because they give the flavor of current work. Many secondary references are included so that the reader has a chance to learn some of the basic material before plunging into research papers.

Finally, it should be noted that the popularity of formal language theory has fluctuated greatly. It is my contention that many of the themes in formal language theory are central to theoretical computer science and that despite these fluctuations formal language theory will continue as a lively and vital area of research.

2. PRELIMINARIES

In this section we establish notation and state some results on two fundamental classes of languages, the regular sets and the recursively enumerable sets.

2.1. We assume as an undefined primitive the notion of a *symbol*. A *string* (*word*, *sentence*) is a finite sequence of symbols and the *empty* word e is the empty sequence. The *length* of a string w is the number of symbols in w; it is denoted by $|w|$. Thus $|e| = 0$ and, if w is a string and a is a symbol, then $|a| = 1$ and $|wa| = |w| + |a|$. The basic operation on strings used here is concatenation: the concatenation xy of strings x and y is the result of juxtaposing x and y, and so $|xy| = |x| + |y|$.

A finite set of symbols is an *alphabet* (*vocabulary*). If Σ is an alphabet, then Σ^* is the set of all strings over Σ, $\Sigma^* = \{a_1 \cdots a_n \mid n \geq 1, \text{ each } a \in \Sigma\} \cup \{e\}$. Any subset of Σ^* is a *language over* Σ. Thus, to say that L is a language is to say that there is some alphabet Σ such that $L \subseteq \Sigma^*$.

We will be concerned with operations on languages. First, there are the Boolean operations: union, intersection, and complementation. Second, there are the Kleene operations: concatenation (the concatenation of languages L_1 and L_2 is $L_1L_2 = \{w_1w_2 \mid w_1 \in L_1 \text{ and } w_2 \in L_2\}$), Kleene $*$ ($L^* = \bigcup_{i=0}^{\infty} L^i$ where $L^0 = \{e\}$, $L^1 = L$, and $L^{n+1} = LL^n$), and Kleene $+$ ($L^+ = \bigcup_{i=1}^{\infty} L^i$).

If Σ is an alphabet, then Σ^* is the *free semigroup with identity* (*free monoid*) generated by Σ. Here the semigroup multiplication is concatenation of strings, an associative binary operation. The empty word e is the identity. Since every $w \in \Sigma^*$ has a unique factorization as a finite sequence of symbols from Σ, Σ^* is the *free* semigroup with identity generated by Σ. For a language $L \subseteq \Sigma^*$, L^* is the subsemigroup (of Σ^*) with identity e generated by L.

Another operation on languages is *reversal*. If Σ is an alphabet and $w \in \Sigma^*$, the reversal w^R of w is defined as follows: $w^R = w$ if $w \in \Sigma \cup \{e\}$; $w^R = ay^R$ if $w = ya$ for $y \in \Sigma^*$, $a \in \Sigma$. If $L \subseteq \Sigma^*$, then the reversal of L is $L^R = \{w^R \mid w \in L\}$.

An important consideration in formal language theory is the study of several types of mappings between languages. The prototype of the mappings considered in this paper is the notion of *homomorphism*. For alphabets Σ and Δ, a homomorphism h from Σ^* to Δ^* is a function $h : \Sigma^* \to \Delta^*$ with the property that for all $x, y \in \Sigma^*$, $h(xy) = h(x)h(y)$. Since each $w \in \Sigma^*$ has a unique factorization as a finite sequence of symbols from Σ, a homomorphism is uniquely determined by defining its values on the symbols in Σ. Thus we are considering homomorphisms between free semigroups.

A homomorphism $h : \Sigma^* \to \Delta^*$ is *nonerasing* if for every $w \in \Sigma^*$, $h(w) = e$ implies $w = e$ (equivalently, $h(a) \neq e$ for every $a \in \Sigma$) and is *length-preserving* if $|h(w)| = |w|$ for every $w \in \Sigma^*$ (equivalently, $|h(a)| = 1$ for every $a \in \Sigma$). Thus a symbol is "erased" if it is mapped to the empty word. Clearly one can view homomorphisms as functions taking strings to strings with no thought of the simple algebraic structure involved.

If $h : \Sigma^* \to \Delta^*$ is a homomorphism and $L \subseteq \Sigma^*$, we write $h(L)$ for $\{h(w) \mid w \in L\}$. If $L \subseteq \Delta^*$, we write $h^{-1}(L)$ for $\{w \in \Sigma^* \mid h(w) \in L\}$. We refer to the mapping h^{-1} from subsets of Δ^* to subsets of Σ^* as an *inverse homomorphism.*

A homomorphism substitutes a string for a symbol. It is of considerable interest in formal language theory to substitute a language for a symbol. To define this notion consider an alphabet Σ. For each $a \in \Sigma$, let Σ_a be an alphabet and let $T(a) \subseteq \Sigma_a^*$. Let $T(e) = \{e\}$ and $T(wa) = T(w)T(a)$ for each $a \in \Sigma$, $w \in \Sigma^*$. Then T is a *substitution on* Σ. If $\mathscr{L}$ is a class of languages such that for each $a \in \Sigma$, $T(a) \in \mathscr{L}$, then T is an $\mathscr{L}$*-substitution.* A class $\mathscr{L}_1$ *is closed under* $\mathscr{L}_2$ *substitution* if for every $L \in \mathscr{L}_1$, when Σ is an alphabet such that $L \subseteq \Sigma^*$ and T is an $\mathscr{L}_2$-substitution on Σ, then $T(L) \in \mathscr{L}_1$.

One of the important themes in the study of formal languages and abstract automata is the characterization of classes of languages and automata by means of operations under which the class of languages is closed. Recall that if θ is an *n*-ary operation on languages, then a class $\mathscr{L}$ of languages is *closed under operation* θ if for every choice $L_1, \ldots, L_n$ of languages in $\mathscr{L}$, $\theta(L_1, \ldots, L_n)$ is in $\mathscr{L}$.

2.2. Now we turn to the notion of "regular sets," the languages recognized by finite-state machines operating as acceptors. The study of finite-state machines as devices to compute functions arises in the theory of switching circuits and the study of logical circuit design. The restriction of this study to machines that compute only characteristic functions is one of the most important areas of the theory of abstract automata.

A *deterministic finite-state acceptor* $D = (K, \Sigma, \delta, q_0, F)$ has a finite set K of *states*, an *input* alphabet Σ, a *transition* function $\delta : K \times \Sigma \to K$, an *initial* state $q_0 \in K$, and a set F of *accepting* states, $F \subseteq K$. The transition function is extended to $\delta^* : K \times \Sigma^* \to K$ by defining $\delta^*(q,e) = q$ and

$$\delta^*(q, ya) = \delta(\delta^*(q, y), a)$$

for every $q \in K$, $a \in \Sigma$, $y \in \Sigma^*$, and since δ^* agrees with δ on $K \times \Sigma$ the notation is abused by referring to δ on $K \times \Sigma^*$ instead of δ^*.

The *language accepted by D* is

$$L(D) = \{w \in \Sigma^* \mid \delta(q_0, w) \in F\}.$$

A language L is a *regular set* (*regular language*) if there is some deterministic finite-state acceptor D such that L(D) = L.

Intuitively, a finite-state acceptor is a machine which has an input tape (see Figure 1). On the input tape there is a read-only "head" which moves across the tape from left to right, reading the contents of successive tape squares. The set of states and the transition function represent the "logic" of the machine and may be viewed as a program with no variables other than a single input variable which takes values read by the input head.

These definitions describe a finite-state acceptor as an extremely simple model of a computer. The regular sets so described may represent sets of numbers or logical predicates or sets of strings with a simple linguistic or syntactic pattern. In compiler construction one constructs finite-state acceptors to recognize portions of the input as part of the lexical analysis. Notice that the construction of such a device requires only a bounded amount of memory.

While finite-state acceptors are quite simple, the class of regular sets is rich in structure. This is shown in part by the following result.

THEOREM 2.1. The class of regular sets is closed under the Boolean operations, the Kleene operations, substitution, reversal, homomorphism, and inverse homomorphism.

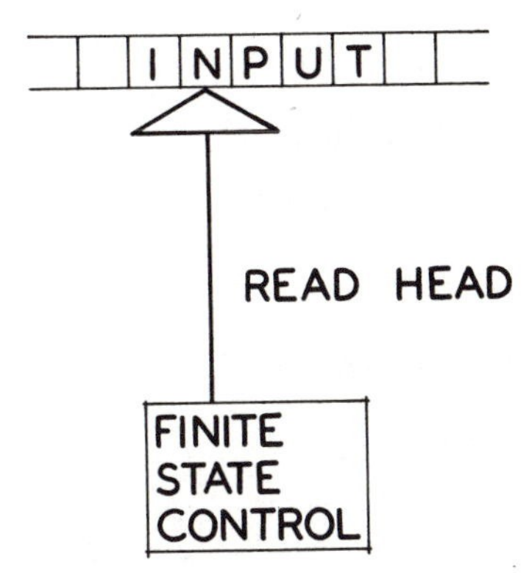

FIG. 1. A finite-state acceptor.

Let us sketch the constructions used in proving that the class of regular sets is closed under the Boolean operations. First, recall that the only available method of specifying a regular set is by specifying a finite-state acceptor which accepts all and only members of that set. For $i = 1, 2$, let

$$M_i = (K_i, \Sigma, \delta_i, q_i, F_i)$$

be a finite-state acceptor. Define

$$\delta : (K_1 \times K_2) \times \Sigma \to K_1 \times K_2$$

by

$$\delta((p_1, p_2), a) = (\delta_1(p_1, a), \delta_2(p_2, a)) \quad \text{for} \quad p_1 \in K_1, p_2 \in K_2, a \in \Sigma.$$

Let

$$F_3 = (K_1 \times F_2) \cup (F_1 \times K_2)$$

and let

$$F_4 = F_1 \times F_2.$$

Then

$$M_3 = (K_1 \times K_2, \Sigma, \delta, (q_1, q_2), F_3)$$

is a finite-state acceptor such that

$$L(M_3) = L(M_1) \cup L(M_2),$$

and

$$M_4 = (K_1 \times K_2, \Sigma, \delta, (q_1, q_2), F_4)$$

is a finite-state acceptor such that

$$L(M_3) = L(M_1) \cap L(M_2).$$

Further, if $F_5 = K_1 - F_1$, then

$$M_5 = (K_1, \Sigma, \delta_1, q_1, F_5)$$

is a finite-state acceptor such that

$$L(M_5) = \Sigma^* - L(M_1).$$

If $M = (K, \Sigma, \delta, q_0, F)$ is a given finite-state acceptor and $w \in \Sigma^*$ is a given input string (see Figure 2), then we can determine whe-

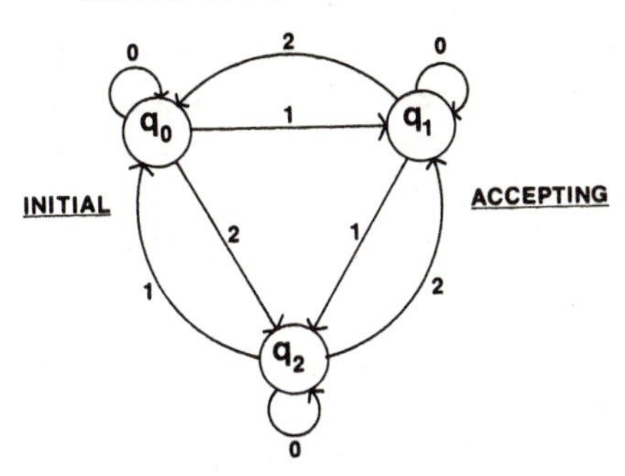

FIG. 2. A finite-state acceptor with initial state q_0 and one accepting state, q_1, such that an input string w in $\{0, 1, 2\}^*$ is accepted if and only if the sum of the digits in w is 1 (mod 3).

ther w is in L(M) by computing $\delta(q_0, w)$ and inspecting F to determine whether $\delta(q_0, w)$ is in F. Further, it is clear that if L(M) is not empty, then there is some string w in L(M) such that $|w| < t$ where t is the number of states in K. Thus, one can determine whether L(M) is empty by checking the finitely many strings in Σ^* with length less than t for membership in L(M). Similarly, one can determine whether L(M) is finite by checking the finitely many strings in Σ^* with length between t and $2t - 1$: L(M) is infinite if and only if there is some w in L(M) such that $t \leq |w| \leq 2t - 1$.

From these comments and the facts about Boolean operations, one can show the following result.

THEOREM 2.2. For each of the following problems, there is an algorithm that provides the solution:

(i) Given a finite-state acceptor M and a string w, does M accept w?
(ii) Given a finite-state acceptor M, is L(M) empty?
(iii) Given a finite-state acceptor M, is L(M) finite?
(iv) Given two finite-state acceptors M_1 and M_2, are $L(M_1)$ and $L(M_2)$ equal?

Let us consider a characterization of the regular sets in terms of closure operations.

THEOREM 2.3. Let Σ be an alphabet. The class of regular sets over Σ is the smallest class containing the finite subsets of Σ^* and closed under union, concatenation, and Kleene*.

The characterization of the regular subsets of Σ^* given in Theorem 2.3 is important since it shows that each regular set can be described by a certain type of formal polynomial, a "regular expression," as well as by a finite-state acceptor.

If Σ is an alphabet, then the class of *regular expressions over* Σ is defined inductively as follows:

(i) (λ) and (0) are regular expressions, where λ, 0, (,) are symbols not in Σ;
(ii) if a is in Σ, then (a) is a regular expression;
(iii) if P and Q are regular expressions over Σ, then so are $(P + Q)$, $(P \cdot Q)$, and (P^*).

The correspondence between regular expressions and regular sets is clear:

(i) (λ) denotes $\{e\}$ and (0) denotes $\varnothing$;
(ii) for $a \in \Sigma$, (a) denotes $\{a\}$;
(iii) if P(Q) is a regular expression denoting the set $\bar{P}(\bar{Q})$, then $(P + Q)$ denotes $\bar{P} \cup \bar{Q}$, $(P \cdot Q)$ denotes $\bar{P}\bar{Q}$, and (P^*) denotes $\bar{P}^*$.

Usually the symbol $\cdot$ and the parentheses are omitted when no ambiguity is introduced.

The fact that a set is regular if and only if it is denoted by a regular expression follows from Theorem 2.3.

Another characterization of the regular sets has become important in the study of the algebraic structure of abstract automata and of formal languages.

Let Σ be an alphabet. A *congruence* relation ρ on Σ^* is an equivalence relation with the property that for any $x, y \in \Sigma^*$, if $x\rho y$, then for all $z, w \in \Sigma^*$, $wxz\rho wyz$. A congruence relation is of *finite index* if it has only finitely many congruence classes.

THEOREM 2.4. Let Σ be an alphabet and let $L \subseteq \Sigma^*$. The following are equivalent:

(i) L is a regular set;
(ii) L is the union of some of the congruence classes of a congruence relation on Σ^* that is of finite index;

(iii) The relation $\equiv$ defined as follows is a congruence relation on Σ^* that is of finite index: for all $x, y \in \Sigma^*$ $x \equiv y$ if and only if for all $w, z \in \Sigma^*$, whenever wxz is in L, then wyz is in L, and conversely.

2.3. At the heart of the study of abstract automata are the basic questions of computability theory: What is an algorithm? What does it mean to say that a function is computable? Logicians have put forth numerous formal models to realize the notion of "algorithm," the model of most interest for the study of automata being the Turing machine.

A *Turing machine* (see Figures 3 and 4) is a device with a finite set of "tape symbols," including a "blank," a finite set of states, a read-write head which operates on a potentially infinite tape, and a finite set of possible operations:

(i) erase a symbol and print a new symbol (overprint);
(ii) change state;
(iii) move right or left one square on the tape or do not move at all;
(iv) halt.

How do Turing machines differ from the finite-state acceptors described in the previous section? First, the head can write (change symbols) as well as read. Second, the head can move left as well as

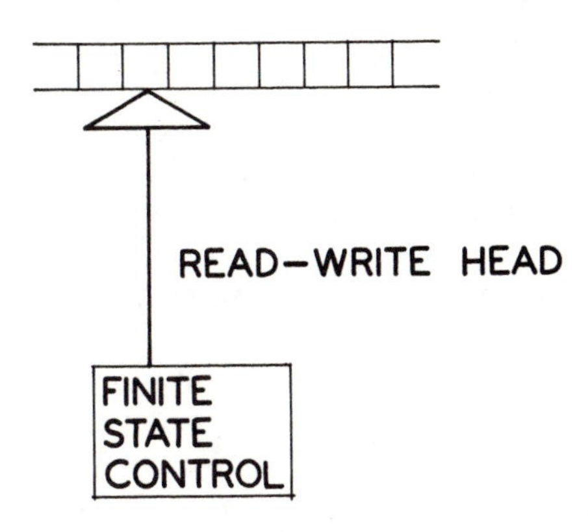

FIG. 3. A Turing machine.

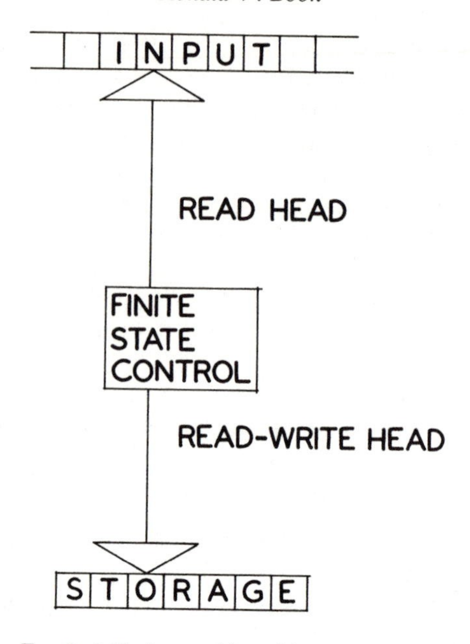

FIG. 4. A Turing machine with one work tape.

right. Third, the head can move beyond the portion of the tape which initially contains the input and can write on initially blank tape squares.

For the sake of the uninitiated reader, one version of Turing machine formalism is presented here.

A *Turing machine* over alphabet Δ is a structure $(K, \Delta, \Gamma, \delta, q_0)$ where K is a finite set of *states*, Δ is a finite alphabet with blank $\beta \in \Delta$, $\Gamma = \Delta \cup \{L, N, R\}$ where L, N, R are three symbols not in $K \cup \Delta$; δ is a function from some subset of $K \times \Delta$, into $\Delta \times \{L, N, R\} \times K$; and $q_0 \in K$ is the initial state. It is assumed that K and Γ are disjoint.

The function δ may be viewed as a finite set of quintuples such that (q, S, T, Z, p) is in the set if and only if $\delta(q, S) = (T, Z, p)$.

An *instantaneous description* (ID) relative to M is a string of the form w_1qw_2 where $w_1, w_2 \in \Delta^*$, $w_2 \neq e$, and $q \in K$.

Let A and B be ID's relative to M. We say A *yields* B and write $A \vdash B$ if for some $S, T \in \Delta - \{\beta\}$, $w, w' \in \Delta^*$, and $q, p \in K$, one of the following occurs:

(1) $\delta(q, S) = (T, N, p)$,
$A = wqSw'$, $B = wpTw'$;
(2) $\delta(q, S) = (T, R, p)$, *either*
$A = wqSw'$, $w' \neq e$, and
$B = wTpw'$, *or* $A = wqS$ and $B = wTp\beta$;
(3) $\delta(q, S) = (T, L, p)$, either
$A = wS'qSw'$, $S' \in \Delta$, and
$B = wpS'Tw'$, or $A = qSw'$ and $B = p\beta Tw'$.

An ID A is a *terminal (halting)* ID if there is no B such that $A \vdash B$.

A *computation of M* is a sequence of ID's $A_0, A_1, \ldots$ such that *either* (1) the sequence is finite, its last member A_n is terminal, and $A_{i-1} \vdash A_i$ for $i = 1, \ldots, n$, or (2) the sequence is infinite and $A_{i-1} \vdash A_i$ for all $i \geq 1$. A computation is *proper* or *improper* according as (1) or (2) is the case. If $A_0, \ldots, A_n$ is proper, then A_n is the *resultant* of the computation beginning with A_0.

For ID's $A_0, A_1, \ldots, A_n$, if $A_{i-1} \vdash A_i$, $i = 1, \ldots, n$, then we write $A_0 \overset{*}{\vdash} A_n$.

It is clear that for any Turing machine $M = (K, \Delta, \Gamma, \delta, q_0)$ all the relevant information about how M will behave is contained in δ. Thus δ may be considered to be the machine itself.

By considering certain instantaneous descriptions of a Turing machine to be initial and interpreting the tape contents of terminal instantaneous descriptions, one sees that a Turing machine computes a (partial) function. It can be shown that the class of all such "Turing computable" functions is precisely the class of functions specified by Post normal systems or by the lambda-calculus of Church or by Markov normal algorithms or by the recursive functions of Gödel, Herbrand, and Kleene.

We shall consider a function to be computable by algorithm if and only if it is computed by a Turing machine that halts on every input. Such a function shall be called *recursive* or *total recursive.* A partial function computed by Turing machine shall be called *partial recursive.*

Here we shall be concerned with the characteristic function of a set. A set is *recursive* if its characteristic function is total recursive.

A set B is *recursively enumerable* if there is a partial recursive function f such that f(b) is defined and equal to 1 when b is in B and f(b) is undefined or is equal to 0 when b is not in B. Thus we may consider Turing machines with certain distinguished accepting states so that a recursive set is the set of inputs accepted by a Turing machine that halts on every input, and a recursively enumerable set is the set of inputs accepted by a Turing machine that may not halt on some of its inputs.

A question or problem or predicate is said to be *decidable (solvable)* if there is an algorithm which will provide the correct answer to every instance of the question; otherwise the question is *undecidable (unsolvable)*. Generally we discuss questions with "yes-no" or "0-1" answers when we consider its decidability. To say that a problem is undecidable is to say that there is no algorithm that will compute its solution on every input. Of course this means that there are infinitely many instances of the problem.

Note that a Turing machine is a finite object. Thus by defining a Turing machine one finitely specifies the set of inputs accepted by the machine even though this set may be infinite, and so one can "name" a recursively enumerable set by specifying a Turing machine that accepts all and only members of that set. Since a Turing machine is a finite object, it can be encoded as a string of symbols in a certain form or as an integer. Such an encoding, usually called a "Gödel numbering," provides an algorithm to map the description of a machine $M = (K, \Delta, \Gamma, \delta, q_0)$ to its encoding and also provides an algorithm to produce the description of M as a set of quintuples from its encoding. This allows one to enumerate the class of all Turing machines or, equivalently, the class of all recursively enumerable sets. Using such an enumeration, one can construct a "universal Turing machine," a Turing machine U such that on inputs e and x, U simulates the computation of the machine with name e on the input encoded by x. Further, using such an enumeration as well as a diagonalization, one can obtain a basic result of computability theory.

THEOREM 2.5. There is no algorithm which when given a description of a Turing machine M and an input x will answer the question "Does M halt when started on input x?"

This question is known as the "Halting Problem for Turing machines" and Theorem 2.5 can be restated as follows: The Halting Problem for Turing machines is undecidable. An equivalent formulation in terms of recursively enumerable sets is as follows: There is no algorithm to answer the question "Is x in L?" for an arbitrary recursively enumerable set L (specified by a Turing machine) and an arbitrary string x. Phrased in this way we say that the "membership problem" for the recursively enumerable sets is undecidable.

There are numerous examples of questions about Turing machines that are undecidable. We list several that occur in various forms in automata theory:

(i) Does a Turing machine halt on all of its inputs? (Equivalently, is a recursively enumerable set actually recursive?)
(ii) The finiteness problem: Does a Turing machine halt on only finitely many inputs? (Is a recursively enumerable set finite?)
(iii) The emptiness problem: Does a Turing machine fail to halt on every input? (Is a recursively enumerable set empty?)
(iv) The equivalence problem: Do two Turing machines accept precisely the same set of strings? (Are two recursively enumerable sets equal?)

Throughout theoretical computer science the question of whether or not a problem is decidable has provided an important theme for study. In this paper we shall use the questions about Turing machines stated above as prototypes for the questions to be asked about the classes of automata, grammars, and languages under investigation. Notice that we have already done this with the class of finite-state acceptors in Theorem 2.2 where the answers turn out to be exactly the opposite of those for Turing machines.

2.4. There are a number of secondary sources that describe finite-state acceptors and regular sets in great detail. In particular, the text by Salomaa [81] is very useful. A fundamental reference in this area is the collection of papers edited by Shannon and McCarthy [88]. Another useful collection (that is unfortunately out of print) is edited by Moore [67]. A particularly important paper for the study of formal languages is that of Rabin and Scott [75].

There are many books in logic which discuss Turing machines

and other formalisms for studying computability as well as the recursive and recursively enumerable sets. Books that place special emphasis on the study of computability include those by Davis [28], Rogers [76], Yasuhara [92], Brainerd and Landweber [17], Hennie [55], and Machtey and Young [63].

3. CONTEXT-FREE GRAMMARS AND LANGUAGES

The study of context-free languages is fundamental to theoretical computer science. Major advances in the use of artificial languages such as programming languages as well as in the study of natural languages came with the realization that formal mathematical machinery was required in order to generate the infinite set of strings of a language. Historically, the notion of a context-free grammar as an important generative structure was developed simultaneously by researchers in programming languages and linguistics.

In this section some of the important features of context-free grammars and languages are described. This development is carried further in Section 4.

3.1. We begin by defining context-free languages as the languages generated by context-free grammars.

A *context-free grammar* is a structure $G = (V, \Sigma, P, S)$ where V is a finite set of symbols called the *alphabet* or the *vocabulary* of the grammar, $\Sigma \subset V$ is the *terminal* alphabet, $S \in (V - \Sigma)$ is the *initial* or *starting symbol* (sometimes called the *axiom* of the grammar), and P is a finite set of ordered pairs, $P \subset (V - \Sigma) \times V^*$. An element of P is a *production* or *rewriting rule* and is written $Z \rightarrow \gamma$ instead of (Z, γ).

The definition of a context-free grammar does not explain how a language is obtained from a grammar; it defines a context-free grammar as a "static" object. To explain the "dynamics" involved in the generation of a language, the notion of "derivation" must be defined.

Let $G = (V, \Sigma, P, S)$ be a context-free grammar. Define a binary relation $\underset{G}{\Rightarrow}$ on V^* as follows: for any $\alpha, \beta \in V^*$ and $Z \rightarrow \gamma \in P$, $\alpha Z \beta \underset{G}{\Rightarrow} \alpha\gamma\beta$. For each $n \geq 0$ define a binary relation $\overset{n}{\underset{G}{\Rightarrow}}$ on V^* as follows: For every $\theta \in V^*$, $\theta \overset{0}{\underset{G}{\Rightarrow}} \theta$; for $\theta_1, \theta_2 \in V^*$, if $\theta_1 \underset{G}{\Rightarrow} \theta_2$, then $\theta_1 \overset{1}{\underset{G}{\Rightarrow}} \theta_2$;

for $\theta_1, \theta_2, \theta_3 \in V^*$, if $\theta_1 \overset{n}{\underset{G}{\Rightarrow}} \theta_2$ and $\theta_2 \overset{1}{\underset{G}{\Rightarrow}} \theta_3$, then $\theta_1 \overset{n+1}{\underset{G}{\Rightarrow}} \theta_3$. Define a binary relation $\overset{*}{\underset{G}{\Rightarrow}}$ on V^* as follows: if $\theta_1, \theta_2 \in V^*$ and $\theta_1 \overset{n}{\underset{G}{\Rightarrow}} \theta_2$ for some $n \geq 0$, then $\theta_1 \overset{*}{\underset{G}{\Rightarrow}} \theta_2$. If $\theta_0, \theta_1, \ldots, \theta_n \in V^*$ and for each $i = 1, \ldots, n$, $\theta_{i-1} \underset{G}{\Rightarrow} \theta_i$, then $\theta_0 \underset{G}{\Rightarrow} \theta_1 \underset{G}{\Rightarrow} \cdots \underset{G}{\Rightarrow} \theta_n$ is a *derivation of length* n *in* G.

The relation $\overset{*}{\underset{G}{\Rightarrow}}$ is the transitive, reflexive closure of the relation $\underset{G}{\Rightarrow}$. We read "$\theta_1$ generates θ_2" for $\theta_1 \overset{*}{\underset{G}{\Rightarrow}} \theta_2$ or "θ_1 generates θ_2 in n steps" for $\theta_1 \overset{n}{\underset{G}{\Rightarrow}} \theta_2$. When no ambiguity can arise, we omit the subscript G and write $\Rightarrow (\overset{n}{\Rightarrow})$ for $\underset{G}{\Rightarrow} (\overset{n}{\underset{G}{\Rightarrow}})$.

If $G = (V, \Sigma, P, S)$ is a context-free grammar, then the *language generated by* G is $L(G) = \{w \in \Sigma^* \mid S \overset{*}{\Rightarrow} w\}$. A string $y \in V^*$ such that $S \overset{*}{\Rightarrow} y$ is a *sentential form of* G.

A language L is a *context-free language* if there is a context-free grammar G such that $L(G) = L$.

It is useful to represent derivations graphically by means of "derivation trees." A tree is a directed acyclic graph with a distinguished node, the root. The root has in-degree 0; all other nodes have in-degree 1 and are accessible from the root. The nodes in a derivation tree are labeled with the symbols used in the derivation. The concept is illustrated in the following examples.

EXAMPLES 3.1.

(a) Let $V = \{a, b, S\}$, $\Sigma = \{a, b\}$, and $P = \{S \to aSb, S \to ab\}$ (see Figure 5). The grammar $G_1 = (V, \Sigma, P, S)$ is such that $L(G_1) = \{a^n b^n \mid n > 0\}$.

(b) Let $V = \{(,), S\}$, $\Sigma = \{(,)\}$, and $P = \{S \to (S), S \to SS, S \to e\}$ (see Figure 6). Let $G_2 = (V, \Sigma, P, S)$. The language $L(G_2)$ is the set of all well-formed strings of correctly balanced parentheses. This language is called the (semi-) Dyck language on one letter.

(c) Let $V = \{a, b, S\}$, $\Sigma = \{a, b\}$, and $P = \{S \to aSa, S \to bSb, S \to e\}$. Let $G_3 = (V, \Sigma, P, S)$. The language $L(G_3) = \{ww^R \mid w \in \{a, b\}^*\}$ is the set of "palindromes" in $\{a, b\}^*$, the set of strings that read the same backwards as forwards.

(d) Let $V = \{(,), [,], S\}$, $\Sigma = \{(,), [,]\}$, and $P = \{S \to (S),$

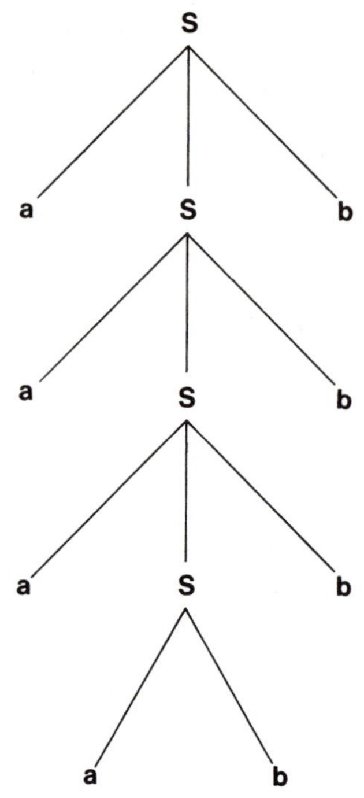

FIG. 5. A derivation tree for the derivation $S \Rightarrow aSb \Rightarrow aaSbb \Rightarrow aaaSbbb \Rightarrow aaaabbbb$ in the grammar G_1 of Example 3.1(a).

$S \rightarrow [S]$, $S \rightarrow SS$, $S \rightarrow e\}$. Let $G_4 = (V, \Sigma, P, S)$. The language $L(G_4)$ is the set of all well-formed strings of two types of parentheses which are nested and correctly balanced. This language is called the (semi-) Dyck language on two letters.

(e) Let $V = \{x, y, z, +, *, (,), S, T, A\}$, $\Sigma = \{x, y, z, +, *, (,)\}$, and $P = \{S \rightarrow A, S \rightarrow (T + T), S \rightarrow (T*T), T \rightarrow (T + T), T \rightarrow (T*T), T \rightarrow A, A \rightarrow x, A \rightarrow y, A \rightarrow z\}$. Let $G_5 = (V, \Sigma, P, S)$. The language $L(G_5)$ is the set of all well-formed expressions over $\{x, y, z\}$ with binary operators $+$ and $*$.

Note that none of the languages in Examples 3.1 is a regular set. On the other hand, one can show that the regular sets are generated by those context-free grammars that are "left-linear": $G = (V, \Sigma, P, S)$ is *left-linear* if each rule is of the form $Z \rightarrow \alpha Y$ where $Z \in V - \Sigma$, $\alpha \in \Sigma^*$, and $Y \in (V - \Sigma) \cup \{e\}$.

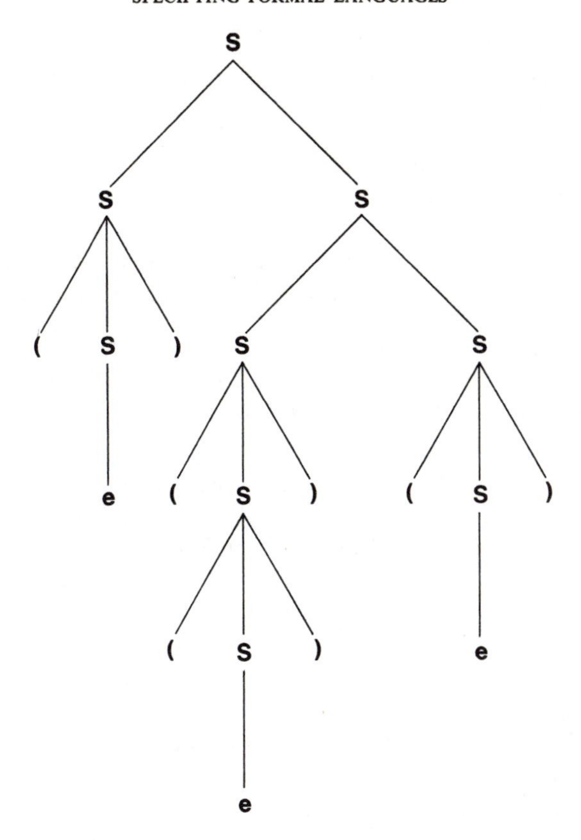

FIG. 6. A derivation tree for the derivation $S \stackrel{*}{\Rightarrow} ()(())()$ in the grammar G_2 of Example 3.1(b).

Representation of derivations by means of derivation trees represents the essential property of being context-free. If two nodes are independent (neither is the descendant of the other), then the subtree rooted at one node represents a derivation that does not depend on the derivation represented by the subtree rooted at the other. This property is represented in terms of derivations by considering "left-to-right" derivations: At every step the leftmost nonterminal symbol is transformed. This derivation corresponds to the construction of a derivation tree by always taking the leftmost possible branch and adjoining the next subtree to the leftmost node of the current frontier if that node is labeled with a nonterminal symbol. Defining this notion formally involves two steps.

(i) Let $G = (V, \Sigma, P, S)$. Define a binary relation $\overset{L}{\Rightarrow}$ on V^* as follows: For any $\alpha, \beta, \gamma \in V^*$, and $Z \in V - \Sigma$, if $\alpha Z \beta \Rightarrow \alpha\gamma\beta$ and $\alpha \in \Sigma^*$, then $\alpha Z \beta \overset{L}{\Rightarrow} \alpha\gamma\beta$.
(ii) Now define $\overset{L,n}{\Rightarrow}$ for each $n \geq 0$ in order to obtain the notion of "left-to-right derivation of length n" and define $\overset{L,*}{\Rightarrow}$ to be the transitive, reflexive closure of $\overset{L}{\Rightarrow}$. (Details are omitted.)

The notion of a left-to-right derivation provides a normal form for derivations of terminal strings from the initial symbol in a context-free grammar. This is seen from the following result.

THEOREM 3.2. Let $G = (V, \Sigma, P, S)$ be a context-free grammar. For any $\rho \in V^*$, $w \in \Sigma^*$, and $n \geq 1$, there is a derivation of w from ρ in G with n steps if and only if there is a left-to-right derivation of w from ρ in G with n steps. Hence, $L(G) = \{w \in \Sigma^* \mid$ there is a left-to-right derivation of w from S in G$\}$.

Proof. It is sufficient to show that if there is a derivation $\rho \Rightarrow \cdots \Rightarrow w$ of length n, then there is a left-to-right derivation $\rho \overset{L}{\Rightarrow} \cdots \overset{L}{\Rightarrow} w$ of length n. The proof is by induction on n.

If $\rho \Rightarrow w$ is a derivation of length 1, then there exist $\alpha, \beta, \gamma \in \Sigma^*$ and $Z \in V - \Sigma$ such that $\rho = \alpha Z \beta$, $w = \alpha\gamma\beta$, and $Z \to \gamma$ is the rewriting rule applied in $\rho \Rightarrow w$. Since Z is the only nonterminal symbol in ρ, the derivation $\rho \Rightarrow w$ is already a left-to-right derivation of length 1.

Assume the result for all $\rho \in V^*$, $w \in \Sigma^*$, and all derivations of length no greater than n (for some $n \geq 1$). Suppose that $\rho_0 \Rightarrow \rho_1 \Rightarrow \cdots \Rightarrow \rho_n \Rightarrow w$ is a derivation of length $n + 1$ with $w \in \Sigma^*$. Now $\rho_1 \Rightarrow \cdots \Rightarrow \rho_n \Rightarrow w$ is a derivation of length n in G so that by the induction hypothesis there is a left-to-right derivation $\rho_1 \overset{L}{\Rightarrow} \theta_2 \overset{L}{\Rightarrow} \cdots \overset{L}{\Rightarrow} \theta_n \overset{L}{\Rightarrow} w$ of length n in G. If $\rho_0 \overset{L}{\Rightarrow} \rho_1$, then $\rho_0 \overset{L}{\Rightarrow} \rho_1 \overset{L}{\Rightarrow} \theta_2 \overset{L}{\Rightarrow} \cdots \overset{L}{\Rightarrow} \theta_n \overset{L}{\Rightarrow} w$ is the desired left-to-right derivation of w from ρ_0 of length $n + 1$. Otherwise, there exist $w_1 \in \Sigma^*$, $y_1, y_2, \gamma \in V^*$, and $Z_1, Z_2 \in V - \Sigma$ such that $\rho_0 = w_1 Z_1 y_1 Z_2 y_2$, $\rho_1 = w_1 Z_1 y_1 \gamma y_2$, and $Z_2 \to \gamma$ is the rewriting rule applied in the derivation $\rho_0 \Rightarrow \rho_1$.

Since Z_1 is the leftmost nonterminal symbol in ρ_1 and $\rho_1 \overset{L}{\Rightarrow} \theta_2$ is the left-to-right derivation, there exists $\beta \in V^*$ such that $\theta_2 = w_1\beta y_1\gamma y_2$ and $Z_1 \to \beta$ is the rewriting rule applied in $\rho_1 \overset{L}{\Rightarrow} \theta_2$. Thus, $\rho_0 = w_1Z_1y_1Z_2y_2 \overset{L}{\Rightarrow} w_1\beta y_1Z_2y_2 \Rightarrow w_1\beta y_1\gamma y_2 = \theta_2$, so that $w_1\beta y_1Z_2y_2 \Rightarrow \theta_2 \overset{L}{\Rightarrow} \cdots \overset{L}{\Rightarrow} \theta_n \overset{L}{\Rightarrow} w$ is a derivation of length n. By the induction hypothesis there is a left-to-right derivation $w_1\beta y_1Z_2 y_2 \overset{L}{\Rightarrow} \Gamma_2 \overset{L}{\Rightarrow} \cdots \overset{L}{\Rightarrow} \Gamma_n \overset{L}{\Rightarrow} w$ of length n in G so that $\rho_0 = w_1Z_1y_1Z_2 y_2 \overset{L}{\Rightarrow} w_1\beta y_1Z_2y_2 \overset{L}{\Rightarrow} \Gamma_2 \overset{L}{\Rightarrow} \cdots \overset{L}{\Rightarrow} \Gamma_n \overset{L}{\Rightarrow} w$ is a left-to-right derivation of length n + 1 that begins with ρ_0 and ends with w. □

3.2. Representation of derivations of context-free grammars by means of derivation trees suggests certain transformations that yield normal forms and grammars. In particular, binary branching derivation trees suggest certain restrictions on the form of the rewriting rules in the grammars.

A context-free grammar G = (V, Σ, P, S) is in *Chomsky Normal Form* if each production in P is of one of the following forms:

$$\left.\begin{array}{l} S \to e \\ Z \to a \\ Z \to Y_1Y_2 \end{array}\right\} a \in \Sigma, \quad Z \in V - \Sigma, \quad Y_1, Y_2 \in V - \Sigma - \{S\}.$$

THEOREM 3.3. From a context-free grammar G_1, one can effectively construct a context-free grammar G_2 in Chomsky Normal Form such that $L(G_2) = L(G_1)$.

If G = (V, Σ, P, S) is a Chomsky Normal Form grammar, then a nonempty string w is in L(G) if and only if there is a derivation of w from S in G of length $2|w| - 1$, and the empty string is in L(G) if and only if $S \to e$ is a rewriting rule of G. Thus we have the following result.

COROLLARY 3.4. There is an algorithm such that given a context-free grammar G and a string w the algorithm determines the answer to the question "Is w in L(G)?" Thus, every context-free language is a recursive set.

In a context-free grammar $G = (V, \Sigma, P, S)$ it is quite common to have certain symbols in $V - \Sigma$ that are "recursive": for some $y_1, y_2 \in V^*$, $y_1 y_2 \neq e$, $Z \overset{*}{\Rightarrow} y_1 Z y_2$. If Z is such that for some nonempty string y, $Z \overset{*}{\Rightarrow} Zy$, then a number of problems can arise. This situation is eliminated when attention is restricted to Greibach Normal Form grammars.

A context-free grammar $G = (V, \Sigma, P, S)$ is in *Greibach Normal Form* if each production in P has one of the following forms:

$$\left.\begin{aligned} &S \to e \\ &Z \to a \\ &Z \to aY_1 \\ &Z \to aY_1Y_2 \end{aligned}\right\} a \in \Sigma, \quad Z \in V - \Sigma, \quad Y_1, Y_2 \in V - \Sigma - \{S\}.$$

THEOREM 3.5. From a context-free grammar G_1 one can effectively construct a context-free grammar G_2 in Greibach Normal Form such that $L(G_2) = L(G_1)$.

Notice that if $G = (V, \Sigma, P, S)$ is a Greibach Normal Form grammar, then a nonempty string w is in L(G) if and only if there is a derivation of w from S in G of length $|w|$.

Given a grammar $G = (V, \Sigma, P, S)$ it may be the case that L(G) is empty or that certain symbols in V or rewriting rules in P are never available for use in derivations of strings in L(G) beginning with S. It is desirable that such symbols and rules be eliminated, and this can be accomplished effectively.

A context-free grammar $G = (V, \Sigma, P, S)$ is *reduced* if either $V = \{S\}$ and $\Sigma = P = \varnothing$ or (i) for each $Y \in V$ there exist $\alpha, \beta \in V^*$ such that $S \overset{*}{\Rightarrow} \alpha Y \beta$ and (ii) for each $Z \in V - \Sigma$ there exists $y \in \Sigma^*$ such that $Z \overset{*}{\Rightarrow} y$.

THEOREM 3.6. From a context-free grammar G_1, one can effectively construct a reduced context-free grammar G_2 such that $L(G_2) = L(G_1)$ and, if G_1 is in Chomsky Normal Form (Greibach Normal Form), then so is G_2.

COROLLARY 3.7. There is an algorithm to determine whether $L(G) = \varnothing$ for a context-free grammar G; that is, the emptiness problem for context-free grammars is decidable.

Consider an arbitrary context-free grammar $G = (V, \Sigma, P, S)$. Let k be the number of symbols in $V - \Sigma$. For any symbol $Z \in V - \Sigma$, consider $S(Z) = \{y \in V^* \mid$ there is a derivation of y from Z of length at most $k\}$. Now if $Y \in V$ is any symbol such that for some $\alpha_1, \alpha_2 \in V^*$, $Z \stackrel{*}{\Rightarrow} \alpha_1 Y \alpha_2$, then there exist $\beta_1, \beta_2 \in V^*$ such that $\beta_1 Y \beta_2$ is in S(Z). This fact is quite useful in proving Theorems 3.3, 3.5, and 3.6 in that it provides a bound on the number of derivations to be considered when testing for certain conditions regarding the rewriting rules of the grammar.

3.3. There is a particularly useful result regarding the "structure" of context-free languages. Its proof depends on simple properties of derivation trees.

THEOREM 3.8. Let L be a context-free language. There exist integers p and q such that every string $w \in L$ satisfying $|w| > p$ may be written as $w = uvxyz$ with $vy \neq e$, $|vxy| \leq q$, and $\{uv^n xy^n z \mid n \geq 0\} \subseteq L$.

Proof. Let $G = (V, \Sigma, P, S)$ be a Chomsky Normal Form grammar such that $L(G) = L$. Let k be the number of nonterminal symbols.

Note that since G is in Chomsky Normal Form, if a derivation starting with any nonterminal symbol has a derivation tree with longest path of length t, then the length of the string generated is at most 2^t, and if the string generated is in Σ^*, then it has length at most 2^{t-1}.

Let $p = 2^{k-1}$ and $q = 2^k$. Suppose that $w \in L(G)$ and $|w| > p$. The longest path in the derivation tree of any derivation of w from S in G has length at least $k + 1$ and so has at least $k + 2$ nodes, $k + 1$ of which are labeled with nonterminal symbols. Thus at least two of these nodes are labeled with the same nonterminal symbol, say $A \in V - \Sigma$ labels nodes n_1 and n_2 with n_2 closer to the leaf than n_1 and with the subpath rooted at n_1 having length at most $k + 1$.

Consider the subtree with root n_1. No path in this tree has length greater than $k + 1$ so that the terminal leaf string has length at most $2^k = q$. Let x be the terminal leaf string generated by the subtree with root at n_2. Let v and y be strings so that vxy is the terminal leaf string generated by the subtree with root at n_1. Thus,

$|vxy| \leq q$. Since G is in Chomsky Normal Form and n_1 and n_2 are in the same path with n_1 above n_2, either $v \neq e$ or $y \neq e$.

Let u and z be strings so that $w = uvxyz$. By considering the portion of the derivation which does not contain the subtree rooted at node n_1, we have $S \overset{*}{\Rightarrow} uAz$. By considering the portion of the derivation from node n_1 to node n_2 we have $A \overset{*}{\Rightarrow} vAy$, and hence for any $n > 0$, $A \overset{*}{\Rightarrow} v^nAy^n$. By considering the portion of the derivation from node n_2, we have $A \overset{*}{\Rightarrow} x$. Thus, $\{uv^nxy^nz \mid n \geq 0\} \subseteq L$.

□

COROLLARY 3.9. For each context-free grammar G, there is an integer $k > 1$ such that L(G) is infinite if and only if L(G) contains a string w such that $k \leq |w| < 2k$. Hence there is an algorithm to determine whether a context-free grammar generates only a finite language, that is, the finiteness problem for context-free grammars is decidable.

Theorem 3.8 is known as the "pumping lemma" for context-free languages and is an example of an "intercalation" theorem. A stronger intercalation theorem for context-free languages is known.

For any set Δ of symbols, any $w \in \Delta^*$ such that $w \neq e$, and any integer i such that $1 \leq i \leq |w|$, the *symbol* a *occurs in the* ith *position of* w if $w = y_1ay_2$ and $|y_1| = i - 1$.

THEOREM 3.10. For each context-free grammar $G = (V, \Sigma, P, S)$, there is an integer $k > 1$ such that for any $w \in L(G)$ with $|w| \geq k$, if any k or more distinct positions in w are designated as distinguished, then there exist $Z \in V - \Sigma$ and $u, v, x, y, z \in \Sigma^*$ such that each of the following conditions is fulfilled:

(i) $S \overset{*}{\Rightarrow} uZz$, $Z \overset{*}{\Rightarrow} vZy$, and $w = uvxyz$;
(ii) x contains at least one of the distinguished positions of w;
(iii) Either both u and v contain distinguished positions, or both y and z contain distinguished positions;
(iv) vxy contains at most k distinguished positions.

These results can be used to show that certain languages are not context-free. For example, none of $\{a^nb^nc^n \mid n > 0\}$, $\{a^nb^mc^nd^m \mid n,$

$m > 0\}$, or $\{wcw^R cw \mid w \in \{a, b\}^*\}$ is context-free but each can be expressed as the intersection of two context-free languages.

COROLLARY 3.11. The class of context-free languages is not closed under intersection.

It should be clear that for an arbitrary context-free grammar G and an arbitrary string w in L(G), there may be more than one derivation of w in G or even more than one left-to-right derivation. Thus we are interested in the "ambiguity" of w in G.

Let $G = (V, \Sigma, P, S)$ be a context-free grammar. A string $w \in L(G)$ is *ambiguous in G* if there exist two distinct left-to-right derivations of w from S in G. The grammar G is an *ambiguous* context-free grammar if there exists a string in L(G) that is ambiguous in G; otherwise, G is *unambiguous*. A context-free language L is an *inherently ambiguous* context-free language if for every context-free grammar G with $L(G) = L$, G is ambiguous; otherwise, L is *unambiguous*.

The following result can be established by using Theorem 3.10.

THEOREM 3.12. There exist inherently ambiguous context-free languages, e.g., $\{a^i b^j c^k \mid i = j \text{ or } j = k\}$.

3.4. At this point the reader will note that the only method of showing that a language is context-free is to exhibit a context-free grammar and to show that the grammar generates the language. To provide another method of showing that a language is context-free as well as to enrich our understanding of this class of languages, we consider closure properties of this class.

THEOREM 3.13. The class of all context-free languages is closed under each of the following operations: union, concatenation, Kleene $*$, intersection with regular sets, inverse homomorphism, reversal, substitution, and arbitrary homomorphic mappings.

The operations given in Theorem 3.13 are not independent; for example, if *L* is any class of languages that contains the regular sets and is closed under substitution, then *L* is closed under union, concatenation, Kleene $*$, and arbitrary homomorphic mappings.

Further, these operations do not characterize "context-free-ness." However, some of these operations can be used to provide such a characterization.

For any $n \geq 1$, let Δ_n be a set of 2n distinct letters, $\Delta_n = \{a_1, \ldots, a_n, \bar{a}_1, \ldots, \bar{a}_n\}$. Let $\sim$ be the congruence on Δ_n^* determined by defining $a_i \bar{a}_i \sim e$ for each $i = 1, \ldots, n$. The *Dyck set* D_n *on* n *letters* is the set $\{w \in \Delta_n^* \mid w \sim e\}$.

Generalizing from D_1 and D_2 in Examples 3.1(b) and 3.1(d), it is clear that for every n, D_n is a context-free language. For any $n \geq 1$, any two Dyck sets on n letters are isomorphic as subsemigroups of free semigroups so that one refers to *the* Dyck set on n letters. Intuitively, D_n is the set of balanced nested strings of matching parentheses of n types.

From the Dyck sets we obtain a characterization of the context-free languages. This result is a version of the "Chomsky-Schützenberger Theorem." It represents an important theme in the mathematical theory of formal languages.

THEOREM 3.14. For each context-free language L there is a regular set R, a nonerasing homomorphism h_1, and a homomorphism h_2 such that $L = h_1(h_2^{-1}(D_2) \cap R)$, where D_2 is the Dyck set on two letters.

Proof. We shall sketch the construction of a regular set R and a homomorphism h_1 such that $h_1(D_t \cap R) = L$ where t is a constant that depends on a grammar generating L. If $\Delta_t = \{a_1, \ldots, a_t, \bar{a}_1, \ldots, \bar{a}_t\}$ and $\Delta_2 = \{a_1, a_2, \bar{a}_1, \bar{a}_2)$, then the homomorphism $h_2 : \Delta_t^* \to \Delta_2^*$ determined by defining $h_2(a_i) = a_1^i a_2$ and $h_2(\bar{a}_i) = \bar{a}_2 \bar{a}_1^i$ for every $i = 1, \ldots, t$ has the property that $h_2^{-1}(D_2) = D_t$. Hence, $h_1(h_2^{-1}(D_2) \cap R) = L$.

Let $G = (V, \Sigma, P, S)$ be a Greibach Normal Form grammar such that $L(G) = L - \{e\}$. For each symbol $Z \in V$, let $\bar{Z}$ be a new symbol and let $\Delta_t = V \cup \{\bar{Z} \mid Z \in V\}$, so that t is the number of symbols in V. Let p and q be two new symbols, $p, q \notin \Delta_t$. Let $G_0 = (\{p, q\} \cup \Delta_t, \Delta_t, P_0, p)$ be the left linear grammar obtained by defining P_0 as follows:

(i) $p \to Sq$ is in P_0;
(ii) for each $Z \in V - \Sigma$, $a \in \Sigma$ such that $Z \to a$ is in P, $q \to a\bar{a}Zq$ is in P_0;

(iii) for each $Z, Y \in V - \Sigma$, $a \in \Sigma$ such that $Z \to aY$ is in P, $q \to a\bar{a}\bar{Z}Yq$ is in P_0;

(iv) for each $Z, Y_1, Y_2 \in V - \Sigma$, $a \in \Sigma$ such that $Z \to aY_1Y_2$ is in P, $q \to a\bar{a}\bar{Z}Y_2Y_1q$ is in P_0;

(v) $q \to e$ is in P_0.

Let $R = L(G_0)$ so that G_0 being a left-linear grammar implies that R is a regular set. Let $h_1 : \Delta_t^* \to \Sigma^*$ be the homomorphism determined by defining $h_1(a) = a$ for $a \in \Sigma$ and $h_1(\bar{a}) = h_1(Z) = h_1(\bar{Z}) = e$ for $a \in \Sigma$, $Z \in V - \Sigma$. By considering left-to-right derivations in G, one can show that $h_1(D_t \cap R) = L - \{e\}$ and if $e \in L$, then $h_1(D_t \cap R') = L$ where $R' = R \cup \{e\}$. By using a technical variation on this construction, h_1 can be made nonerasing. □

Since each Dyck set is a context-free language and the class of context-free languages is closed under inverse homomorphism, homomorphic mappings, and intersection with regular sets, the following result is immediate.

COROLLARY 3.15. The class of context-free languages is the smallest class containing D_2 and closed under homomorphism, inverse homomorphism, and intersection with regular sets.

It should be noted that in Theorem 3.14, if D_2 is replaced by D_1, then we cannot obtain all the context-free languages.

3.5. There is a result of mathematical interest regarding the number of occurrences of individual letters in the strings making up a context-free language.

For any $k > 0$, let $N^{(k)}$ be the set of all k-tuples of natural numbers, so that $N^{(k)}$ is closed under addition by coordinates and under scalar multiplication.

A subset Q of $N^{(k)}$ is *linear* if there exist $\alpha, \beta_1, \ldots, \beta_m \in N^{(k)}$ such that $Q = \{\alpha + n_1\beta_1 + \cdots + n_m\beta_m \mid n_i \in N\}$, and a subset of $N^{(k)}$ is *semi-linear* if it is a finite union of linear sets.

Let Σ be a finite alphabet, say $\Sigma = \{a_i \mid 1 \le i \le k\}$. The *Parikh* *mapping* ψ_k of Σ^* onto $N^{(k)}$ is defined as follows:

$$\psi_k(e) = (0, \ldots, 0);$$

$$\psi_k(a_i) = (Z_{i1}, \ldots, Z_{ik}) \quad \text{where } Z_{ij} = 0 \text{ for } j \neq i \text{ and } Z_{ii} = 1;$$

$$\psi_k(b_1 \ldots b_n) = \sum_{j=1}^{n} \psi_k(b_j) \qquad \text{for } n \geq 1, \text{ each } b_j \in \Sigma.$$

THEOREM 3.16. If L is a context-free language and $L \subseteq \Sigma^*$ where Σ is a finite alphabet, then the image of L under the Parikh mapping $\psi_{\#(\Sigma)}$ is a semi-linear set.

Of course there exist languages that are not context-free but do have a semi-linear image under the Parikh mapping, e.g., $\{a^n b^n c^n \mid n \geq 1\}$.

3.6. It has been noted that certain questions about context-free grammars are decidable, e.g., the question "For a context-free grammar G, is L(G) empty?" However, some other important questions are undecidable. To show that a question about context-free grammars is undecidable, one of two basic techniques is usually employed: reduction to the halting problem (or some other undecidable question) for Turing machines or reduction to the Correspondence Problem.

Let us consider what is involved in reducing a question about context-free grammars to the halting problem for Turing machines. We begin by considering a Turing machine M which has one tape and one read-write head that operates on that tape. Without loss of generality, assume that the computation of machine M on an input string w halts if and only if M accepts w. A finite computation of M on input w can be represented by a sequence of "instantaneous descriptions," strings that describe M's tape contents and finite-state control. Such a sequence $ID_0, ID_1, \ldots, ID_n$ of instantaneous descriptions represent an accepting computation of M on w if ID_0 represents M's initial configuration on input w, ID_n represents an accepting (halting) configuration of M, and for each $j = 1, \ldots, n$, ID_j represents M's tape contents and finite-state control after the transition function acts on the instantaneous description ID_{j-1}. Let Δ be the alphabet containing all the symbols used by M as well as symbols representing the states in M's finite state control. Let # be a symbol not in Δ. For each input string w, one can construct a context-free grammar G (depending only on M and w) such that L(G) is the set of all strings y in $(\Delta \cup \{\#\})^*$ such that y is not $\# ID_0 \# ID_1 \# \ldots \# ID_n \#$ where $ID_0, ID_1, \ldots, ID_n$ represents an accepting computation of M on input w. Thus each y in L(G) either is not of the form of such an encoding of an accepting computation or is of the form of such an encoding but contains a "mistake"; that is, if $y = \# x_0 \# x_1 \# \ldots \# x_n \#$ where each x_i is a string in Δ^* that encodes a configuration of M, then either x_0 is not of the form of an initial configuration of M on w or x_n is not of the form of an accepting configuration of M or for some j, $1 \leq j \leq n$, the string x_j is not of the form of the configuration obtained by applying M's transition function to the configuration represented by x_{j-1}. Due to our assumptions concerning M the computation of M on w halts if and only if M accepts w. Thus, M accepts w if and only if $L(G) = (\Delta \cup \{\#\})^*$. Now $(\Delta \cup \{\#\})^*$ is a regular set and hence L(G) is a context-free language. Thus the question "For context-free grammars G_1 and G_2, is $L(G_1)$ equal to $L(G_2)$?" is undecidable, for otherwise the halting problem for Turing machines would be decidable.

THEOREM 3.17. The equivalence problem for context-free grammars is undecidable.

Variations on the technique described above can be used to obtain the following results.

THEOREM 3.18. Each of the following questions is undecidable:

(a) For a context-free grammar G, is L(G) regular?
(b) For a context-free grammar G and a regular set R, is L(G) = R?
(c) For a context-free grammar G and a finite alphabet Σ, is $\Sigma^* - L(G)$ empty?
(d) For a context-free grammar G, is L(G) co-finite?
(e) For a context-free grammar G, is G ambiguous?

(f) For a context-free grammar G, is L(G) inherently ambiguous?
(g) For context-free grammars G_1 and G_2, is $L(G_1) \cap L(G_2)$ empty?
(h) For context-free grammars G_1 and G_2, is $L(G_1) \cap L(G_2)$ finite?
(i) For context-free grammars G_1 and G_2, is $L(G_1) \subseteq L(G_2)$?

3.7. The concept of a context-free grammar and the language it generates originates with Chomsky [19]–[25], who attempted to develop a reasonable mathematical model for the description of natural language. The theory developed initially through the work of Chomsky and of Bar-Hillel [7]–[9]. Around 1960 it was discovered that the formal description languages used by Backus to specify certain aspects of programming languages were precisely the context-free languages. What has become known as Backus-Naur Form (BNF) was used in the syntactic definition of the language ALGOL 60 [68].

The textbook by Ginsburg [37] is a fairly complete treatment of the theory of context-free languages circa 1966. The books by Salomaa [82] and by Hopcroft and Ullman [58] discuss the class of context-free languages as well as other classes of languages. The textbooks by Aho and Ullman [3], [94] and by Lewis, Rosenkrantz, and Stearns [61] describe those aspects of the theory of context-free languages that are of great use in compiler design.

There are several papers which are particularly useful in tracing the development of the theory of context-free languages. These papers contain some of the results noted in this section. Theorem 3.5 is due to Greibach [45], Theorem 3.8 to Bar-Hillel, Perles, and Shamir [9], Theorem 3.10 to Ogden [71], Theorem 3.14 to Chomsky and Schützenberger [25], and Theorem 3.16 to Parikh [72]. A different viewpoint of the specification of context-free languages is due to Nivat [69].

4. PUSHDOWN STORE ACCEPTORS

Two characterizations of the context-free languages were established in the last section: First, a language was defined to be context-free if it was generated by a context-free grammar; second, it was shown that a language is context-free if and only if it can be represented as $h_1(h_2^{-1}(D_2) \cap R)$ where h_1 and h_2 are homomorphisms, D_2 is the Dyck set on two letters, and R is a regular set. In this section we provide a characterization in terms of a class of abstract automata, the "pushdown store acceptors."

A "pushdown store acceptor" is an automaton with an input tape, a finite set of states, and a last-in first-out data structure: a pushdown store. A pushdown store may be regarded as a one-way infinite tape whose contents can be changed only at one end, the "top." The information obtained from the pushdown store in a single step is the top symbol on the tape. Reading a symbol from the pushdown store automatically erases ("pops") it from the tape. Symbols can be added only at the top and only a bounded number can be added ("pushed down") in any single step. The acceptors have an input tape with a head that moves across the tape from left to right, reading the contents of successive tape squares but never writing. Depending on the current state, the input symbol being scanned, and the symbol on the top of the pushdown store, the transition function determines whether or not a new input symbol is to be read, what the next state is to be, and how the pushdown store is to be altered.

Pushdown store acceptors may be viewed as extensions of finite-state acceptors: add a pushdown store as auxiliary storage to a finite-state acceptor. However, there are several other differences.

A finite-state acceptor reads input from left to right and reads a new input symbol at every step. Any abstract automaton (or Turing machine) with this property is said to operate in *real time*. Notice that if an automaton operates in real time, then a computation on an input string of length n has at most n steps. A pushdown store acceptor need not read a new input at each step; it may perform a sequence of transitions that only change state and manipulate the pushdown store. It can be shown that for each pushdown store acceptor there is a constant k such that a computation on an input string of length n has at most kn steps. Any abstract automaton (or Turing machine) with this property is said to operate in *linear time*.

There is an extremely important difference between the way that a finite-state acceptor operates and the way that a pushdown store acceptor operates. In the definition of finite-state acceptor (and of

Turing machines) given in Section 2, at each step of the computation there is exactly one transition that the machine can make. This means that the machine is "deterministic." However, when studying abstract automata, one considers the "nondeterministic" mode of operation in which there is a finite number of possible transitions and the automaton must "guess" the correct choice of transitions. In this way, nondeterministic automata do not faithfully model the behavior of actual computing machines but do form a mathematical construct that plays an important role in the study of automata and formal languages and of computational complexity. In particular, nondeterministic pushdown store acceptors characterize the class of context-free languages while the deterministic pushdown store acceptors do not.

4.1. We begin the formal definitions by defining deterministic pushdown store acceptors.

A *deterministic pushdown store acceptor* $D = (K, \Sigma, \Gamma, \delta, q_0, F)$ (see Figure 7) has a finite set K of states, an input alphabet Σ, a pushdown alphabet Γ, a (partial) transition function

$$\delta : K \times (\Sigma \cup \{e\}) \times (\Gamma \cup \{e\}) \to K \times \Gamma^*,$$

an initial state $q_0 \in K$, and a set F of accepting states. The transition function is restricted so that for each $q \in K$ and $Z \in \Gamma \cup \{e\}$, either (a) $\delta(q, e, Z)$ is undefined and for each $a \in \Sigma$, $\delta(q, a, Z)$ is defined, or (b) $\delta(q, e, Z)$ is defined and for each $a \in \Sigma$, $\delta(q, a, Z)$ is undefined.

In the definition of a deterministic pushdown store acceptor, the transition function is defined in such a way that, based on the current state and the current "top" of the pushdown store, either a new input symbol is read (part (a)) or a transition involving only change of state and manipulation of the pushdown store is specified (part (b)). This is an essential feature of the definition of a deterministic machine.

Again note that the definition specifies a pushdown store acceptor as a static object. To explain the dynamics of how such an acceptor computes is to explain how the instructions encoded in the transition function are applied to the input tape and the push-

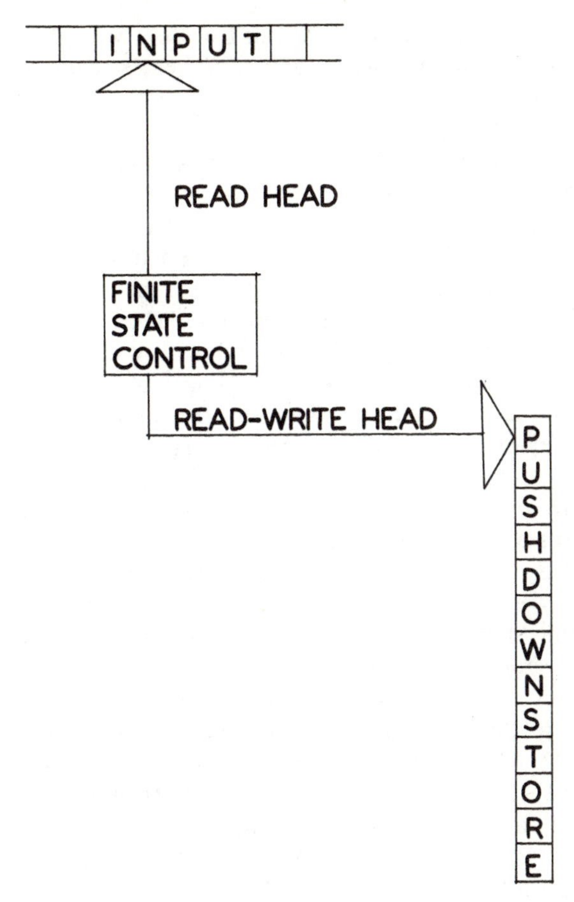

FIG. 7. A pushdown store acceptor.

down store. With a finite-state acceptor it was sufficient to extend the transition function to input strings instead of individual input symbols, but here it is necessary to define "instantaneous descriptions" and a "yield" relation between instantaneous descriptions.

If $D = (K, \Sigma, \Gamma, \delta, q_0, F)$ is a pushdown store acceptor, then an *instantaneous description* of D is an element of $K \times \Sigma^* \times \Gamma^*$. An *initial* instantaneous description is any element of $\{q_0\} \times \Sigma^* \times \{e\}$. Define a binary relation $\vdash$ (read: "yields" or "yields in one step") on the set of instantaneous descriptions of D as follows: for p,

$q \in K$, $a \in \Sigma \cup \{e\}$, $u, v \in \Gamma^*$, $Z \in \Gamma \cup \{e\}$, $w \in \Sigma^*$, if $\delta(q, a, Z) = (p, u)$, then $(q, aw, Zv) \vdash (p, w, uv)$. Let $\overset{*}{\vdash}$ denote the transitive reflexive closure of $\vdash$.

An instantaneous description (q, w, y) is interpreted as a pushdown store acceptor configuration with current state q, input string w, and pushdown store contents y; w represents the input remaining to be processed, and if $w \neq e$, then the leftmost symbol is the currently scanned input symbol; y represents the current pushdown store contents, and if $y \neq e$, then the leftmost symbol of y is the symbol on the "top" of the pushdown store. If $(q, aw, Zy) \vdash (p, w, uy)$, then from state q, while reading $a \in \Sigma \cup \{e\}$ (i.e., either reading $a \in \Sigma$ as input or ignoring the input if $a = e$) with $Z \in \Gamma \cup \{e\}$ on the top of the pushdown store (i.e., the pushdown store is not empty and the symbol $Z \in \Gamma$ is the contents of the top tape square or the pushdown store is empty and $Z = e$), the acceptor goes to state p and replaces Z on the top of the pushdown store with u. If $u = e$, then this transition "pops" Z from the pushdown store; if $u \neq e$, then this transition erases Z and writes u in place of Z, "pushing down" the rightmost $|u| - 1$ symbols of u into the pushdown store (one symbol per tape square) so that the leftmost symbol (top) of the store contains the leftmost symbol of u.

There are three notions of "acceptance" by a pushdown store acceptor $D = (K, \Sigma, \Gamma, \delta, q_0, F)$. Define $L(D) = \{w \in \Sigma^* \mid$ there exists $q \in F$ such that $(q_0, w, e) \overset{*}{\vdash} (q, e, e)\}$, $T(D) = \{w \in \Sigma^* \mid$ there exist $q \in F$ and $u \in \Gamma^*$ such that $(q_0, w, e) \overset{*}{\vdash} (q, e, u)\}$, and $N(D) = \{w \in \Sigma^* \mid$ there exist $q \in K$ such that $(q_0, w, e) \overset{*}{\vdash} (q, e, e)\}$.

For a pushdown store acceptor D, L(D) represents "acceptance by final state and empty store," T(D) represents "acceptance by final state," and N(D) represents "acceptance by empty store."

It is easy to show that for any deterministic pushdown store acceptor D_1, one can construct a deterministic pushdown store acceptor D_2 with the property that $L(D_1) = N(D_2) = T(D_2)$, and that for such a D_1 one can construct a D_3 with the property that $N(D_1) = L(D_3) = T(D_3)$. However, the language $L = \{a\}^* \cup \{a^n b^m \mid n \geq m \geq 0\}$ is such that there is a deterministic pushdown store acceptor D_1 such that $T(D_1) = L$ but there is no acceptor D_2 such that $N(D_2) = L$ or $L(D_2) = L$. Thus, in order to describe the largest class of languages as "deterministic context-free," we choose "acceptance by final state" as the method of acceptance to specify these languages.

A language L is *deterministic context-free* if there is a deterministic pushdown store acceptor D such that $T(D) = L$.

At this point we have not justified the use of the term "deterministic context-free" since we have not shown that T(D) is a context-free language when D is a deterministic pushdown store acceptor. However, this will be an immediate corollary of the characterization of the context-free languages as the languages accepted by nondeterministic pushdown store acceptors.

EXAMPLES 4.1.

(a) Let $\Sigma = \{a, b, c\}$ and $L = \{wcw^R \mid w \in \{a, b\}^*\}$. For $K = \{q_0, q_1, q_2\}$, $\Gamma = \{a, b\}$, $F = \{q_2\}$, and the transition function δ given below, $D = (K, \Sigma, \Gamma, \delta, q_0, F)$ is a deterministic pushdown store acceptor such that $T(D) = L$.

$$\begin{aligned}
\delta(q_0, a, e) &= (q_0, a)\\
\delta(q_0, b, e) &= (q_0, b)\\
\delta(q_0, c, e) &= (q_1, e)\\
\delta(q_0, a, a) &= (q_0, aa)\\
\delta(q_0, a, b) &= (q_0, ab)\\
\delta(q_0, b, a) &= (q_0, ba)\\
\delta(q_0, b, b) &= (q_0, bb)\\
\delta(q_0, c, a) &= (q_1, a)\\
\delta(q_0, c, b) &= (q_1, b)\\
\delta(q_1, a, a) &= (q_1, e)\\
\delta(q_1, b, b) &= (q_1, e)\\
\delta(q_1, e, e) &= (q_2, e)
\end{aligned}$$

From the initial state q_0, D reads a string $w \in \{a, b\}^*$ and copies it onto the pushdown store. When D first reads c, D transfers into state q_1 and then attempts to match the remaining input with the contents of the pushdown store. The input is accepted if and only if the computation ends in state q_2, and D moves into state q_2 if and only if D is in state q_1 with the pushdown store empty (so that the input read before the c is the reversal of the input read after the c).

(b) Let $\Sigma = \{(,), [,]\}$ and let $L \subseteq \Sigma^*$ be the Dyck set on two letters (see Example 3.1(d)). For $K = \{q_0, q_1\}$, $\Gamma = \{(, [\}$, $F = \{q_0\}$, and the transition function given below, $D = (K, \Sigma, \Gamma, \delta, q_0, F)$ is a deterministic pushdown store acceptor such that $T(D) = L$.

$$\begin{aligned}
\delta(q_0, (, e) &= (q_1, ()\\
\delta(q_0, [, e) &= (q_1, [)\\
\delta(q_1, (, () &= (q_1, (()\\
\delta(q_1, [, () &= (q_1, [()\\
\delta(q_1, (, [) &= (q_1, ([)\\
\delta(q_1, [, [) &= (q_1, [[)\\
\delta(q_1,), () &= (q_1, e)\\
\delta(q_1,], [) &= (q_1, e)\\
\delta(q_1, e, e) &= (q_0, e)
\end{aligned}$$

Deterministic pushdown store acceptors "can be made to halt," that is, for every such D_1 one can construct D_2 such that $T(D_2) = T(D_1)$, and on every input string D_2's computation halts. Thus one can show that the complement of a deterministic context-free language is also a deterministic context-free language. This is not a property of the class of all context-free languages, for otherwise the class of context-free languages would be closed under intersection. Hence there are context-free languages that are not deterministic context-free; for example, $L_1 = \{ww^R \mid w \in \{a, b\}^*\}$ and $L_2 = \{a^p b^q c^r \mid p \neq q \text{ or } q \neq r\}$. Finally note that deterministic pushdown store acceptors "can be made to operate in linear time" but that the language $\{w_1 \# w_2 \# \dots \# w_n \# c^{n-i} \# w_i^R \mid n \geq 1,\ 0 \leq i \leq n-1,$ each $w_j \in \{a, b\}^*\}$ is a deterministic context-free language that cannot be accepted in real time by a deterministic pushdown store acceptor (or multitape Turing machine).

4.2. Now we turn to the study of nondeterministic pushdown store acceptors.

A *nondeterministic pushdown store acceptor*

$$D = (K, \Sigma, \Gamma, \delta, q_0, F)$$

has a finite set K of states, an input alphabet Σ, and pushdown alphabet Γ, a transition function δ: $K \times (\Sigma \cup \{e\}) \times (\Gamma \cup \{e\}) \to$ (finite subsets of $K \times \Gamma^*$), an initial state q_0, and a set F of accepting states.

The notion of instantaneous description is defined just as in the deterministic case. For a nondeterministic pushdown store acceptor

$D = (K, \Sigma, \Gamma, \delta, q_0, F)$, define a binary relation $\vdash$ on the set of instantaneous descriptions of D as follows: for $p, q \in K$, $a \in \Sigma \cup \{e\}$, $Z \in \Gamma \cup \{e\}$, $u, v \in \Gamma^*$, $w \in \Sigma^*$, if $(p, u) \in \delta(q, a, Z)$ then $(q, aw, Zv) \vdash (p, w, uv)$. Let $\vdash^*$ denote the transitive reflexive closure of $\vdash$.

The definition of a nondeterministic pushdown store acceptor differs from that of a deterministic acceptor in that the transition function specifies a finite set of possible next moves. To represent all possible computations of a nondeterministic pushdown store acceptor on a given input string, a "computation tree" of instantaneous descriptions is used. A single computation is represented by a path in this tree starting at the root and ending at a leaf.

The three definitions of acceptance given for deterministic pushdown store acceptors are valid for nondeterministic pushdown store acceptors. However, in the case of the nondeterministic model, the three methods of acceptance are of equal power: $\{L(D) \mid D$ is a nondeterministic pushdown store acceptor$\}$ = $\{T(D) \mid D$ is a nondeterministic pushdown store acceptor$\}$ = $\{N(D) \mid D$ is a nondeterministic pushdown store acceptor$\}$.

It must be emphasized that a nondeterministic pushdown store acceptor D accepts a given input string w if and only if there exists an accepting computation in the computation tree of D and w. Some computations of D on w may not end in an accepting configuration even though D accepts w. For D to reject w, all computations of D on w must end in nonaccepting computations. Generally, a nondeterministic pushdown store acceptor D cannot tell whether it rejects an input string w; to determine this, one must "deterministically simulate" all of D's computations and construct the computation tree for D and w.

EXAMPLES 4.2.

(a) Let $\Sigma = \{a, b\}$ and $L = \{ww^R \mid w \in \{a, b\}^*\}$. For $K = \{q_0, q_1, q_2\}$, $\Gamma = \{a, b\}$, $F = \{q_2\}$, and the transition function δ given below, $D = (K, \Sigma, \Gamma, \delta, q_0, F)$ is a nondeterministic pushdown store acceptor such that $T(D) = L(D) = N(D) = L$.

$$\delta(q_0, a, e) = \{(q_0, a)\}$$
$$\delta(q_0, b, e) = \{(q_0, b)\}$$
$$\delta(q_0, a, a) = \{(q_0, aa), (q_1, e)\}$$

$$\begin{aligned}
\delta(q_0, a, b) &= \{(q_0, ab)\}\\
\delta(q_0, b, a) &= \{(q_0, ba)\}\\
\delta(q_0, b, b) &= \{(q_0, bb), (q_1, e)\}\\
\delta(q_1, b, b) &= \{(q_1, e)\}\\
\delta(q_1, a, a) &= \{(q_1, e)\}\\
\delta(q_1, e, e) &= \{(q_2, e)\}
\end{aligned}$$

D reads input symbols from Σ and writes these symbols on the pushdown store while in state q_0. The only use of nondeterminism is the "guess" that half the input has been read and so it is time to transfer control to state q_1 and match the remaining input with the contents of the pushdown store.

(b) Let $\Sigma = \{a, b, c\}$ and $L = \{a^p b^q c^r \mid p \neq q \text{ or } q \neq r\}$. For $K = \{q_i \mid i = 0, \ldots, 9\}$, $\Gamma = \{1\}$, $F = \{q_5\}$, and the transition function δ given below, $D = (K, \Sigma, \Gamma, \delta, q_0, F)$ is a nondeterministic pushdown store acceptor such that $T(D) = L(D) = L$.

$$\begin{aligned}
\delta(q_0, e, e) &= \{(q_1, e), (q_6, e)\}\\
\delta(q_1, a, e) &= \{(q_1, 1)\}\\
\delta(q_1, a, 1) &= \{(q_1, 11)\}\\
\delta(q_1, b, 1) &= \{(q_2, e)\}\\
\delta(q_2, b, 1) &= \{(q_2, e)\}\\
\delta(q_2, b, e) &= \{(q_3, e)\}\\
\delta(q_2, c, 1) &= \{(q_4, e)\}\\
\delta(q_3, b, e) &= \{(q_3, e)\}\\
\delta(q_3, c, e) &= \{(q_4, e)\}\\
\delta(q_4, c, e) &= \{(q_4, e)\}\\
\delta(q_4, c, 1) &= \{(q_4, e)\}\\
\delta(q_4, e, e) &= \{(q_5, e)\}\\
\delta(q_4, e, 1) &= \{(q_4, e)\}\\
\delta(q_6, a, e) &= \{(q_6, e)\}\\
\delta(q_7, b, e) &= \{(q_7, 1)\}\\
\delta(q_7, b, 1) &= \{(q_7, 11)\}\\
\delta(q_7, c, 1) &= \{(q_8, e)\}\\
\delta(q_8, c, 1) &= \{(q_8, e)\}\\
\delta(q_8, c, e) &= \{(q_4, e)\}\\
\delta(q_8, e, 1) &= \{(q_9, e)\}\\
\delta(q_9, e, 1) &= \{(q_9, e)\}\\
\delta(q_9, e, e) &= \{(q_5, e)\}
\end{aligned}$$

In this case D initially guesses that $p \neq q$ (by going into state q_1) or that $q \neq r$ (by going into state q_6). Once this guess is made, D simply checks to see whether the guess is correct.

Now we sketch the proof of the characterization of the context-free languages as the languages accepted by the nondeterministic pushdown store acceptors.

Recall that a language L is context-free if and only if there is a Greibach Normal Form grammar $G = (V, \Sigma, P, S)$ such that $L(G) = L$. From G construct a nondeterministic pushdown store acceptor $D = (K, \Sigma, \Gamma, \delta, q_0, F)$ as follows: $K = \{q_0, q_1, q_2\}$, $\Gamma = \{\$\} \cup (V - \Sigma)$ where \$ is a symbol not in $V - \Sigma$, $F = \{q_2\}$, and δ is given by (i)–(iii).

(i) $\delta(q_0, e, e) = \begin{cases} \{(q_1, S\$), (q_2, e)\} & \text{if } S \to e \text{ is in P,} \\ \{(q_1, S\$)\} & \text{otherwise.} \end{cases}$

(ii) For each $a \in \Sigma$ and $Z \in V - \Sigma$, $\delta(q_1, a, Z) = \{(q_1, \gamma) \mid \gamma \in (V - \Sigma)(V - \Sigma) \cup (V - \Sigma) \cup \{e\}$ and $Z \to a\gamma$ is in $P\}$.

(iii) $\delta(q_1, e, \$) = \{(q_2, e)\}$.

Recall that w is in L(G) if and only if there is a left-to-right derivation of w from S in G. The nondeterministic pushdown store acceptor D is constructed so that its accepting computations simulate the left-to-right derivations of G. At any point in a computation, the symbol on the top of the pushdown store corresponds to the leftmost nonterminal symbol in the string in the corresponding derivation.

This construction yields the following result.

THEOREM 4.3. For every context-free grammar G, one can construct a nondeterministic pushdown store acceptor D such that $L(D) = T(D) = N(D) = L(G)$.

In the construction of the nondeterministic pushdown store acceptor D from the Greibach Normal Form grammar G, the use of the symbol \$ as an "endmarker" for the pushdown store allows one to claim that $L(D) = T(D) = N(D)$. Thus in the next theorem we shall be concerned only with L(D).

We wish to show that if $D = (K, \Sigma, \Gamma, \delta, q_0, F)$ is a nondeterministic pushdown store acceptor, then L(D) is a context-free language.

To accomplish this we show how to construct a context-free grammar G so that the left-to-right derivations of strings in L(G) correspond to the accepting computations of D.

Let $\Delta = (K \times \Gamma \times K) \cup (K \times \{e\} \times K)$. We view the elements of Δ as individual symbols and we assume that $\Sigma \cap \Delta = \emptyset$. Let S be a new symbol, $S \notin \Sigma \cup \Delta$, and let $V = \Sigma \cup \Delta \cup \{S\}$. Define the set P of productions of G as follows:

(i) $\{S \to (q_0, e, p) \mid p \in F\} \subseteq P$;

(ii) For every choice $q, q_1, \ldots, q_{t+1} \in K$, $a \in \Sigma \cup \{e\}$, $Z \in \Gamma \cup \{e\}$, $Y_1, \ldots, Y_t \in \Gamma$, $t \geq 1$ such that $\delta(q, a, Z)$ contains $(p, Y_1 \ldots Y_t)$ where $p = q_{t+1}$, $(q, Z, p) \to a(q_1, Y_1, q_2)(q_2, Y_2, q_3) \cdots (q_t, Y_t, q_{t+1})$ is in P;

(iii) For every $q, p \in K$, $a \in \Sigma \cup \{e\}$, $Z \in \Gamma \cup \{e\}$ such that $\delta(q, a, Z)$ contains (p, e), $(q, Z, p) \to a$ is in P;

(iv) For every $p \in F$, $(p, e, p) \to e$ is in P.

The symbols (p, Z, q) "encode" three pieces of information. The first coordinate represents the current state of a computation of D. The second coordinate represents the symbol on the top of D's pushdown store if the store is not empty ($Z \neq e$) or is the empty word e if the store is empty. The third coordinate represents a "guess" of the state D will have reached when, in a subcomputation starting in state p with Z on the top of the store, this square is emptied for the first time.

The productions in P allow for the simulation by a left-to-right derivation in G of a computation of D. By induction on the length n of the computation, it can be shown that for any $w \in \Sigma^*$, $q \in K$, and $Y_1, \ldots, Y_t \in \Gamma$, $(q_0, w, e) \vdash^* (q, e, Y_1 \ldots Y_t)$ is a computation of length n in D if and only if for every choice $q_1, \ldots, q_{t+1} \in K$ with $q_1 = q$, $(q_0, e, q_{t+1}) \overset{*}{\Rightarrow} w(q_1, Y_1, q_2) \ldots (q_t, Y_t, q_{t+1})$ is a left-to-right derivation of length n in G. Similarly, for any $w \in \Sigma^*$ and $p \in F$, $(q_0, w, e) \vdash^* (p, e, e)$ is a computation in D if and only if $S \Rightarrow (q_0, e, p) \overset{*}{\Rightarrow} w(p, e, p) \Rightarrow w$ is a left-to-right derivation in G.

This construction leads to the following result.

THEOREM 4.4. If D is a nondeterministic pushdown store acceptor, then from D one can construct a context-free grammar G such that L(G) = L(D).

From Theorems 4.3 and 4.4 we have our characterization of the context-free languages as the languages accepted by nondeterministic pushdown store acceptors. Further, the constructive interchange between grammars and acceptors preserves the decidability and undecidability of questions about context-free languages whether these languages are specified by grammars or by acceptors.

4.3. The importance of the pushdown store as a data structure useful in natural language processing was recognized very early (see Kuno [60]). Papers by Chomsky [22], Schützenberger [85], and Evey [32] demonstrated the usefulness of the formal model in studying context-free languages. Hartmanis [52] used pushdown store acceptors to show the undecidability of certain questions about context-free languages. Ginsburg and Greibach [39] did a careful analysis of deterministic pushdown store acceptors.

The use of a pushdown store when studying multitape Turing machines is illustrated in Book and Greibach [15].

Again, the text of Ginsburg [37] is a source bringing together much of the early results on the study of context-free languages by means of pushdown store acceptors. The texts by Aho and Ullman [3], [94] and by Lewis, Rosenkrantz, and Stearns [61] and the dissertation by Brosgol [18] illustrate how deterministic pushdown store machines can be used in compiler construction.

The languages accepted by deterministic pushdown store acceptors are deterministic context-free languages. Grammars to generate these languages are the LR(k) grammars of Knuth [95].

5. PARSING

As noted in Section 4, the class of context-free languages is precisely the class of languages accepted by nondeterministic pushdown store acceptors. For applications in programming and compiling, it is necessary to have an algorithm for recognition and parsing of context-free grammars. Not only must one be able to determine whether or not a given string is generated by a given context-free grammar (recognition) but also when the string is so generated it is necessary to find a derivation tree for it (parsing). Clearly, these tasks can be performed by transforming the given grammar into some kind of standard form, say Chomsky Normal

Form, and then enumerating all derivations from the initial symbol of a certain length (in the case of Chomsky Normal Form, derivations of length $2|w|-1$ where w is the string in question). However, it is quite clear that such a procedure is hopelessly time-consuming and it is necessary to find efficient techniques for recognition and parsing.

A great deal of effort has been expended in the development and analysis of algorithms for context-free recognition and parsing. In this section, one recognition algorithm is sketched.

Let $G = (V, \Sigma, P, S)$ be a context-free grammar in Chomsky Normal Form. Recall that $e \in L(G)$ if and only if $S \to e$ is a production in P, and that all other rules are of the form $Z \to a$ or $Z \to Y_1 Y_2$ with $a \in \Sigma$, $Z \in V - \Sigma$, and $Y_1, Y_2 \in V - \Sigma - \{S\}$. Given a nonempty string $w \in \Sigma^*$, say $w = a_1 \ldots a_n$, a "recognition matrix" for w is as follows: If $V - \Sigma$ contains q symbols, say $V - \Sigma = \{Z_1, Z_2, \ldots, Z_q\}$ with $Z_1 = S$, then let

$$T = [t(i, j, k)]_{i, j = 1}^{n},\ {}_{k=1}^{q}$$

be the three-dimensional binary matrix defined by $t(i, j, k) = 1$ if and only if the substring $a_j \ldots a_{j+i-1}$ can be generated from Z_k. In such a recognition matrix, $t(n, 1, 1) = 1$ if and only if $a_1 \ldots a_n \in L(G)$.

A recognition algorithm will result from an algorithm to construct the recognition matrix T given $a_1 \ldots a_n$ and G. Let us consider a simple iterative algorithm that constructs T. First, note that the $i = 1$ plane is easily constructed by inspecting the set P of productions since $t(1, j, k) = 1$ if and only if $Z_k \to a_j$ is a production in P. Second, note that all other entries are obtained by considering those productions in P of the form $A \to BC$ where $A, B, C \in V - \Sigma$. Thus, supposing that all values $t(i, j, k)$ have been computed for $r < i$ where i is some integer greater than 1 and all j, k, $1 \leq j, k \leq n$, compute $t(i, j, k)$ by taking the Boolean sum

$$\sum_{Z_k \to Z_{k_1} Z_{k_2} \in P} \ \sum_{s=1}^{i-1} t(s, j, k_1) t(i - s, j + s, k_2).$$

One can see that $t(i, j, k) = 1$ if and only if for some production $Z_k \to Z_{k_1} Z_{k_2} \in P$ and some s between 1 and $i - 1$, both $t(s, j, k_1)$

and $t(i - s, j + s, k_2) = 1$. Restating this in terms of the definition of T, $Z_k \overset{*}{\Rightarrow} a_j \dots a_{j+i-1}$ if and only if there exist $Z_{k_1}, Z_{k_2} \in V - \Sigma$ such that $Z_k \to Z_{k_1} Z_{k_2} \in P$ and for some s, $1 \leq s \leq i - 1$, $Z_{k_1} \overset{*}{\Rightarrow} a_j \dots a_{j+s-1}$ and $Z_{k_2} \overset{*}{\Rightarrow} a_{j+s} \dots a_{j+i-1}$.

Often the recognition matrix T is modified to be a two-dimensional matrix R whose entries are subsets of $V - \Sigma$: Z_k is in the entry R(i, j) if and only if T(i, j, k) = 1. For a fixed grammar G, the modified matrix R can be computed from $a_1 \dots a_n$ in a number of steps that is proportional to n^3, and so the question "Is w in L(G)?" can be answered in that amount of time. When the answer is "yes," then a parse can be obtained by searching the matrix.

There are many variations on the recognition algorithm described above, some depending on the form of the grammar, some on the type of machine used in implementing the algorithm, some on the way the matrix is represented. One of the more recent variations employs a different representation of the recognition matrix, a reduction of the computation of the transitive closure of a binary relation to that of Boolean matrix multiplication, and "fast" matrix multiplication techniques to obtain a running time on the order of $n^{2+\varepsilon}$.

A wide variety of parsing techniques are discussed in Aho and Ullman [3], [94] and in Lewis, Rosenkrantz, and Stearns [61]. A careful analysis of the Cocke-Kasami-Younger algorithm presented in this section can be found in Graham and Harrison [51].

6. VARIATIONS ON CONTEXT-FREE LANGUAGES

We have noted three characterizations of the context-free languages: generation by context-free grammars, acceptance by nondeterministic pushdown store acceptors, and "algebraic" representation in terms of a "generator," the Dyck set on two letters. We now consider both extensions and restrictions of the context-free languages based on these three methods of specification. Some of the resulting classes of languages have received considerable attention in the literature.

6.1. Let us consider restrictions of the context-free languages. One approach puts restrictions on the form of the rules of a grammar.

A context-free grammar $G = (V, \Sigma, P, S)$ is *linear context-free* if each production in P is of the form $Z \to \alpha Y \beta$ where $Z \in V - \Sigma$, $\alpha, \beta \in \Sigma^*$, and $Y \in (V - \Sigma) \cup \{e\}$. A language L is a *linear context-free language* if there is a linear context-free grammar G such that $L(G) = L$.

The class of linear context-free languages is a rich class of languages that plays an important role in the algebraic theory of languages as well as in the study of languages specified by multi-tape Turing machines. Every regular set is a linear context-free language, the languages $\{ww^R \mid w \in \{a, b\}^*\}$ and $\{a^n b^n \mid n \geq 0\}$ are linear context-free but not regular, and the Dyck sets are context-free but not linear context-free.

The linear context-free languages have a simple algebraic characterization. Consider a linear context-free grammar $G = (V, \Sigma, P, S)$. Let the productions in P be enumerated as $r_1, \ldots, r_m$, and let $T = \{t_1, \ldots, t_m\}$ be a set of m new symbols. Let $R = \{t_{j_1} \ldots t_{j_n} \mid n \geq 1$, each $t_{j_i} \in T$, there is a derivation $\gamma_0 \Rightarrow \gamma_1 \Rightarrow \cdots \Rightarrow \gamma_n$ in G such that $\gamma_0 = S$ and $\gamma_n \in \Sigma^*$ and for each $i = 1, \ldots, n$, r_{j_i} is the production applied in the step $\gamma_{i-1} \Rightarrow \gamma_i\}$. Let $h_1, h_2 : T^* \to \Sigma^*$ be the homomorphisms determined as follows: For r_i in P, if r_i is $Z \to \alpha Y \beta$, where $Y \in (V - \Sigma) \cup \{e\}$, then $h_1(t_i) = \alpha$ and $h_2(t_i) = \beta$.

It is clear that if $t_{j_1} \ldots t_{j_n}$ is in R, then there is a derivation $S \Rightarrow \gamma_1 \Rightarrow \cdots \Rightarrow \gamma_n$ in G with $\gamma_n \in \Sigma^*$ and $h_1(t_{j_1} \ldots t_{j_n})h_2(t_{j_n} \ldots t_{j_1}) = \gamma_n$. Clearly every string in L(G) can be so represented. Thus, $L(G) = \{h_1(y)h_2(y^R) \mid y \in R\}$. It is easy to see that R is a regular set; in fact, R is the language generated by the left linear grammar $G' = ((V - \Sigma) \cup T, T, P', S)$ where $P' = \{Z \to t_i Y \mid$ the production r_i in P is $Z \to \alpha Y \beta$ where $Y \in V - \Sigma\} \cup \{Z \to t_i \mid$ the production r_i in P is $Z \to \delta$ where $\delta \in \Sigma^*\}$.

The construction sketched above yields one part of the following characterization.

THEOREM 6.1. A language L is linear context-free if and only if there exist homomorphisms h_1 and h_2 and a regular set R such that $L = \{h_1(w)h_2(w^R) \mid w \in R\}$.

Using Theorem 6.1 and the fact that the class of linear context-free languages is closed under certain simple operations, one obtains the following result.

THEOREM 6.2. The class of linear context-free languages is the smallest class containing $\{ww^R \mid w \in \{a, b\}^*\}$ and closed under homomorphism, inverse homomorphism, and intersection with regular sets.

It should be noted that the class of linear context-free languages is not closed under concatenation or under Kleene $*$. For example, $\{a^n b^n a^m b^m \mid n, m \geq 1\}$ is the concatenation of two linear languages but is not itself linear context-free.

Now let us consider a class of restricted nondeterministic pushdown store acceptors that accept precisely the linear context-free languages. These acceptors are restricted by allowing the pushdown store to make only one change from writing (pushing) to erasing (popping) during any computation. Thus, the read-write head on the top of the pushdown store is allowed to make exactly one "turn" or "reversal."

It is easy to see that if L is a linear context-free language, then there is a nondeterministic "one-turn" pushdown store acceptor D with the property that $L(D) = L$. We use the representation $\{h_1(w)h_2(w^R) \mid w \in R\}$ for some regular set R and homomorphisms h_1 and h_2. The pushdown store acceptor D reads the first part of its input as $h_1(w)$, nondeterministically guessing a string w and storing w on its pushdown store while checking that w is in R by simulating a deterministic finite-state acceptor for R in its finite-state control. Then D empties its pushdown store while reading the remainder of its input and checks whether this string is indeed $h_2(w^R)$.

To show that the language accepted by a nondeterministic one-turn pushdown store acceptor is a linear context-free language, one can use a "nondeterministic finite-state transducer," that is, a finite-state acceptor that is nondeterministic and that produces output. The string written on the pushdown store before the pushdown store makes its turn is the output of a nondeterministic finite-state transducer, and the input read after the turn is made is the output of another transducer whose input is the contents of the pushdown store. The class of relations represented by such transducers is closed under composition. If the set of input strings to a nondeterministic finite-state transducer is restricted to a regular set R, then the output is expressible as $h_1(h_2^{-1}(R) \cap R')$ where R' is another regular set and h_1 and h_2 are homomorphisms. Applying Theorems 6.1 and 6.2 in this situation yields the following result.

THEOREM 6.3. A language L is linear context-free if and only if it is accepted by a nondeterministic one-turn pushdown store acceptor.

In characterizing the linear context-free languages we restricted the behavior of pushdown store acceptors by restricting the way that the pushdown store's read-write head could move. Another type of restriction that one can make is to restrict the pushdown store's alphabet to a single letter. The resulting acceptor is called a "one-counter acceptor." The class of languages accepted by such acceptors can be characterized as the smallest class containing the Dyck set on one letter and closed under homomorphism, inverse homomorphism, and intersection with regular sets.

The class of one-counter languages and the class of linear context-free languages are not comparable: $\{ww^R \mid w \in \{a, b\}^*\}$ is not a one-counter language and the Dyck set on one letter is not linear context-free.

If one considers pushdown store acceptors that are counters but also are restricted to be one-turn, then one obtains the smallest class of languages containing $\{a^n b^n \mid n \geq 0\}$ and closed under homomorphism, inverse homomorphism, and intersection with regular sets.

While numerous other subclasses of the context-free languages have been studied, the classes described here are ubiquitous in formal language theory and the methods used in defining and characterizing them are typical of the restrictions imposed throughout the subject.

6.2. Now we turn to the specification of classes of languages that are not all context-free. We begin by considering specification by a generative structure.

A *rewriting system* $G = (V, \Sigma, P, S)$ has an alphabet V, a terminal alphabet $\Sigma \subset V$, an initial symbol $S \in V - \Sigma$, and a finite set P of productions (rewriting rules) of the form

$$w_1 Z_1 w_2 Z_2 \ldots w_t Z_t w_{t+1} \rightarrow w_1 y_1 w_2 y_2 \ldots w_t y_t w_{t+1}$$

where each $w_i \in \Sigma^*$, each $Z_i \in V - \Sigma$, each $y_i \in V^*$, and there is some i such that $y_i \neq Z_i$. Define a binary relation $\Rightarrow$ on V^* as follows: for $\alpha, \beta \in V^*$, if $\rho \rightarrow \theta$ is in P, then $\alpha\rho\beta \Rightarrow \alpha\theta\beta$. Let $\overset{*}{\Rightarrow}$ denote the transitive-reflexive closure of $\Rightarrow$. The *language generated by* G is $L(G) = \{w \in \Sigma^* \mid S \overset{*}{\Rightarrow} w\}$.

The notion of a rewriting system is very general. Productions may rewrite more than one symbol per step and the new parts of the string (i.e., y_i's) may depend upon more than the single corresponding symbol Z_i. It is useful to consider restrictions on the form of the productions in a rewriting system. One restriction is to allow only one symbol to be rewritten in a step. With this restriction, rewriting systems may also be viewed as generalizations of context-free grammars.

This important extension of the notion of context-free grammar is obtained by adding "context," that is, a symbol Z may be rewritten as γ only if Z occurs with the string α on its immediate left and the string β on its immediate right. Thus, a "context-sensitive" rewriting rule has the form $\alpha Z\beta \rightarrow \alpha\gamma\beta$, and we distinguish between the case where "erasing" is allowed, i.e., where γ may be the empty word, and where it is not.

A *context-sensitive* (*without erasing*) *grammar* is a structure $G = (V, \Sigma, P, S)$ where V is an alphabet, $\Sigma \subset V$ is the terminal alphabet, $S \in V - \Sigma$ is the initial symbol, and P is a finite set of productions (rewriting rules) of the form $\alpha Z\beta \rightarrow \alpha\gamma\beta$ where $\alpha, \beta, \gamma \in V^*$, $\gamma \neq e$, and $Z \in V - \Sigma$. If the restriction that $\gamma \neq e$ is dropped, then the grammar is *context-sensitive with erasing.*

Since context-sensitive grammars (with or without erasing) are rewriting systems, the notions of derivation and of language generated by the grammar can be applied.

In a context-sensitive (without erasing) grammar $G = (V, \Sigma, P, S)$, there are no rules that decrease length: if $\rho \rightarrow \theta$ is in P, then $|\rho| \leq |\theta|$. It is clear that if L is a context-free language and $e \notin L$, then there is a context-sensitive grammar G such that $L(G) = L$. The following strategy has been used to extend the definition to include languages containing the empty word. Let $G = (V, \Sigma, P, S)$ be such that $S \rightarrow e$ is in P (so that $e \in L(G)$), $P - \{S \rightarrow e\}$ has no erasing productions, and S does not occur on the right-hand side of any production in P (so that erasing cannot occur by $\alpha Z\beta \Rightarrow \alpha\gamma S\delta\beta \Rightarrow \alpha\gamma\delta\beta$). Grammars with such a restriction are sometimes called "extended" context-sensitive.

A language L is *context-sensitive* (*extended context-sensitive*) if there is a context-sensitive (extended context-sensitive) grammar G such that L(G) = L.

While the restriction that a production of a rewriting system rewrite only one symbol per step forces one to consider only context-sensitive with erasing grammars, there is no weakening of generative capacity. A language is generated by a context-sensitive with erasing grammar if and only if it is generated by an arbitrary rewriting system. A language is context-sensitive if and only if it is generated by a rewriting system in which each production has its right-hand side at least as long as its left-hand side.

The hierarchy of grammars that goes from context-sensitive with erasing to context-sensitive to context-free to left linear context-free (also called finite-state) is known as the Chomsky hierarchy of grammars. The corresponding hierarchy of classes of languages corresponds to the four classes of languages that have been dominant in the theory of formal languages: the recursively enumerable sets, which are the languages generated by the context-sensitive with erasing grammars; the context-sensitive (extended context-sensitive) languages; the context-free languages; and the regular sets, which are the languages generated by left linear context-free grammars.

In Section 2 the class of recursively enumerable languages was defined to be the class of languages accepted by unrestricted Turing machines. We have noted above that the class of recursively enumerable languages is the class of languages generated by context-sensitive with erasing grammars. There are numerous algebraic characterizations of this class but many of those of interest in the study of formal languages and abstract automata stem from characterizations using automata. We will describe some of these characterizations as they arise in our survey of automata.

The class of context-sensitive (or extended context-sensitive) languages can be characterized by certain restricted Turing machines. A *linear-bounded automaton* is a Turing machine that is restricted in such a way that in every computation the machine visits only those tape squares upon which the input is originally written. A language is accepted by a nondeterministic linear-bounded automaton if and only if it is context-sensitive. This provides an automata-theoretic characterization of the context-sensitive languages.

The class of linear-bounded automata has a "universal Turing machine theorem"; that is, there exists a linear-bounded automaton U that will accept suitably encoded inputs M and x where M is a linear-bounded automaton and x is an input accepted by M. From this fact one obtains an algebraic characterization of the context-sensitive languages in the following form.

THEOREM 6.4. There is a context-sensitive language L_0 with the property that for every context-sensitive language L there exist a nonerasing homomorphism h_1, a homomorphism h_2, and a regular set R such that $L = h_1(h_2^{-1}(L_0) \cap R)$.

Both the class of recursively enumerable sets and the class of context-sensitive languages are closed under many of the operations arising in the study of the context-free languages. In particular, both of these classes are closed under union, intersection, and nonerasing homomorphism. The class of recursively enumerable sets is not closed under complementation but it is not known whether the class of context-sensitive languages is closed under complementation. Every context-sensitive language is a recursive set and every recursively enumerable set is the homomorphic image of a context-sensitive language so that the class of context-sensitive languages is not closed under arbitrary homomorphic mappings.

Generalizations of grammars and rewriting systems have been obtained in several ways. One method is to regulate the way productions are applied, say by taking subsets of the set of productions and imposing an order upon the members of each subset. Thus a specified sequence of productions must be applied one after the other, and until the sequence is exhausted, no other productions can be applied. Another method of generalizing these systems has been to allow context to occur anywhere in the string, not necessarily contiguous to the symbol being rewritten.

Within the wide variety of studies of generative systems in the literature, one of the most fruitful in terms of mathematical explorations has been the study of "developmental systems and languages," often called "L-systems." While originally created to model certain phenomena in developmental biology, it has become one of the most vigorous parts of formal language theory. In the traditional notion of a rewriting system, productions are applied sequentially, one production applied at each step. In developmental systems productions are applied in parallel. At each step every symbol that can be rewritten must be rewritten.

The simplest type of developmental system is an "OL system." An *OL system* $G = (V, P, x)$ has an alphabet V, a finite set P of context-free productions, and a string $x \in V^* - \{e\}$, the axiom. A binary relation $\Rightarrow$ on V^* is defined as follows:

If $Z_1, \ldots, Z_k \in V$ and for each $i = 1, \ldots, k$, $Z_i \rightarrow \gamma_i$ is in P, then $Z_1 \ldots Z_k \Rightarrow \gamma_1 \ldots \gamma_k$. The transitive reflexive closure of $\Rightarrow$ is $\overset{*}{\Rightarrow}$. The language generated by G is $L(G) = \{w \in V^* \mid x \overset{*}{\Rightarrow} w\}$.

Thus an OL system has no distinguished terminal symbols and productions are applied in parallel.

Consider the OL system $G = (\{a\}, \{a \rightarrow aa\}, a)$. In this case $L(G) = \{a^{2^n} \mid n \geq 0\}$. Now G has only one production but L(G) is an infinite language that is not context-free. However a rewriting system with only one production generates either a singleton set or the empty set.

There are many variations on the notion of an OL system and both languages and sets of infinite sequences have been studied. Some of the language-theoretic results have been applied in the biological setting that motivated the original study and this area has had serious impact on mathematical studies of growth and form.

6.3. The method of specifying classes of languages that has proved most fruitful for defining new classes is that of acceptance by abstract automata. Starting with the definition of pushdown store acceptors, we can illustrate some of the methods used to define new classes of automata.

Recall that a pushdown store acceptor has an input tape that is read in only one direction, a finite set of states, a pushdown store that provides auxiliary storage, and a transition function that describes how the states change in accordance with the input symbol currently read and the information obtained from the auxiliary storage—in this case, the symbol on the top of the pushdown store. Further, both deterministic and nondeterministic modes of operation are considered. Thus we see that several parameters are involved: (i) the method of reading input; (ii) the type of auxiliary storage; (iii) the mode of operation, (iv) the number of read heads on the input tape. In the case of automata that read the input only in one direction, it is possible to allow still another parameter: (v) whether the transition function may force the acceptor to read a new input symbol at each step, that is, to operate in real time.

Consider a type of automaton obtained by varying one of these parameters. An auxiliary storage tape that has the basic form of a pushdown store but which can be read below the top without erasing is called a *stack*. As with a pushdown store, a stack can be altered only at the top, that is, the read-write head can erase the contents of a tape square only at the top and it can write only at the top, but the read-write head of a stack can visit the interior of the stack in the read-only mode. A *one-way stack acceptor* has an input tape that is read from left to right, a finite set of states, an auxiliary storage tape that is a stack, a transition function, and a set of accepting states. Since the interior of a stack may be read without erasing, a deterministic stack acceptor can accept languages such as $\{a^n b^n c^n \mid n \geq 1\}$ and $\{wcw \mid w \in \{a, b\}^*\}$ that are not context-free.

The class of languages accepted by one-way nondeterministic (deterministic) stack acceptors has many of the same positive and negative closure properties as the class of context-free (deterministic context-free) languages. There is an intercalation theorem for one-way stack acceptors that allows one to show that $\{a^{n^2} b^{n^2} c^{n^2} \mid n \geq 1\}$ is a language not accepted by such an automaton and hence that the class of languages accepted by one-way nondeterministic (or deterministic) stack acceptors is not closed under intersection.

Since a pushdown store acceptor is a one-way stack acceptor, all of the questions about nondeterministic pushdown store acceptors that are undecidable (e.g., Is $L_1 = L_2$? Is L regular?) are also undecidable when asked about one-way nondeterministic stack acceptors. The class of languages accepted by one-way nondeterministic stack acceptors is a class of recursive sets and this class is closed under arbitrary homomorphic mappings. Hence, the emptiness problem is decidable. From the intercalation theorem one can show that finiteness is decidable. As in the case of deterministic pushdown store acceptors, the decidability of the equivalence problem for one-way deterministic stack acceptors (i.e., Are $L(M_1)$ and $L(M_2)$ equal?) is open.

There are restrictions on the definition of one-way stack acceptors that yield classes of automata whose power of acceptance is incomparable to that of the nondeterministic pushdown store acceptors. One such restriction is to force the stack to be "nonerasing," that is, the top symbol of the stack may be replaced by a symbol and symbols may be pushed down, but the length of the stack cannot decrease. The language $\{a^{n^2} \mid n \geq 1\}$ is accepted (by final state) by a one-way deterministic nonerasing stack acceptor, and the Dyck set on two letters is not accepted by any one-way nondeterministic nonerasing stack acceptor. Another restriction is that the stack's contents cannot be altered once the read-write head visits the interior of the stack for the first time; the resulting stack is called a "checking" stack and a one-way deterministic checking-stack acceptor can accept the language $\{a^n b^n c^n \mid n \geq 1\}$.

Just as with pushdown store acceptors, one can restrict the stack alphabet to be a single letter, thereby obtaining a (nonerasing, checking) "stack counter." With any type of stack acceptor one may further consider the number of times the stack's read-write head changes direction, obtaining "finite-turn" ("reversal-bounded") stack acceptors.

Having taken any of those variations of a stack as auxiliary storage, one may consider the different classes of acceptors obtained by varying the parameters (i), (iii), (iv), and (v). In each case the languages so specified are recursive sets.

It is usually the case that a class of languages specified by deterministic acceptors is closed under complementation if the acceptors "can be made to halt," that a class of languages specified by nondeterministic acceptors is closed under union, that a class of languages specified by nondeterministic acceptors with one-way input is closed under nonerasing homomorphism, and that a class of languages specified by acceptors with two-way input is closed under intersection. If no restrictions are placed on the finite set of states, then usually the class of languages specified is closed under intersection with regular sets.

If an acceptor is allowed two separate storage structures, e.g., two pushdown stores or two counters or two stacks, then the resulting acceptor may have the full power of a Turing machine unless some other restriction is imposed. For example, the running time may be restricted by bounding it by a recursive function of the size of the input. The study of abstract automata operating with restrictions on such computational resources as running time or amount of storage space fall into the area of computational complexity which will be briefly described in Section 7.

6.4. Many of the classes of languages studied in the literature share certain positive closure properties. In a number of cases, the proofs that the classes are closed under these operations are remarkably similar. This has led to the development of a theory emphasizing the study of classes of languages defined by specific collections of closure properties, of classes of languages characterized as the smallest class containing a given base and closed under certain operations, and of the relationship between closure properties of classes of languages and characteristic properties of the behavior of abstract automata. This theory is mathematical and much of its impact is on the theory of abstract automata and the theory of formal languages as theories that are part of the mathematical foundations of computer science.

An *abstract family of languages* (AFL) is a class of languages containing at least one nonempty language and closed under nonerasing homomorphism, inverse homomorphism, intersection with regular sets, union, concatenation, and Kleene $+$. A *semi-AFL* is defined in the same way as an AFL except that closure under concatenation and Kleene $+$ is not required. An AFL (semi-AFL) is *full* if it is closed under arbitrary homomorphism. An AFL (semi-AFL) is (full) *principal with generator* L if it is the smallest (full) AFL (semi-AFL) containing the language L.

By drawing upon the many classes of languages specified by grammars or automata for examples, the various subsets of the defining AFL operations have been studied in terms of their relative independence. The class of context-free languages provides an example of a full principal AFL (one possible generator is the Dyck set on two letters). The class of linear context-free languages is a full principal semi-AFL that is not an AFL (since it is not closed under concatenation and Kleene $+$). The class of languages accepted by one-way nondeterministic acceptors that operate in real time and that have a finite number of counters as auxiliary storage (each such acceptor has a fixed finite number of counters, but there is no bound on the number of counters that the acceptors in the class may possess) is an AFL that is not full and is not principal; it is closed under intersection and under those substitutions T such that for any symbol a, T(a) is in the class and T(a) does not contain e.

In the study of abstract automata and formal languages, many types of acceptors have been defined in terms of the parameters described above. The classes of acceptors that specify AFL are "abstract families of acceptors."

A *storage schema* (Γ, I, f, g) has a nonempty set Γ of *storage symbols*, a nonempty set I of *instructions*, a partial function f from $\Gamma^* \times I$ to $\Gamma^* \cup \{\varnothing\}$ that determines how storage is to be altered at each step of an acceptor's computation, and a partial function g from Γ^* to the finite subsets of Γ^* that specifies what information can be obtained from the storage in one step of an acceptor's computation.

The storage manipulation (write) function f is restricted so that for each γ in $\cup\{g(\gamma) \mid \gamma \in \Gamma^*\}$ there is an identity element 1_γ in Γ with the property that $f(\gamma', 1_\gamma) = \gamma'$ for all γ such that γ is in $g(\gamma')$. This condition provides a uniform procedure for allowing the storage to remain unchanged while the acceptor manipulates the finite-state control or reads input.

The storage information (read) function g is restricted so that the empty storage configuration is distinguished from all other storage configurations.

For a storage schema (Γ, I, f, g) the *abstract family of acceptors* (AFA) defined by (Γ, I, f, g) is the class of all acceptors $(K, \Sigma, \delta, q_0, F)$ where K is a finite set of states, Σ is an input alphabet, $q_0 \in K$ is the initial state, $F \subseteq K$ is the set of accepting states, and the transition function δ is a function from $K \times (\Sigma \cup \{e\}) \times g(\Gamma^*)$ into the finite subsets of $K \times I$ such that $\{\gamma \mid \delta(q, a, \gamma) \neq \varnothing$ for some $q \in K$, $a \in \Sigma \cup \{e\}\}$ is finite.

An AFA is a storage schema together with all acceptors with finite-state control and a finite number of input symbols and auxiliary storage specified by the storage schema. Each acceptor has an initial state and a set of accepting states.

The class of nondeterministic pushdown store acceptors defined in Section 4 is an AFA. The storage schema determines that the top of the pushdown store is read at each step of a computation by defining $g(Z_1 \dots Z_n) = Z_1$ for $n \geq 1$, each $Z_i \in \Gamma$, and $g(e) = e$. The storage schema determines that only the top of the pushdown store can be altered. The definitions of instantaneous description and the "compute" relation for nondeterministic pushdown store acceptors provide examples of how these notions are defined for AFA.

For each AFL $\mathscr{L}$ there exists an AFA such that the languages accepted by final state and empty store by those acceptors that (i) are specified by this AFA, and (ii) operate in such a way that there is a fixed finite bound on the number of consecutive transitions that do not read new input, are precisely all and only the languages in $\mathscr{L}$. If the AFL is full, then there is an AFA with the same properties except that condition (ii) may be omitted. Conversely, in just this way every AFA specifies an AFL and every AFA specifies a full AFL when condition (ii) is omitted.

The formal definition of AFA is cumbersome but the concept has given rise to some intuitive descriptions of the behavior of automata that have been quite fruitful in suggesting both techniques and results. The reader should consider the nondeterministic pushdown store acceptors as typifying the concept of AFA.

Since only nondeterministic acceptors with one-way input are specified by the definition of AFA, this theory does not provide a framework for a comprehensive classification of abstract automata studied in the literature. However, aspects of this theory have been very useful in studying classes of languages specified by various automata.

Many questions about classes of languages have been studied in forms of AFL (and AFA): AFL closed under substitution or reversal or intersection, AFL such that every language is semi-linear, AFL characterized as the smallest AFL containing some collection of "bounded" languages (a language L is *bounded* if there is some $k \geq 1$ and some strings $y_1, \dots, y_k$ such that $L \subseteq \{y_1\}^* \dots \{y_k\}^*$), etc. Perhaps the most fruitful application of this theory has been to the study of subclasses of the class of context-free languages and to AFL and semi-AFL that have many of the structural properties of the class of context-free languages. A rich algebraic theory has evolved through the study of operators that produce infinite hierarchies of classes of languages when applied without limit to a base class.

Some of the characterizing properties of AFL and semi-AFL and some of the important examples of AFL and semi-AFL play an essential role in the study of automata-based computational complexity. A different abstract specification of certain classes of languages relating to the basic questions of automata-based computational complexity is described in Section 7.

6.5. The text by Ginsburg [37] offers background on subclasses of context-free languages but for a broader treatment of classes of languages one should see the texts by Salomaa [82] and by Hopcroft and Ullman [58]. Certain motivation for some of this development can be found in Chomsky [21], [23]. The study of L-systems is enjoying great popularity among those working in formal language theory. The text by Herman and Rozenberg [57] provides motivation and much of the basic groundwork. Stack acceptors were introduced by Ginsburg, Greibach, and Harrison [40] and much of the early work on stack languages is summarized in Hopcroft and Ullman [58].

The primary reference on the notion of abstract family of languages is the memoir by Ginsburg, Greibach, and Hopcroft [41] while a secondary reference is the book by Ginsburg [38]. Examples of the rich algebraic theory that has been developed can be found in papers by Goldstine [42]-[44] and by Greibach [46], [49], [50].

7. COMPLEXITY CLASSES OF FORMAL LANGUAGES

The study of computational complexity is currently the most active area within theoretical computer science (see the article by Preparata in this Study for a general discussion of this subject). This area can be approached in different ways: an axiomatic approach, "abstract" complexity, that is a part of recursive function theory; the study of classes of languages accepted (or functions computed) by various models of computation with restricted resources and the relationships between these classes; and the analysis of specific algorithms and the complexity of specific concrete functions or problems in terms of restricted classes of algorithms.

In this section we focus on classes of formal languages specified by abstract automata with restricted computational resources.

7.1. In studying computational complexity one of the first tasks is to define the concept of computational difficulty (or complexity). Such a definition requires a method of specifying or representing algorithms as well as measure of cost that is applicable to that method of specification. For example, if one wishes to define computational difficulty by means of running times of programs, then one must specify the class of programs allowed in terms of their structure and atomic operations (or in terms of a formal programming language) and how individual steps are to be counted, i.e., what a step is and how much time a step takes.

One goal of a theory of computational complexity is to provide a measure (or even a definition) of the "intrinsic" complexity of a function to be computed or of a problem to be solved. Thus one would like to define the complexity of a function in a way that is independent of the method of specification and the applicable measure of difficulty. This goal suggests the need for formal comparisons between different models of computation and their various measures of complexity.

Two quite different types of measures have been studied. One is a "static" measure. The standard example of a static measure is the size of a program (i.e., the number of statements), a parameter that does not depend on the input. On the other hand, a "dynamic" measure such as the running time of a program does depend on the size of the input and thus describes the behavior of computations of the program instead of its structure.

By studying these questions in an abstract framework based on recursive function theory, it has been shown that there exist functions with no intrinsic dynamic complexity: For every program that computes the function, there is another program that computes the same function but runs faster on infinitely many inputs. By studying certain concrete problems, it has been shown that there exist recognition problems whose minimum running times are invariant (up to a certain factor) under wide changes of model.

In the study of formal languages and abstract automata, it appears that the most important questions concern recognition problems that are known to be not only recursive but even primitive recursive (in fact, subelementary). We restrict the discussion here to a description of the basic questions of automata-based computational complexity, focusing on multitape Turing machines as the basic model of computation and placing recursive bounds on the computational resources of time and space.

7.2. The study of automata-based computational complexity focuses on the study of abstract computing devices, their organization and their computational power. This work is aimed at understanding the dependence of computational difficulty on the properties of the computing devices on which the computation is performed. There are many models of computation and many complexity measures that have been studied, but the most influential and widely studied model is the multitape Turing machine with the running time of a computation (i.e., the number of steps in a computation) and the amount of space used in a computation (i.e., the number of memory cells visited) as complexity measures.

Given a model and a measure, one wishes to characterize the power of that model with respect to bounds on the measure. This leads to the study of hierarchies of classes of languages accepted based on hierarchies of recursive functions that bound the measure. In the case of multitape Turing machines, the possibility of hierarchies based on the number of tapes has also been considered.

With the mode of operation as a parameter, the question of the equivalence of the deterministic and nondeterministic modes arises. With the assumption that the nondeterministic mode is usually more powerful (at least when considering classes defined by recursive bounds on the time or space used by multitape Turing machines), the question takes a slightly different form: When using the deterministic mode of operation, what is the additional cost of recognizing a language originally specified by a resource-bounded machine operating in the nondeterministic mode?

Given a model of computation, one may consider tradeoffs between measures. In the case of Turing machines, one wishes to know how much space (time) is necessary to recognize a language originally specified by a time-bounded (space-bounded) Turing machine.

While there are a number of other questions to be considered when studying automata-based complexity, the themes just indicated play a dominant role. We turn to a brief survey of the known (partial) answers to these questions.

First consider time-bounded computation. In this case we often restrict attention to machines that have an input tape that is read from left to right, finite-state control, and some finite number of auxiliary storage tapes. Without further restriction such machines have the power of ordinary Turing machines and so the class of languages accepted by such machines is the class of all recursively enumerable sets. However a recursive bound on the running time forces the language accepted by the machine to be a recursive set. For this discussion such a recursive bound will be monotone increasing and a "real-time countable" function: A function f is *real-time countable* if there is a deterministic multitape Turing machine M such that upon input of length n, M runs for exactly f(n) steps and then halts.

For any real-time countable function f, let DTIME(f) (NTIME(f)) be the class of languages accepted by those deterministic (nondeterministic) multitape Turing machines whose running times are bounded by f. If f(n) = n, then DTIME(f) is the class of *real-time definable* languages and NTIME(f) is the class of *quasi-real-time* languages. The class of languages accepted by those deterministic (nondeterministic) Turing machines whose running times are bounded by a polynomial in the length of the input is denoted by P (NP), so that $P = \bigcup_{k \geq 1} \mathrm{DTIME}(n^k)$ (and $NP = \bigcup_{k \geq 1} \mathrm{NTIME}(n^k)$).

The classes DTIME(f) and NTIME(f) have been defined without restricting the number of auxiliary storage tapes used by the machines. In the case of nondeterministic machines, it is sufficient to restrict attention to machines with only two auxiliary storage tapes (where one may be a pushdown store and the other a stack). In the case of deterministic machines that operate in real time, the number of storage tapes cannot be so restricted: For every integer $k \geq 1$, there is a language that is accepted in real time by a deterministic Turing machine with $k + 1$ auxiliary storage tapes but not by any deterministic machine with only k storage tapes that operates in real time. For other time bounds it is not known whether additional storage tapes provide additional computing power. However, if one is willing to sacrifice time, then one can restrict the number of storage tapes: every language in DTIME(f) can be accepted by a deterministic machine M_1 with only two storage tapes that operate in time f(n) log f(n) and by a machine M_2 with only one storage tape that operates in time $(f(n))^2$.

Given a time bound f, how much larger must g be to ensure that DTIME(g) – DTIME(f) is not empty or that NTIME(g) – NTIME(f) is not empty? Using the method of "enumerate and diagonalize" just as it is used in showing that there are recursively enumerable sets that are not recursive, one can show that if $\lim_{n\to\infty} f(n) \log f(n)/g(n) = 0$, then there is a language in DTIME(g) that is not in DTIME(f). It is known that one can do more in linear time than in real time: there is a deterministic context-free language that is not in DTIME(n). Thus, the case of real-time computation is quite special since

$$\bigcup_{c>0} \mathrm{DTIME}(cn) \neq \mathrm{DTIME}(n).$$

However for any running time f such that $\lim_{n\to 0} n/f(n) = 0$, $\bigcup_{c>0} \mathrm{DTIME}(cf) = \mathrm{DTIME}(f)$.

In the nondeterministic case things are different. For any time bound f, $\mathrm{NTIME}(f) = \bigcup_{c>0} \mathrm{NTIME}(cf)$. Also, if $\lim_{n\to\infty} f(n)/g(n) = 0$ and $\lim_{n\to\infty} \sup f(n+1)/g(n) < \infty$, then $\mathrm{NTIME}(f) \subsetneq \mathrm{NTIME}(g)$.

As noted in the discussion in the previous sections, there are classes of automata for which the deterministic and nondeterministic modes of operation have the same power of computation (e.g., finite-state acceptors, unrestricted Turing machines) and there are classes of automata for which the nondeterministic mode of operation provides more computational power than the deterministic mode (e.g., pushdown store acceptors, one-way stack acceptors). For any time bound f it is easy to see that

$$\mathrm{DTIME}(f) \subseteq \mathrm{NTIME}(f) \subseteq \bigcup_{c>0} \mathrm{DTIME}(2^{cf}).$$

It is not known whether these inclusions are strict for any time bounds at all. Nor is it known whether there is a time bound f such that

$$\bigcup_{k\geq 1} \mathrm{DTIME}(f^k) = \bigcup_{k\geq 1} \mathrm{NTIME}(f^k).$$

While these problems have been studied in automata theory for many years, they have received renewed attention due to the interest today in the study of P and NP and the many attempts to show that P ≠ NP. While many of those studying P and NP do so from the standpoint of concrete complexity, there are a number of questions about formal languages and automata that are closely related. For example, (i) a language L is in NTIME(n) if and only if there exist three deterministic context-free languages L_1, L_2, and L_3 and a nonerasing homomorphism h such that

$$h(L_1 \cap L_2 \cap L_3) = L$$

and (ii) $P = NP$ if and only if P is an AFL if and only if $\text{NTIME}(n) \subseteq P$.

Now let us consider the effect of placing bounds on the amount of space used in a computation. Just as with time bounds, a recursive bound on the amount of space used in a Turing machine's computations forces the language accepted by the machine to be a recursive set. For this discussion such a recursive bound will be monotone increasing and a "tape-constructible" function: a function f such that $f(n) \geq \log n$ is *tape-constructible* if there is a deterministic multitape Turing machine M such that upon input of length n, M marks exactly f(n) tape squares on the some one distinguished storage tape and then halts, while visiting no more than f(n) tape squares on any of its storage tapes.

In the case of space-bounded machines, we shall consider multitape machines with a distinguished input tape that can be read in both directions (that is, an input tape with a read-only head that can move both left and right). When the space bound f(n) is such that $f(n) \geq n$, the ability to read input in both directions during a computation provides no additional computational power. When f(n) does not grow as fast as n (e.g., $f(n) = \log n$), the situation can be quite different with the ability to read input in both directions providing a great deal of additional computational power.

For any tape-constructible function f, let DSPACE(f) (NSPACE(f)) be the class of languages accepted by those deterministic (respectively, nondeterministic) multitape Turing machines whose work space is bounded by f. In the case $f(n) = n$, the machines are called (deterministic or nondeterministic) linear-bounded automata.

The classes DSPACE(f) and NSPACE(f) are defined without restricting the number of auxiliary storage tapes used by the machines. In both the deterministic and nondeterministic cases, it is sufficient to restrict attention to machines with only one auxiliary storage tape.

Given a space bound f, how much larger must g be to ensure that DSPACE(g)−DSPACE(f) is not empty or that NSPACE(g)−NSPACE(f) is not empty? Again using the method of "enumerate and diagonalize," one can show that if $\lim_{n \to \infty} f(n)/g(n) = 0$, then there is a language in DSPACE(g) that is not in DSPACE(f), so that DSPACE(f) $\subsetneq$ DSPACE(g). Combining this result with certain results on "simulation" and "translation," one can show that if $\lim_{n \to \infty} (f(n))^2/g(n) = 0$, then there is a language in NSPACE(g) that is not in NSPACE(f) so that NSPACE(f) $\subsetneq$ NSPACE(g).

When we compare the deterministic and nondeterministic modes of operation for space-bounded computation, again we do not know whether there is a space-bound f such that DSPACE(f) = NSPACE(f). However, it is known that for any space-bound f, NSPACE(f) $\subseteq$ DSPACE(f^2). It is not known whether this inclusion is strict or whether f^2 can be replaced by $f^{1+\epsilon}$ for some $\epsilon < 1$ or by a function such as $f(n) \log f(n)$.

In the case of the space-bound $f(n) = n$, the question of whether DSPACE(n) and NSPACE(n) are equal has received a good deal of attention. The class NSPACE(n) is exactly the class of context-sensitive (without erasing) languages and the question of equality of DSPACE(n) and NSPACE(n) is usually referred to as the "LBA problem" since Turing machines operating in space-bound n are equivalent to linear-bounded automata. Thus, this question is often stated as follows: Is every context-sensitive language accepted by a deterministic linear-bounded automaton?

Since for any space bound f, NSPACE(f) $\subseteq$ DSPACE(f^2), we see that $\bigcup_{k \geq 1}$ DSPACE(n^k) = $\bigcup_{k \geq 1}$ NSPACE(n^k). This class is often referred to as PSPACE.

A good deal of effort has been directed toward the general question of equality of DSPACE(f) and NSPACE(f) by focussing on the special case where $f(n) = \log n$. It is known that DSPACE (log n) = NSPACE(log n) if and only if for all space bounds f, DSPACE(f) = NSPACE(f) (recall that we required a tape-constructible function f to be such that $f(n) \geq \log n$). By using certain encoding techniques and by considering DSPACE(log n) as a class of languages and examining its closure properties, the following statements have been shown to be equivalent: (i) DSPACE (log n) = NSPACE(log n); (ii) DSPACE(log n) is closed under Kleene +; (iii) every linear context-free language is in DSPACE (log n); (iv) every language accepted by a nondeterministic one-counter acceptor is in DSPACE(log n).

7.3. In order to compare classes of languages, it is not always necessary to examine the entire class. Determining the inherent complexity of some one language in the class may provide enough information to determine the complexity of the entire class. For example, to determine whether P equals NP it is sufficient to determine whether one specific language is in P.

A statement in the propositional calculus is in *conjunctive normal form* if it is a conjunction of clauses where each clause is a disjunct of propositional variables and negations of variables. A statement in conjunctive normal form is *satisfiable* if there is an assignment of truth values to the variables under which the entire statement is "true," that is, under which every clause is "true." The set of statements in conjunctive normal form that are satisfiable can be represented as a language L_0 when the propositional variables are encoded as strings over an alphabet.

It is easy to see that L_0 is in NP: A nondeterministic Turing machine (in fact, a nondeterministic one-way stack acceptor) can "guess" an assignment of truth values to the variables in a given statement and then determine whether the assignment yields the value "true" for the statement. All known deterministic algorithms for recognizing L_0 take exponential time; but L_0 has the property that L_0 is in P if and only if $P = NP$. This result is established by showing that for every language L in NP there is a function f_L such that f_L can be computed by a deterministic Turing machine in polynomial time and $f_L^{-1}(L_0) = L$. Thus one can reduce the question "Is w in L?" to the question "Is $f_L(w)$ in L_0?" Suppose that L_0 is in P. Then there is some $k \geq 1$ and some deterministic Turing machine M_0 recognizing L_0 and running in time n^k. For an arbitrary language L in NP, one can construct a deterministic Turing machine M_L such that on input w, M_L first computes $f_L(w)$ and then simulates M_0 on $f_L(w)$, so that M_L accepts w if and only if M_0 accepts $f_L(w)$. Thus, M_L recognizes L and since f_L can be computed in polynomial time, say time n^t, M_L runs in time $|w|^t + |w|^{tk}$. Hence, L is in P.

Any language L in NP with the property that $\{f^{-1}(L) \mid f$ can be computed in polynomial time by a deterministic Turing machine} = NP is called NP-*complete.* Many problems from logic, combinatorial mathematics, operations research, and the theory of automata have been shown to be NP-complete when suitably represented as formal languages.

The discussion above can be summarized as follows.

THEOREM 7.1. (a) Some NP-complete language is in P if and only if P = NP.

(b) The set of conjunctive normal form statements that are satisfiable is NP-complete.

Generally if $\mathscr{L}$ is a class of languages and $\mathscr{C}$ is a class of functions such that $\mathscr{C}^{-1}(\mathscr{L}) \subseteq \mathscr{L}$, then a language L_0 in $\mathscr{L}$ is $\mathscr{L}$-*complete* (or *complete for* $\mathscr{L}$ if $\mathscr{C}^{-1}(\{L_0\}) = \mathscr{L}$.

We shall consider other important classes of languages and problems (or languages) that are complete for these classes.

Recall from Section 2 that questions about regular sets are easily decidable if the sets are specified by deterministic finite-state acceptors. When a regular set is specified by a regular expression, it is usually the case that a deterministic finite-state acceptor is constructed from the expression before testing membership, finiteness, etc. Here we consider the complexity of questions about regular expressions.

If E is a regular expression, let L(E) be the set of strings denoted by E. For any alphabet Σ, let IN(Σ) = {L(E) | E is a regular expression over Σ such that L(E) $\neq \Sigma^*$}.

THEOREM 7.2. The language IN({0,1}) is complete for NSPACE(n), and DSPACE(n) = NSPACE(n) if and only if IN({0, 1}) is in DSPACE(n).

Theorem 7.2 says that testing whether a regular expression over {0, 1} does not denote the set {0, 1}* can be done nondeterministically using linear space and that this test can always be made deterministically in linear space if and only if every context-sensitive language (that is, every language in NSPACE(n)) can be accepted by a deterministic linear-bounded automaton.

The proof of Theorem 7.2 can be outlined as follows. Given a nondeterministic Turing machine M with exactly one tape and one read-write head such that during a computation only that portion of the tape where the input is originally written is used, and given an input string w, construct a regular expression E over an alphabet Σ such that w is accepted by M if and only if $L(E) \neq \Sigma^*$. There is an algorithm to compute E from M and w that operates in linear time so that the length of the expression E is bounded by a constant multiple of the length of w. The regular expression E has the form $E_1 + E_2 + E_3$ where (i) the regular expression E_1 denotes the set of strings that cannot encode accepting computations of M on w because they do not begin with the initial instantaneous description of M on w, (ii) the regular expression E_2 denotes the set of strings that cannot encode accepting computations of M because they do not end with an accepting instantaneous description, and (iii) the regular expression E_3 denotes the set of strings that cannot encode accepting computations of M because they are not of the form of a sequence of instantaneous descriptions such that each follows from the previous one. Recall from Section 3 that the set of "noncomputations" of a one-tape one-head Turing machine is a context-free language; this fact was used to show that the question "For a context-free grammar G, is L(G) equal to Σ^*?" is undecidable. Here the fact that the linear-bounded automaton M visits no more than $|w|$ tape squares in its computations on w allows one to construct the regular expression E_3. Note that since w is given and M is a linear-bounded automaton, each instantaneous description in any of M's computations on w is of length $|w| + 1$. The fact that it is decidable whether M accepts w is translated to the fact that it is decidable whether E is equivalent to Σ^*, and Theorem 7.2 shows that this decision problem "costs" nondeterministic linear space.

It is easy to see that there is a function f that transforms any regular expression E over Σ to a regular expression f(E) over $\{0, 1\}$ such that $L(f(E)) \neq \{0, 1\}^*$ if and only if $L(E) \neq \Sigma^*$; f simply encodes symbols from Σ as strings in $\{0, 1\}^*$. The function f can be computed in linear time by a deterministic Turing machine so that for any regular expression E, the expression f(E) has length bounded by a constant multiple of the length of E. Thus, "Is L(E) not equal to Σ^*?" and "Is L(f(E)) not equal to $\{0, 1\}^*$?" both have the same space complexity.

An algorithm that runs in linear time uses at most linear space. Hence, IN($\{0, 1\}$) is in DSPACE(n) if and only if DSPACE(n) = NSPACE(n).

For any alphabet Σ, Σ^* is a regular set; so Theorem 7.2 shows that the question of inequivalence of regular expressions requires at least nondeterministic linear space. In fact, this question is also complete for NSPACE(n).

There are restrictions of the problem of inequivalence of regular expressions that are of interest. In particular, if the regular expressions contain no occurrence of $*$ or if the alphabet is simply $\{0\}$, then the question of inequivalence is NP-complete.

All of the questions considered so far have been directed toward nondeterministic computation: One obtains either an answer "yes" (or "accept") or no answer at all. Thus we see that NP is closed under complementation if and only if the set of conjunctive normal form statements that are not satisfiable is in NP, and the class of context-sensitive languages, NSPACE(n), is closed under complementation if and only if $\{E \mid E$ is a regular expression over $\{0, 1\}$ such that $L(E) = \{0, 1\}^*\}$ is context-sensitive. These techniques have also been used to compare time and space.

Now we consider briefly the complexity of questions about context-free languages. As noted in Section 5, the membership question for a language specified by a Chomsky Normal Form grammar is solvable in time $n^{2+\varepsilon}$, that is, every context-free language is in DTIME($n^{2+\varepsilon}$). Also, it is known that every context-free language is in DSPACE($(\log n)^2$) and that there is a linear context-free language L such that L is in DSPACE(log n) if and only if DSPACE(log n) = NSPACE(log n).

There is a "hardest" context-free language, a language L_0 such that for any time-bound f, L_0 is in DTIME(f) if and only if every context-free language is in DTIME(f), and for any space-bound g, L_0 is in DSPACE(g) (NSPACE(g)). This language, a nondeterministic version of the Dyck set D_2, is complete for the class of context-free languages, where the function used to "reduce" an arbitrary context-free language to L_0 is a homomorphism. That is, $\{h^{-1}(L_0), h^{-1}(L_0 \cup \{e\}) \mid h$ is a homomorphism$\}$ is the class of all context-free languages.

A class of languages closed under inverse homomorphism and intersection with regular sets is a *cylinder*. The smallest cylinder containing a given language L_0 is a *principal* cylinder with generator L_0. If $\mathscr{L}$ is a principal cylinder with generator L_0, then the time or space complexity of the membership problem for a language in $\mathscr{L}$ is determined by the complexity of the membership problem for L_0. Thus the class of context-free languages is a principal cylinder, as are DSPACE(n^k), NSPACE(n^k), and NTIME(n^k) for any k. On the other hand, the class of deterministic context-free languages and the class of linear context-free languages are not principal cylinders, nor are P and NP.

Returning to the context-free languages, recall that the emptiness problem is decidable. It can be shown that this question (when suitably encoded) is complete for P; in this case, the functions used are computed by deterministic Turing machines that use at most log n space.

7.4. Borodin [16] has provided an excellent overview of the study of computational complexity. The text by Aho, Hopcroft, and Ullman [4] touches upon many of the topics studied in this field. A pioneering paper by Hartmanis and Stearns [54] still plays an important role in the study of automata-based complexity.

The real-time definable languages were studied by Rosenberg [78] and the quasi-real-time languages by Book and Greibach [15]. The hierarchy for deterministic real-time machines based on the number of tapes was suggested by Rabin [74] and established by Aanderaa [1]. The hierarchy for deterministic machines based on running times was established by Hartmanis and Stearns [54] and improved by Hennie and Stearns [56], while for nondeterministic machines Seiferas [86] strengthened the initial results of Cook [27].

The importance of the "P = ?NP" question was pointed out by Cook [26] and underscored by Karp [59]. Some connections between language theory and the "P = ?NP" question can be found in Book [11].

Classes specified by space bounds have been extensively studied. Hierarchy results can be found in Hartmanis and Stearns [54], Stearns, Hartmanis, and Lewis [89], and Seiferas [86], [87]. Savitch [83] established the best result on deterministic simulation of nondeterministic machines known to date. Translations between classes specified by different space bounds can be found in Savitch [83], [84] and Book [12]–[14]. Problems relating to "DSPACE (log n) = ?NSPACE(log n)" have been studied by Sudborough [90] and Monien [66].

The inherent complexity of questions about regular expressions was explored by Meyer and Stockmeyer [65] and extended by

Hartmanis and Hunt [53]. Greibach [47] established the existence of a hardest context-free language. Properties of cylinders and principal cylinders can be found in Autebert [5] and Boasson and Nivat [10].

The paper by Wrathall [97] is an excellent example of the use of language theory in studying computational complexity.

8. CONCLUDING REMARKS

Some of the topics described in this paper can be treated from quite different points of view. For example, Eilenberg [29] presents certain topics in formal language theory as part of algebra. But in addition, it must be noted that there are a great many topics that simply cannot be covered here. Some of these are considered part of the core of formal language theory and some are considered as areas to which ideas from language theory can be successfully applied. A very few such topics are listed here.

In studies of the semantics of programming languages and of program correctness and verification, program schemes are quite useful. For an introduction to this topic and how it relates to language theory, see the papers by Manna [64], Luckham, Park and Paterson [62], Friedman [33], [34], Garland and Luckham [36], and Nivat [70]. The text by Greibach [48] is quite useful in learning about this area.

In addition to studying languages as sets of strings, one often wishes to study the "structure" of these strings, e.g., the derivation tree of a derivation from a context-free grammar. Thus the study of tree automata and tree languages has developed from a topic useful in mathematical logic to an important topic within theoretical computer science. Thatcher [91] provides an introduction to this topic and the interested reader should also see the papers by Engelfriet [30], [31], Perrault [73], Rosen [77], and Rounds [80], and the dissertation by Baker [6].

Formal grammars can be used as a mechanism to describe the generating of graphs and other two-dimensional objects. In this way they play an important role in the development of programming languages useful for picture-processing systems and for syntactic pattern recognition. In this area the books by Rosenfeld [79] and Fu [35] are of value.

Other areas ot computer science and applied mathematics which have found useful tools in formal language theory are described in a collection of papers edited by Yeh [93].

Papers in formal language theory are published in a wide variety of journals, but in particular in the *Journal of Computer and System Sciences, Mathematical Systems Theory,* and *Theoretical Computer Science* with occasional papers in the *Journal of the Association for Computing Machinery, SIAM Journal of Computing, Acta Informatica,* and *Information and Control.* Many of the results are reported in various symposia but particularly in the annual IEEE Symposium on Foundations of Computer Science (formerly, the IEEE Symposium on Switching and Automata Theory) and in the International Colloquium on Automata, Languages, and Programming, sponsored by the European Association for Theoretical Computer Science. Papers in formal language theory have also been presented at the annual ACM Symposium on Theory of Computing and the Conference on Information Science and Systems.

Since this paper was first written, a number of important publications in formal language theory have appeared. Harrison [98] has written an introduction to many aspects of the field. Greibach [99] has written a scholarly history of the early development of formal language theory that illustrates how it developed from computational linguistics into a part of theoretical computer science. A book by Berstel [100] develops the theory of context-free languages from the algebraic standpoint by using the notion of rational transduction. Rozenberg and Salomaa [101] provide a mathematical development of L-systems, and Wood [102] shows how abstract notions of a grammar are related to L-systems. Salomaa [103] has brought together a number of combinatorial properties of formal languages in a delightful way. A symposium on formal language theory was held in December 1979 with the goal of putting past work into perspective and reporting on the status of open problems. The Proceedings [104] contains the texts of thirteen invited lectures which speak to these goals.

REFERENCES

1. S. Aanderaa, "On k-tape versus (k − 1)-tape real time computation," in *Complexity of Computation*, R. Karp, ed., SIAM-AMS Proc., VII, Amer. Math. Soc., 1973, pp. 75–96.
2. A. Aho, ed., *Currents in the Theory of Computing*, Prentice-Hall, Englewood Cliffs, N. J., 1973.

3. A. Aho and J. Ullman, *The Theory of Parsing, Translation, and Compiling* (two vols.), Prentice-Hall, Englewood Cliffs, N. J., 1972.
4. A. Aho, J. Hopcroft, and J. Ullman, *The Design and Analysis of Computer Algorithms*, Addison-Wesley, Reading, Mass., 1976.
5. J. Autebert, "Non principalité du cylindre des langages à compteur," *Math. Systems Theory*, **11** (1978), 157–167.
6. B. Baker, "Tree Transductions and Families of Tree Languages," Ph.D. dissertation, Harvard University, 1973.
7. Y. Bar-Hillel, *Language and Information*, Addison-Wesley, Reading, Mass., 1964.
8. Y. Bar-Hillel, C. Gairman, and E. Shamir, "On categorical and phrase-structure grammars," *Bull. Res. Council Israel, Sec. F*, **9** (1960), 155–166. Also in [7], 99–115.
9. Y. Bar-Hillel, M. Perles, and E. Shamir, "On formal properties of simple phrase structure grammars," *Z. Phonetik, Sprachwissenschaft und Kommunikations forschung*, **14** (1961), 143–172. Also in [7], 116–150.
10. L. Boasson and M. Nivat, "Le cylindre des langages linéaires," *Math. Systems Theory*, **11** (1978), 147–155.
11. R. Book, "On languages accepted in polynomial time," *SIAM J. Computing*, **1** (1972), 281–287.
12. ———, "Comparing complexity classes," *J. Comput. System Sci.*, **9** (1974), 213–229.
13. ———, "Tally languages and complexity classes," *Info. Control*, **26** (1974), 186–193.
14. ———, "Translational lemmas, polynomial time, and $(\log n)^j$-space," *Theoretical Comput. Sci.*, **1** (1975), 215–226.
15. R. Book and S. Greibach, "Quasi-realtime languages," *Math. Systems Theory*, **4** (1970), 97–111.
16. A. Borodin, "Computational complexity; theory and practice," in [2], 35–89.
17. W. Brainerd and L. Landweber, *Theory of Computation*, Wiley, New York, 1974.
18. B. Brosgol, "Deterministic Translation Grammars," Ph.D. dissertation, Harvard University, 1973.
19. N. Chomsky, "Three models for the description of language," *IRE Trans. Info. Theory*, **IT2** (1956), 113–124.
20. ———, *Syntactic Structures*, Mouton, The Hague, 1957.
21. ———, "On certain formal properties of grammars," *Info. Control*, **2** (1959), 137–167.
22. ———, "Context-free grammars and pushdown storage," *M.I.T. Res. Lab. Electron. Quart. Prog. Report*, 65, 1962.
23. ———, "Formal properties of grammars," in *Handbook of Math. Psychology*, D. Luce, R. Bush, and E. Galanter, eds., Wiley, New York, 1963, pp. 323–418.
24. N. Chomsky and G. Miller, "Finite state languages," *Info. Control*, **1** (1958), 91–112.
25. N. Chomsky and M. Schützenberger, "The theory of context-free languages," in *Computer Programming and Formal Systems*, P. Braffort and D. Hirschberg, eds., North-Holland, Amsterdam, 1963, pp. 118–161.

26. S. Cook, "The complexity of theorem-proving procedures," *Proc. Third ACM Symp. Theory Comput.*, 1971, pp. 187–192.
27. S. Cook, "A hierarchy for nondeterministic time complexity," *J. Comput. System Sci.*, **7** (1973), 343–354.
28. M. Davis, *Computability and Unsolvability*, McGraw-Hill, New York, 1958.
29. S. Eilenberg, *Automata, Languages, and Machines*, Academic Press, New York, 1974.
30. J. Engelfriet, "Bottom-up and top-down tree transformations—A survey," *Math. Systems Theory*, **9** (1975), 198–231.
31. J. Engelfriet, "Top-down tree transducers with regular look-ahead," *Math. Systems Theory*, **10** (1977), 289–303.
32. R. Evey, "Theory and Application of Pushdown Store Machines," Ph.D. dissertation, Harvard University, 1963.
33. E. Friedman, "The inclusion problem for simple languages," *Theoretical Comput. Sci.*, **1** (1976), 297–316.
34. E. Friedman, "Equivalence problems for deterministic context-free languages and monadic recursion schemes," *J. Comput. System Sci.*, **14** (1977), 344–359.
35. K. Fu, *Syntactic Methods in Pattern Recognition*, Academic Press, New York, 1974.
36. S. Garland and D. Luckham, "Program schemes, recursion schemes, and formal languages," *J. Comput. System Sci.*, **7** (1973), 119–160.
37. S. Ginsburg, *The Mathematical Theory of Context-free Languages*, McGraw-Hill, New York, 1966.
38. S. Ginsburg, *Algebraic and Automata-Theoretic Properties of Formal Languages*, North-Holland, Amsterdam, 1975.
39. S. Ginsburg and S. Greibach, "Deterministic context-free languages," *Info. Control*, **9** (1966), 620–648.
40. S. Ginsburg, S. Greibach, and M. Harrison, "Stack automata and compiling," *J. Assoc. Comput. Mach.*, **14** (1967), 172–201.
41. S. Ginsburg, S. Greibach, and J. Hopcroft, *Studies in Abstract Families of Languages*, Mem. Amer. Math. Soc., no. 87, 1969.
42. J. Goldstine, "Substitution and bounded languages," *J. Comput. System Sci.*, **6** (1972), 9–29.
43. ———, "Independent sets of one-letter languages," *J. Comput. System Sci.*, **10** (1975), 351–369.
44. ———, "Continuous operations on languages," *Math. Systems Theory*, **11** (1977), 1–8.
45. S. Greibach, "A new normal form for context-free phrase structure grammars," *J. Assoc. Comput. Mach.*, **12** (1965), 42–52.
46. ———, "Simple syntactic operators on full semi-AFLs," *J. Comput. System Sci.*, **6** (1972), 30–76.
47. ———, "The hardest context-free language," *SIAM J. Comput.*, **2** (1973), 304–310.
48. ———, *Theory of Program Structures: Schemes, Semantics, and Verification*, Lecture Notes in Computer Science, vol. 36, Springer-Verlag, 1975.
49. ———, "Control sets on context-free grammar forms," *J. Comput. System Sci.*, **15** (1977), 35–98.

50. ———, "One-way finite visit automata," *Theoretical Comput. Sci.* (to appear).

51. S. Graham and M. Harrison, "Parsing of general context-free languages," in *Advances in Computers*, 14, M. Rubinoff and M. Yovits, eds., 1976, pp. 77–185.

52. J. Hartmanis, "Context-free languages and Turing machine computations," in *Math. Aspects of Computer Science*, J. Schwartz, ed., Proc. Symposia in Applied Mathematics, vol. 19, Amer. Math. Soc., 1967, pp. 42–51.

53. J. Hartmanis and H. Hunt, "The LBA problem and its importance in the theory of computing," in *Complexity of Computation*, R. Karp, ed., SIAM-AMS Proc., vol. 7, Amer. Math. Soc., pp. 1–26.

54. J. Hartmanis and R. Stearns, "On the computational complexity of algorithms," *Trans. Amer. Math. Soc.*, **117** (1965), 285–306.

55. F. Hennie, *Introduction to Computability*, Addison-Wesley, Reading, Mass., 1977.

56. F. Hennie and R. Stearns, "Two tape simulation of multitape machines," *J. Assoc. Comput. Mach.*, **13** (1966), 533–546.

57. G. Herman and G. Rozenberg, *Developmental Systems and Languages*, North-Holland, Amsterdam, 1975.

58. J. Hopcroft and J. Ullman, *Formal Languages and Their Relation to Automata*, Addison-Wesley, Reading, Mass., 1969.

59. R. Karp, "Reducibility among combinatorial problems," in *Complexity of Computer Computations*, R. Miller and J. Thatcher, eds., Plenum Press, New York, 1972, pp. 85–104.

60. S. Kuno, "Computer analysis of natural languages," in *Math. Aspects of Computer Science*, J. Schwartz, ed., Proc. Symposia in Applied Mathematics, vol. 19, Amer. Math. Soc., 1967, pp. 52–110.

61. P. Lewis, D. Rosenkrantz, and R. Stearns, *Compiler Design Theory*, Addison-Wesley, Reading, Mass., 1976.

62. D. Luckham, D. Park, and M. Paterson, "On formalized computer programs," *J. Comput. System Sci.*, **4** (1970), 220–249.

63. M. Machtey and P. Young, *An Introduction to the General Theory of Algorithms*, North-Holland, Amsterdam, 1978.

64. Z. Manna, "Program schemas," in [2], pp. 90–142.

65. A. Meyer and L. Stockmeyer, "The equivalence problem for regular expressions with squaring requires exponential space," *Proc. 13th IEEE Symp. Switching and Automat Theory*, 1972, pp. 125–129.

66. B. Monien, "Transformational methods and their application to complexity problems," *Acta Informatica*, **6** (1976), 95–108.

67. E. Moore, ed., *Sequential Machines: Selected Papers*, Addison-Wesley, Reading, Mass., 1964.

68. P. Naur, ed., "Report on the algorithmic language ALGOL-60," *Comm. ACM*, **3** (1960), 299–314.

69. M. Nivat, "Transduction des Langages de Chomsky," Ph.D. dissertation, Université de Paris, 1967. Also appears in *Ann. Inst. Fourier, Grenoble*, **18** (1968), 339–456.

70. ———, "On the interpretation of polyadic recursive schemes," *Symp. Mathematica*, vol. 15, Academic Press, New York, 1975.

71. W. Ogden, "A helpful result for proving inherent ambiguity," *Math. Systems Theory*, **2** (1968), 191–194.

72. R. Parikh, "On context-free languages," *J. Assoc. Comput. Mach.*, **13** (1966), 570–581.
73. C. Perrault, "Intercalation lemmas for tree transducer languages," *J. Comput. System Sci.*, **13** (1976), 246–277.
74. M. Rabin, "Real-time computation," *Israel J. Math.*, **1** (1963), 203–211.
75. M. Rabin and D. Scott, "Finite automata and their decision problems," *IBM J. Res. Develop.*, **3** (1959), 114–125.
76. H. Rogers, *Theory of Recursive Functions and Effective Computability*, McGraw-Hill, New York, 1967.
77. B. Rosen, "Tree-manipulating systems and Church-Rosser theorems," *J. Assoc. Comput. Mach.*, **20** (1973), 160–187.
78. A. Rosenberg, "Real-time definable languages," *J. Assoc. Comput. Mach.*, **14** (1967), 645–662.
79. A. Rosenfeld, *Picture Processing by Computer*, Academic Press, New York, 1969.
80. W. Rounds, "Mappings and grammars on trees," *Math. Systems Theory*, **4** (1970), 257–287.
81. A. Salomaa, *Theory of Automata*, Pergamon Press, Oxford, 1969.
82. ———, *Formal Languages*, Academic Press, New York, 1973.
83. W. Savitch, "Relationships between nondeterministic and deterministic tape complexities," *J. Comput. System Sci.*, **4** (1970), 177–192.
84. ———, "A note on multihead automata and context-sensitive languages," *Acta Inform.*, **2** (1973), 249–252.
85. M. Schützenberger, "Context-free languages and pushdown automata," *Info. Control*, **6** (1963), 246–264.
86. J. Seiferas, "Nondeterministic Time and Space Complexity Classes," Ph.D. dissertation, M.I.T., 1974.
87. J. Seiferas, "Techniques for separating space complexity classes," *J. Comput. System Sci.*, **14** (1977), 73–99.
88. C. Shannon and J. McCarthy, eds., *Automata Studies*, Princeton University Press, 1956.
89. R. Stearns, J. Hartmanis, and P. Lewis, "Hierarchies of memory limited computations," *Proc. 6th IEEE Symp. Switching Circuit Theory and Logical Design*, 1965, pp. 179–190.
90. I. Sudborough, "A note on tape-bounded complexity classes and linear context-free languages," *J. Assoc. Comput. Mach.*, **22** (1975), 499–500.
91. J. Thatcher, "Tree automata: An informal survey," in [2], pp. 143–172.
92. A. Yasuhara, *Recursive Function Theory and Logic*, Academic Press, New York, 1971.
93. R. Yeh, *Applied Computation Theory: Analysis, Design, and Modeling*, Prentice-Hall, Englewood Cliffs, N. J., 1976.
94. A. Aho and J. Ullman, *Principles of Compiler Design*, Addison-Wesley, Reading, Mass., 1977.
95. D. Knuth, "On the translation of languages from left to right," *Information and Control*, **8** (1965), 607–639.
96. ———, Computer science and its relation to mathematics, *Amer. Math. Monthly*, **81** (1974), 323–342.

97. C. Wrathall, "Rudimentary predicates and relative computation," *SIAM J. Computing*, **7** (1978), 194–209.
98. M. Harrison, *Introduction to Formal Language Theory*, Addison-Wesley, Reading, Mass., 1978.
99. S. A. Greibach, "Formal languages: origin and directions," *Annals of the History of Computing* **3** (1981), 14–41.
100. J. Berstel, *Transductions and Context-Free Languages*, Teubner Studienbucher, Stuttgart, 1979.
101. G. Rozenberg and A. Salomaa, *The Mathematical Theory of L-Systems*, Academic Press, New York, 1980.
102. D. Wood, *Grammar and L Forms: An Introduction*, Lecture Notes in Computer Science 91, Springer-Verlag, 1980.
103. A. Salomaa, *Jewels of Formal Language Theory*, Computer Science Press, 1981.
104. R. Book (ed.), *Formal Language Theory: Perspectives and Open Problems*, Academic Press, New York, 1980.

FORMAL ANALYSIS OF COMPUTER PROGRAMS

Terrence W. Pratt

The goal of this chapter is to provide a brief excursion into some of the problems and results in the formal analysis of computer programs. Most of the work of interest is relatively recent, and much is of a tentative and exploratory nature. As with most developing research areas, there is substantial disagreement over even what the central questions are and how they should be approached. In such a situation, it seems of little value to present any of the current formal theories in depth, as their value may be quite transitory. Instead, an alternative structure seems more appropriate: an approach that begins with an exposition of the problems themselves, followed by a survey of some of the formal approaches to the solution of these problems that are under development. Of course the range of problems to be solved is considerable. For this survey, four problems have been chosen as the basis for discussion, four problems that seem to encompass a substantial part of the work of interest. For the reader whose interest is aroused by any of the topics raised, the concluding section suggests further readings.

This work was supported in part by the National Science Foundation under grant DCR 75-16858.

1. PROGRAMS AND PROGRAMMING LANGUAGES

A computer is a tool ordinarily used to perform quickly and accurately some complex or tedious computations. One has a set of *input data,* upon which a well-defined sequence of operations is to be performed to compute some *output data* of interest. In order to specify the structure of the input and output data and the computation to be performed, a *program* is written according to the rules of some *programming language.* The programming language is simply a predefined notation for the specification of programs.

Let us take a particular problem and a program for its solution. Suppose we wish to determine, to within a given accuracy, the quotient of two positive real numbers, where the quotient is between zero and one. The input data for the program is a set of three numbers, representing the dividend, divisor, and the desired accuracy. The output data from the program is a list of the input data followed by a single real number, the quotient. Figure 1 gives one program for making such a computation. The algorithm is one developed by Wensley [25]. The program is written in the programming language PASCAL, i.e., the notation used to express the algorithm is that of the PASCAL programming language [11].

2. FOUR CENTRAL PROBLEMS IN THE ANALYSIS OF PROGRAMS

The example program is neither particularly complex nor particularly useful, yet some of the central issues in the analysis of programs are apparent even here. The four that concern us are the following:

a. *What does the notation mean?* The first problem in constructing a detailed and precise analysis of the program lies in the notation, the programming language. The program is to be taken as representing a set of instructions for a computer, but what are the instructions, and what exactly is the computer being instructed to do? For example, we see in Figure 1 the program segment:

while $E \leq D$ **do**

begin

⋮

end

We might guess that this construct in the PASCAL language has the customary meaning: The statements between the **begin** and **end** are to be repeated indefinitely until the condition $E \leq D$ becomes false. But we have no assurance that this is the meaning, and, even if it is, there are many details that must be filled in before the meaning has been made entirely precise: What is the meaning of D and E, when is the condition $E \leq D$ tested for the first time, etc.?

It might seem that the problem of specifying the meaning of such programming language constructs would have been solved long

```
program Division(input,output);
    var R,M,N,E: real;
    function Quot(P,Q,E:real): real;
        var A,B,D,Y: real;
        begin A := 0;
              B := Q/2;
              D := 1;
              Y := 0;
              while E ≤ D do
                  begin
                      if P ≥ A + B then begin Y := Y + D/2;
                                              A := A + B
                                        end;
                      B := B/2;
                      D := D/2
                  end;
              Quot := Y
        end;
    begin read (M, N, E);
          R := Quot (M, N, E);
          write (M, N, E, R)
    end.
```

FIG. 1. A PASCAL program to compute quotients.

ago; after all, there are hundreds of programming languages in use—surely the notations they provide must be well defined! Oddly, all but the simplest of these languages are not defined unambiguously in an intelligible way. The development of a means to provide formal, unambiguous, and readable definitions for such languages has been and remains a central problem in the analysis of programs.

b. *What does the program compute*? A program always defines a function. It has a fixed domain of possible sets of input data and a range of possible sets of output data, and each input data set is mapped uniquely into an output data set. When one is given a program to analyze, the first question, almost inevitably, is to ask what function the program computes. Again, it might seem that the answer should be trivial: Isn't it easy to look at the program and determine the function it represents, or at least ask the person who wrote the program? Again the answer is no, and not only because the notation itself tends to be ambiguous: The program quickly becomes too complex to be easily described or understood in terms of the function it computes. Even the programmer who wrote the program is likely to be uncertain as to just what function the program actually does compute. Immediately we are led to the closely related question:

c. *Is the program correct*? Presumably the program is intended to compute some specified function, i.e., given a certain input data set, it should produce the appropriate output data set. But does it do so for the entire range of input data sets? Again it might seem, and certainly would be hoped, that the answer would be readily obtainable. But again, the truth is far different—most programs, when originally designed and written, are incorrect, but the fact is hidden deep in the complexity of the program structure. Most of the errors are detected only after a tedious, time-consuming, and essentially *ad hoc* testing procedure; some are never detected at all. A central issue in the formal analysis of programs is that of providing methods for determining the correctness of programs directly.

d. *Is there a more "efficient" program to compute the same function*? A final issue of considerable importance is that of finding better versions of the program—programs that represent the same function but that have fewer instructions or are less costly in some other measure. Here we confront the question of whether two programs compute the same function, i.e., after we modify the program, how can we be sure that the new version still computes the same function as the original? And if it does, how can we be sure that it is really better than the original?

All of these problems need and are amenable to formal treatment.

3. DEFINITION OF PROGRAMMING LANGUAGES

Let us now take up the problems mentioned in the previous section one by one, considering the problems and difficulties encountered in the application of mathematics to each. The first problem is that of providing precise and complete definitions of the programming languages used to write programs. Clearly, without such definitions efforts at careful analysis of programs to determine correctness or efficiency must surely fail. First, a bit of history will clarify why current programming languages lack precise definitions.

Programming languages such as FORTRAN, COBOL, ALGOL, and PL/I (to name just four of the many hundreds in existence) developed primarily as solutions to a practical problem, as simple expedients. The practical problem was, and still is, that computers are very difficult to use directly. A typical computer has built into it the capability to carry out programs of instructions written in a primitive "machine language." Typically the instructions in a machine language consist only of patterns of binary digits, e.g., the binary sequence 0011001110000000001010101111001 may serve as an instruction to add two numbers stored in the internal memory of the computer and store the result in a third storage location in the same memory. Early users of computers wrote programs directly in such machine languages. However, the task was extremely tedious and error prone. It was soon recognized that better "languages" for writing programs were essential. Here the general purpose nature of computers was brought to bear: It was clear that, once a better notation for writing programs was devised, a single program (written in machine language) could then be executed by the computer whose function would be to *translate* the program from the new notation into instructions in the machine language of the computer. The translated program could then, in a second step, be executed by the computer to produce the desired computation. Any competent programmer could devise a notation, a "programming language," that would be suitable for his purposes. He could then write the appropriate translation program to translate his new programming language into machine language, and immediately his subsequent programming could be entirely in the new programming language—he never again, or perhaps only occasionally, would need to go back to the difficulties of programming in machine language. Once the basic concept of a *compiler* program, a program to translate a program in one notation into a program in a machine language (or another, already available, programming language), was grasped and accepted, numerous programming languages were designed and "implemented" on different computers. Subsequently, the need to move programs between different computers and the need for programmers to communicate with each other has led to some standardized languages, such as FORTRAN and COBOL, for which compiler programs are available on most computers.

For our purposes, the central fact about these programming languages is the *ad hoc* method of their design and definition. If we ask, What is the meaning of this or that statement in a program?—i.e., If I write this down, what will it cause the computer to do?—a precise answer is generally not to be found except in the following way: The computer will perform those instructions produced by the compiler when it translates the given statement into machine language. Since the actions the computer will take for each instruction are completely determined by its internal structure (the manner in which its circuits are connected), and the instructions produced by the compiler are completely determined by its internal structure (i.e., by the actions generated by instructions of the compiler program when it is executed by the computer), the "meaning" in terms of actions to be taken by the computer is completely and unambiguously defined. Unfortunately, it is quite impossible to make use of this definition of the language as given by the structure of the computer and the compiler; both are entirely too complicated to be comprehensible. The computer may have millions of circuits and the compiler may contain thousands of instructions. Thus, like the physicist who knows all the laws of motion yet still cannot catch the ball, we know the computation invoked by a program is completely and precisely determined, yet we still cannot predict what the program will do. Clearly the definition of a programming language by a compiler and computer is unacceptable; its precision is of little value.

The value of a precise, complete, and intelligible definition of a programming language was recognized early. The development of adequate mathematical tools for constructing such definitions has been slow, however. Large early successes were obtained in the development of a formal theory of the syntax of programming languages (a topic discussed in another chapter), most notably in the theory of context-free grammars and their derivatives. Thus it has long been possible to give a precise, complete, and intelligible definition of what it is allowable to *write* in a program of a given language, which constructs are allowed, and which are not, how the statements are punctuated, etc. What has not been possible in any widely accepted form is precise definition of the *meaning* of each statement or other syntactic construct.

A typical programming language may have from ten to fifty or more different types of statements, declarations, expressions, etc. The meaning of each of these constructs is usually closely tied to the context in which it occurs, so that the same statement appearing in different contexts represents a different set of computational steps to be performed. Perhaps the most difficult aspect of defining the semantics of a programming language lies in capturing the notion of the "computational context" in which each statement is executed.

Formal definitions of the semantics of programming languages have been based on a broad range of approaches. Unfortunately, space constraints preclude even an example definition of a simple language. A brief sketch of three different approaches must suffice to convey the flavor of some of the most promising approaches. A final critique suggests some of the difficulties with these approaches. None has been accepted as a method for practical definition of any large class of actual programming languages. Informal methods remain the standard for new language definitions.

The "Abstract Machine" Approach. The most straightforward approach to the formal definition of programming languages is based on the notion of an *abstract machine* (sometimes termed a *virtual computer* or *executing automaton*). An abstract machine consists of:

1. a set of *states*, each of which is a complex formal structure containing representations of programs and data;
2. a set of *primitive operations* on states (state transformations) that map one state into a "next" state;
3. a *transition rule* that maps any state into a primitive operation, representing the next operation to be applied, or into the *halt* operation; and
4. a subset of states that serve as *initial states.*

To complete the definition of a programming language, an abstract machine must be augmented by:

5. a *formal language*, consisting of character strings representing syntactically valid programs in the language, usually defined by a formal grammar of some sort; and
6. a *translation function* mapping each valid program into an initial state of the abstract machine.

The formal language defines the set of valid programs in the language. For each of these valid programs the function that it computes is determined as follows:

1. First the program is mapped into an initial state of the machine, by means of the translation function.

2. From this initial state the transition rule and primitive operations of the abstract machine define a sequence of state transitions, first from the initial state to a next state and then from that state to another following state, etc. Ultimately either the *halt* operation is produced, and thus the sequence ends, or the sequence may continue without halting. If the sequence does end, then the final state in the sequence represents the result of program execution. The initial state is assumed to contain the input data; the final state then contains the output data.

The Vienna Definition Language. The most widely known example of the abstract machine approach is the formal system known as the "Vienna Definition Language" (VDL). This approach was developed by the Vienna Laboratory of IBM for use in the formal definition of the language PL/I. It has been used in definitions of PL/I, ALGOL 60, and BASIC, among others.

In a VDL definition of a programming language, each state of the abstract machine is represented by a finite *tree* with labeled arcs and nodes. The "state tree" that exists at any point during a computation sequence contains complete information about the state of the computation at that point: It contains all the data structures used by the program, the program text itself, various "housekeeping" data structures needed to keep track of data names and attributes, etc., and a "control" subtree that contains the next operation to be executed at any point together with any pending operations. All this information is represented in the form of subtrees within the overall state tree.

The primitive operations in a VDL abstract machine are defined as tree transformations that modify the state tree in appropriate ways. The transition rule is straightforward: The next operation to be applied, given a current state tree, is the operation that labels a terminal node of the control subtree in this current state. Usually the control subtree has only a single terminal node, so that the next operation is uniquely specified, but it is possible for there to be multiple terminal nodes, in which case the abstract machine becomes nondeterministic: More than one sequence of state transitions is possible depending on which of the possible next operations is chosen for application to the current state.

The initial states of the abstract machine are defined as follows: An initial state tree has a number of invariant components, primarily representing elements of the state that are initially empty. The major variant component is the program text subtree. This subtree contains the program that is to be executed, represented as an *abstract syntax tree* derived from the original program. The abstract syntax tree of a program is essentially the parse tree of the program with most of the nonessential syntactic elements deleted, such as punctuation, and a number of implicit specifications added, such as default attributes for data structures.

The set of valid programs in the language is specified by a context-free grammar, and the translation from valid program to the corresponding abstract syntax tree is defined by a translation function that maps the parse trees defined by the context-free grammar into the appropriate abstract syntax trees.

The Scott-Strachey Approach. A second formal approach to programming language definition attempts to directly describe the meaning of a program in terms of the function it computes. The foundations of this approach are due largely to Dana Scott and Christopher Strachey. Concisely summarized, this approach is based on the following constructs that differ from those used in the abstract machine approach:

1. The computation invoked by a particular program is still described in terms of a sequence of states, but the states now contain only the data being manipulated, not the program itself.
2. States are constructed from base sets, functions on these base sets, higher-order functions over these functions, etc. A lattice structure is imposed on these sets, and the functions defined must all be "continuous" in an appropriate sense on these domains.
3. The meaning of the various statement types in the language is defined directly in terms of functions mapping states into states. These definitions typically will involve general recursive equations of the form $f = F(f)$. A unique solution for such systems of recursive equations is provided by a theory of least fixed points, generalizing the classical recursion results of Kleene [12].
4. The overall meaning of a program is defined recursively, as the application of the function defined by the first statement in the program to the initial state, followed by the application of the remainder of the program to the resulting state.
5. Because the meaning of a program is defined directly in terms of its syntactic structure, as the recursive composition of the functions defined by each of the constituent statements of the program in the written sequence, transfers of control from one statement to another within the program (e.g., by **goto** or **exit** statements) cause considerable difficulty. A rather complex formal structure called a "continuation" is introduced to surmount this difficulty.

The resulting definition of a programming language in this formalism is a set of recursive equations over a set of syntactic and semantic domains (base sets). The set of valid programs in the language is defined by a set of productions (equations) that essentially form a context-free grammar for the language.

The Hoare "Axiomatic" Approach. In this approach, the precise definition of any sort of detailed "state" of a computation is avoided entirely. The meaning of a statement in the language is defined in terms of axioms and rules of inference, which allow one to describe properties of the computational state before (or after) execution of a given statement, given known properties of the state after (or before) execution of the statement.

Two examples will illustrate the technique:

1. *Assignment axiom*: Assume a simple language with assignment statements of the form

$$Y := F(X,Y,\dots)$$

whose intended meaning is that the current values of the variables X, Y, ... referenced in the right-hand side are taken as arguments for the function F, whose value is then assigned as the new value of the variable Y. An axiom that formalizes this meaning would be:

$$\{p(X,F(X,Y,\dots),\dots)\}\, Y := F(X,Y,\dots)\{p(X,Y,\dots)\}$$

which is read: If p is a predicate (propositional formula) specifying relationships between the values of the variables in the current state, and if $p(X,Y,\dots)$ is true after execution of the assignment statement, $Y := F(X,Y,\dots)$, then we may deduce that $p(X,F(X,Y,\dots),\dots)$ was a true statement about the relationships between the values of the variables before execution of the assignment statement.

2. *"While" statement inference rule.* Assume a simple language with **while** statements of the form:

while B **do** S

where B is a predicate as above and S is a list of statements. The intended meaning is that the predicate B is to be evaluated using the values of the variables in the current state, and if its value is true, then the statement list S is to be executed to produce a new state, and the entire process repeated until the value of B in the current state is false. The meaning of this statement would be formally defined by the rule:

$$\frac{\{p \wedge B\}S\{p\}}{\{p\}\ \textbf{while}\ B\ \textbf{do}\ S\ \{p \wedge \neg B\}}$$

which is read: If we can establish, for predicates p and B, that whenever both p and B are true before execution of the statement list S then p is true after execution as well, then we may conclude that whenever p alone is true before execution of the **while** statement, **while** B **do** S, that after execution both p and $\neg B$ must be true. Predicate p is termed the *loop invariant.*

Summary of the Problem of Programming Language Definition. None of the techniques for formal definition of programming languages has been entirely successful. Various reasons may be advanced as to why this is so. On the one hand, existing programming languages are, almost without exception, far too irregular semantically to admit simple definitions. This irregularity is largely a result of the *ad hoc* method of their original definition. Thus they are in some sense intrinsically difficult to describe formally. Equally important in contributing to the lack of success of formal definitional techniques has been deep disagreement as to the criteria for evaluating these techniques. Each of the approaches mentioned above has been fairly successful according to some desiderata and has failed according to others.

A brief summary of some of the objections will suffice to illustrate the difficulties, without really doing justice to any of them:

1. The abstract machine approach introduces a complex extra layer of definition. One cannot directly understand the meaning of a particular statement in a program. Instead, the statement must first be traced through a complex translation into a sequence of operations for the abstract machine, and then the effect of these operations must be observed. In addition, the abstract machine approach seems to "overspecify" the language semantics by specifying details about exactly how particular constructs are to be handled, even though this specification is irrelevant to the computation of the output data.

2. The Scott-Strachey approach avoids much of the detail of the state structure. The meaning of a statement is defined directly in terms of its effect on the current state; there is no translation step. However, the definitions produced are complex and rather obscure, due in part to fundamental problems with description of common programming language constructs, particularly transfers of control and changes in the meaning of variable names.

3. The Hoare axiomatic approach also provides direct semantic definitions for statements and other syntactic constructs but suffers even more severely from difficulties with description of some common programming language constructs, particularly transfers

of control, the meaning of variable names, and data structures with shared storage.

4. THE FUNCTION REPRESENTED BY A PROGRAM

Any program defines a mapping from a family of input data sets into a family of output data sets, and thus any program defines a function (of course the range may be the empty set if the program never terminates for any input). It is natural to ask, given a program, What function does it compute? Assume now that the programming language has been carefully defined so that we may ignore any questions of ambiguity in the notation. If we knew what function the program did in fact compute (or represent), then we could determine the correctness of the program, i.e., whether it computed the function we desired. The next section takes up this problem, the problem of proving a program correct. However, it is worthwhile to probe the facile observation that any program represents a function before moving to the correctness problem.

There are a number of characteristics of the functions represented by programs that distinguish them from most of the functions ordinarily seen in elementary mathematics. First, the functions computed are defined, almost without exception, in terms of cases. The family of input data sets is partitioned by the program into a set of disjoint cases, and for each case the program computes the output data set in a different way. Of course the definition of a function in terms of different cases is familiar from mathematics. What is noteworthy is the number of cases treated: Even a relatively simple program with no loops is likely to consider hundreds of different cases. If the program contains a loop, then the number of cases treated is potentially unbounded.

Given a program, it is straightforward to determine the cases that the program discriminates and the computation invoked in each case. Each different execution path through the program corresponds to one case. Along each path, part of the computation is concerned with choosing the path itself (i.e., with computing values that are used at the next branch point to determine the path along which execution should continue), and thus this part of the computation actually discriminates the case into which the input data falls. This part of the program is usually termed the *control struc-*

ture of the program. The remainder of the computation along the path is concerned with computing the appropriate output data. If one writes down for each path the control part of each path and the associated computation of the output, then a description of the function computed may be derived.

The division program of Figure 1 provides a good example. Figure 2 shows a partial description of the cases treated by this program and the output data computed in each case. Since this program contains a loop, the number of cases treated is potentially infinite, i.e., there are an infinite number of different paths through the program.

Case	Output Computed
$E > 1$	$R := 0$
$1/2 < E \leq 1 \wedge P \geq Q/2$	$R := 0 + 1/2 = 1/2$
$1/2 < E \leq 1 \wedge P < Q/2$	$R := 0$
$1/4 < E \leq 1/2 \wedge P \geq (3/4)Q$	$R := 0 + 1/2 + (1/2)/2 = 3/4$
$1/4 < E \leq 1/2 \wedge Q/2 \leq P < (3/4)Q$	$R := 0 + 1/2 = 1/2$
$1/4 < E \leq 1/2 \wedge Q/4 \leq P < Q/2$	$R := 0 + (1/2)/2 = 1/4$
$1/4 < E \leq 1/2 \wedge 0 \leq P < Q/4$	$R := 0$
$1/8 < E \leq 1/4 \wedge P \geq (7/8)Q$	$R := 0 + 1/2 + (1/2)/2 + ((1/2)/2)/2 = 7/8$
$1/8 < E \leq 1/4 \wedge (3/4)Q \leq P < (7/8)Q$	$R := 0 + 1/2 + (1/2)/2 = 3/4$
⋮	

FIG. 2. The function computed by the program of Figure 1.

The assumption that the division program may treat an unbounded number of cases is not realistic, of course. We do not expect the value of E, the accuracy of the quotient, to be arbitrarily small on any real computer. But even if we allow E to be no smaller than 2^{-32} (certainly not unrealistic on many computers), the number of cases treated by the program, although now finite, is so large as to effectively preclude any case-by-case description or analysis.

A second source of complexity in describing the functions computed by programs lies in the complexity of the control structure computations that discriminate the cases treated by the program. These case discriminations almost always involve complex determinations about the *sequence* in which the input data is presented. Thus the various pieces of input data are not treated independently, but, rather, the appearance of a certain piece of input data early in the sequence causes a change in the way a later piece of data is interpreted. Typically a substantial part of the control structure of a program serves to save the "context" created by early data in the input sequence so that it may be used to control the interpretation of the later input data. Thus, if we wish to describe the function computed by a program in terms of the cases treated, then the description of the cases must include these sequential interdependencies in the input data.

Yet a third source of complexity lies in the presence of paths through a program that can never be traversed for any choice of input data. There are no such paths in the division program of Figure 1 because for any possible execution path, there is in fact some choice of input data that will cause execution to take that path (ignoring the problem of a minimum value for E). However, few programs of even moderate size have this property. Consider the program of Figure 3, known as the *91-function.* The definition of this function is:

For integer input value X, the value of the function is

$$X - 10 \text{ if } X > 100, \text{ and } 91 \text{ otherwise.}$$

```
program Example (input, output);
    var X :integer;
    function F91(X :integer) :integer;
        var A,B :integer;
        begin A: = X;
            B: = 1;
            while (A ≤ 100) or (B ≠ 1) do

                              if A ≤ 100 then begin A: = A + 11;
                                                    B: = B + 1 end
                                         else begin A: = A − 10;
                                                    B: = B − 1 end;

            F91: = A − 10

        end;
    begin read(X);
        write(X,F91(x))
    end.
```

FIG. 3. The *91-function.*

Note that the program contains many paths that can never be traversed for any choice of input value. Moreover, the control structure of the program effectively conceals the simplicity of the function computed. It is difficult to determine that the description above is in fact an accurate description of the function computed from inspection of the program itself.

Let us now return to the original problem: Given a program, can we describe the function that it computes? The answer is that, while we can do so, the description is likely to be almost as complex as the program itself. Even if the function has a simple definition, as with the *91-function* above, the program structure may effectively hide the fact. But far more likely the program will treat such a large number of cases, interrelated in such complex ways, that no simple definition is possible. In fact, in many cases, the *program itself* may be essentially the simplest way we can think of to describe the function precisely. Thus, although a program always computes a function, we cannot presume in general that such a function admits of a definition that is substantially simpler than the program itself.

5. PROVING PROGRAM CORRECTNESS

Once a program has been written, the next problem is whether it is "correct," i.e., whether the program performs the intended computation when it is executed. The traditional approach has been almost entirely *ad hoc*; the program is simply tested by being executed with some example data sets, and the resulting output data is inspected. If the output data is that desired for the given input data in all the test cases, then the program is presumed correct.

The difficulties with program testing as a means of determining correctness are easily seen, given our discussion of the preceding section. Each execution path through the program determines a separate computation. In principle each of the possible paths would need to be tested with each of the possible input data sets before one could be sure that the program was correct. Of course the number of cases involved in even a simple program is usually far too large to allow such exhaustive testing.

Program testing, then, is an inadequate means of showing that a program is correct. Not only is testing inadequate in theory, but also in practice: most computer programs beyond the very smallest contain errors, even programs that have been thoroughly tested and used for some years. These latent errors are an enormous practical problem, and better means of assuring the correctness of programs have been eagerly sought.

How can one formalize the notion of a program being "correct"? Informally, the program is correct if it computes the desired function, i.e., if, for each of the desired range of input data sets, the program produces the desired output data set when executed. We need, then, a precise definition of the function that the program is to compute. Given such a function definition, and given that the meaning of the programming language has been precisely defined so that the function that the program computes is also known, then the program is correct if the function that it computes is in fact the given function.

Although this concept of program correctness is straightforward, one is immediately mired in difficulties in attempting to apply the concept to actual programs written in real programming languages. Both the problems of the preceding sections come to the fore: First, we usually do not have a precise, complete, and intelligible definition of the programming language; second, even if we did, the function computed by the program is likely to be extremely complex to specify. If the specification of the function desired is itself almost as complex as the program in question, how can we be sure that the specification is correct? Moreover, if we are to specify the function desired, we must have a precise notation for making this specification. This notation must itself be defined.

The general problem may be restated: We have two notations for defining functions, a *specification language* and a *programming language*. Given a specification written in the specification language and a program written in the programming language, the problem is to determine whether both define the same function. If so, we may say that the program is "correct," i.e., that it is a correct encoding of the specification. Alternatively, we might say that the specification is a "correct" description of what the program computes. Of course both program and specification may be incorrect in the intuitive sense, in that neither actually represents the intent of the person writing them.

Formal Correctness Proofs of Programs. A number of approaches have been studied to the problem of proving programs correct. The most widely known is the *inductive assertions* approach developed first by Floyd [7]. References to other methods are given in the final section below.

In the inductive assertions approach, the usual specification language used is the predicate calculus and ordinary mathematical notation. One gives an *input predicate* defining the domain of the function represented by the program to be proved correct and an *output predicate* defining the relationship between the input data values and the output data values, i.e., defining the function computed. In addition a *loop invariant* must be provided for each loop in the program. A loop invariant is a predicate that is satisfied whenever the loop is entered from outside during program execution and that is also satisfied after each subsequent traversal of the loop (i.e., after each execution of the statements within the loop). From these predicates and a precise definition of the meaning of the statements in the language, it is possible to prove that the program is in fact correct, i.e., that whenever the input data satisfies the input predicate, the output data computed by the program will satisfy the output predicate.

As an example of the technique, consider the division algorithm of Figure 1. Since the main program simply calls the function *Quot* and prints the result, we should concentrate attention on function *Quot*. If the three input values to *Quot* are P, Q, and E, then an appropriate input predicate would be:

$$(0 \leq P < Q) \wedge (E > 0)$$

and the corresponding output predicate would be:

$$P/Q - E < R \leq P/Q$$

where R is the result computed by *Quot*. These two predicates define the function that we wish the program to compute.

The next step requires that we provide a loop invariant suitable for use in the proof. This is the most difficult and creative step, for the appropriate loop invariant is often hard to find. An appropriate loop invariant, a predicate that will be true whenever execution reaches the test $E \leq D$ at the beginning of the **while** loop is:

$$A = Q \times Y$$
$$B = Q \times (D/2)$$
$$D = 2^{-k} \text{ for some integer } k \geq 0$$
$$P/Q - D < Y \leq P/Q.$$

The input and output predicates serve to specify the function that the program is to compute. To prove the program correct, i.e., to prove that the program does in fact compute this function, we proceed by proving three lemmas, or "verification conditions":

LEMMA 1. If the input data satisfies the input predicate, and we begin to execute the program, then when we first reach the **while** loop, the loop invariant is satisfied.

LEMMA 2. If the loop invariant is satisfied and the condition for executing the loop is also satisfied by the current values of the program variables, then after execution of the statements within the loop the loop invariant is again satisfied (for either of the two paths through the loop).

LEMMA 3. If the loop invariant is satisfied but the condition for executing the loop is not satisfied by the current values of the program variables, then after executing the statements following the loop through the final statement in *Quot* the output value satisfies the output predicate.

To formally state the lemmas requires a formal definition of the programming language, so that we have a precise conception of the meaning of "execution" of the statements in the program. Of course, we lack such a definition of PASCAL, the language used to write the program of Figure 1. However, since the program is very simple, it will suffice for this example to assume that the meanings of assignment statements and **while** statements are given by the axioms stated in Section 3. The lemmas to be proved may then be written:

LEMMA 1. If the three values read by the *Read* operation and transmitted as parameters to *Quot* are the numbers p, q, and e, and

$(0 \le p < q) \wedge (e > 0)$ then (taking the loop invariant and substituting the values assigned to each of the variables in the preceding assignments in the program of Figure 1):

$$\begin{aligned}&0 = q \times 0\\&q/2 = q \times (1/2)\\&1 = 2^{-k} \text{ for } k = 0\\&p/q - 1 < 0 \le p/q.\end{aligned}$$

LEMMA 2. If at the beginning of the **while** loop, we know that:

$$\begin{aligned}&E \le D\\&A = Q \times Y\\&B = Q \times (D/2)\\&D = 2^{-k} \text{ for some integer } k \ge 0\end{aligned}$$

and

$$P/Q - D < Y \le P/Q$$

then it is also true that (again substituting new values assigned during execution of the loop):

if $P \ge A + B$ then

$$\begin{aligned}&A + B = Q \times (Y + D/2)\\&B/2 = Q \times ((D/2)/2)\\&D/2 = 2^{-k} \text{ for some integer } k \ge 0\\&P/Q - D/2 < Y + D/2 \le P/Q\end{aligned}$$

or if $P < A + B$ then

$$\begin{aligned}&A = Q \times Y\\&B/2 = Q \times ((D/2)/2)\\&D/2 = 2^{-k} \text{ for some integer } k \ge 0\\&P/Q - D/2 < Y \le P/Q.\end{aligned}$$

From this lemma we can conclude that if the loop invariant is satisfied by one set of values for P, Q, E, A, B, D, Y, and the loop is executed, then the new values also satisfy the loop invariant.

LEMMA 3. If at the beginning of the **while** loop we know that

$$\begin{aligned}&E > D\\&A = Q \times Y\\&B = Q \times (D/2)\\&D = 2^{-k} \text{ for some integer } k \ge 0\end{aligned}$$

and

$$P/Q - D < Y \leq P/Q$$

then

$$P/Q - E < Y \leq P/Q.$$

Proving these three lemmas would suffice to show that the program computes the desired results for inputs in the specified domain, provided that execution ever terminates at all. To complete the "proof of the program" we must also show that program execution always terminates for any input data satisfying the input predicate. Such a proof of termination is usually given separately. For the example program, execution can fail to terminate only if the predicate in the **while** statement were to be satisfied on entry to the loop and always thereafter, for some set of valid input data, no matter how many times the loop was executed. A simple argument suffices to show such a situation impossible for this program. However, in general, proof of termination is nontrivial. This completes the proof of correctness of the division program. As an interesting exercise the reader might enjoy "proving" the 91-function (Figure 3) using the specification given in the accompanying text.

The conclusion provides references to further, more complete, discussions of the formal theory of correctness proofs for programs. Note how closely the problem of proving correctness is tied to the problems discussed in the two preceding sections: We must have both a good useable formal definition of the programming language and a way of specifying the function the program should compute in order to state and prove the conditions for the program to be correct.

6. EQUIVALENCE AND TRANSFORMATION OF PROGRAMS

The fourth and final problem of this discussion is also one that has generated a number of formal studies: the problem of optimizing or improving a program through transformation of its internal structure. The need for such program transformation arises in many applications. The most common case is probably the need to improve the efficiency of a program, either through reduction of its storage requirements or through reduction of its execution time. Another set of useful transformations improve the internal structure of a program to make it more intelligible.

The formal problem is easily stated. Given a programming language L, consider all the programs that may be written in L, the set R_L. There is a natural equivalence relation definable on this set:

DEFINITION: If P and Q are programs in R_L, then $P \simeq Q$ iff P and Q represent the same function, i.e., iff P and Q are defined over the same domain of input data sets and for each data set in this domain, they produce identical output data sets.

Each of the equivalence classes will, in general, contain many programs, since there will be many different ways to compute the same function.

A *transformation* on programs in R_L is a mapping from R_L into R_L. A transformation, T, is *valid* if for any program P, $P \simeq T(P)$, i.e., if the original and transformed programs both compute the same function. Ordinarily the transformations of greatest interest are the valid transformations that also produce an improved program according to some measure of program structure or performance.

A general method for determining the equivalence of two programs in the set R_L (for any given language L) would be a great help. That is, we would like to have a rule for deciding, given any two programs P and Q in R_L, whether $P \simeq Q$. Unfortunately a general solution to this equivalence problem is not possible: The question is undecidable for any programming language L beyond the most trivial (see, e.g., Constable and Muchnick [3]).

Optimizing Transformations. Fortunately it is not necessary to have a general solution to the program equivalence problem in order to develop valid transformations on programs. The largest class of useful transformations is the class of "optimizing" transformations, transformations that decrease the execution time or storage requirements of a program. A few examples will suffice to illustrate the sorts of transformations involved.

1. *Moving constant computations out of loops.* Consider the program segment:

```
⋮
read(Z)
⋮
while X ≠ Y do
   begin
      ⋮
      Y: = sin(X) + square-root(Z)/5
      X: = ...
      ⋮
   end
⋮
```

Assuming that Z is not assigned a new value elsewhere within the loop, then the computation of

$$square\text{-}root(Z)/5$$

is repeated on each execution of the loop, and yet its value is always the same. If, as is often the case in large programs, the loop is executed many thousands or millions of times, the repetition of this "constant" computation may use a substantial amount of execution time. Such a computation may be moved outside of the loop and computed only once, thus decreasing the execution time of the program. An equivalent optimized program segment would be:

```
⋮
read(Z)
   ⋮
T: = square root(Z)/5
while X ≠ Y do
```

```
begin
    ⋮
    Y: = sin(X) + T
    ⋮
end
⋮
```

A formal proof that the transformed program is functionally equivalent to the original program requires a formal definition of the programming language. Even informally there are a number of subtleties that must be considered before one may determine exactly the conditions under which the transformation above is valid. Most notably, the function *square-root* may change the values of variables in the program indirectly, through so-called "side effects". In such a case, even though the value returned by the function is constant, the transformation may not produce a functionally equivalent program.

2. *Eliminating computation of repeated subexpressions.* Consider the program segment:

```
⋮
A[2*I + J]: = R + B[K,2*I + J]
⋮
R: = (2*I + J)/W
⋮
```

The expression $2*I + J$ appears in three places in the segment. If this segment is within a loop which is executed many thousands of times, then much execution time would be saved if the value of $2*I + J$ could be computed only once each time through the loop and this value saved and used in each of the three occurrences instead of being recomputed at each occurrence. The more efficient program would then be:

$$\vdots$$

$$TEMP: = 2*I + J$$

$$A[TEMP]: = R + B[K,TEMP]$$

$$\vdots$$

$$R: = TEMP/W$$

$$\vdots$$

Under what conditions is this transformation valid? Again a precise definition of the meaning of the programming language is needed. It must be proved that the value of the expression is the same at each occurrence when computed as in the original program. Obviously the values of variable I and J must be known to be the same when execution reaches each of the three occurrences of the expression. Thus determining the validity of application of this transformation requires tracing the flow of the computation and the possible changes in values of the variables that may occur.

Many other optimizing transformations have been studied. Aho and Ullman [1] provide a survey of these results.

Transformations That Improve Program Structure. A second class of transformations of interest are those that make the structure of a program more intelligible. Much of this work has been motivated by the problem of changing "unstructured" programs, containing many statement labels and **goto** statements transferring control to these labels, into "structured" programs that have no labels and **goto** statements. The formal foundations for such transformations are well established. Some of the earliest work was that of Bohm and Jacopini [2], who were able to show that for every program in a simple programming language there was a functionally equivalent program that required only simple statement sequences and **while** statement loops. In particular **goto** statements and statement labels were not necessary. Later work has clarified the situations in which such transformations are possible without the introduction of new variables and conditional branching.

Knuth's survey [13] provides a thorough introduction to this work.

7. SUGGESTIONS FOR FURTHER READING

Many methods for the formal definition of programming languages have been developed. The papers in Engeler [6] and Rustin [21], particularly the survey by Wegner [24], are a useful starting point. See also the earlier survey by DeBakker [4]. An integrated treatment of various formal approaches is found in a paper by Hoare and Lauer [9]. The Vienna Definition Language is described in Wegner [23] and Lee [16]. The original work by the IBM Vienna Laboratory is summarized in Lucas and Walk [17]. Other approaches that utilize abstract machines are found in the directed graph models of the author [20] and Landin's work based on the lambda calculus [14]. Tennent [22] provides a survey of the Scott-Strachey approach; see Milne and Strachey [19] for a more complete discussion. The axiomatic approach to language definition is described by Hoare in a series of papers [8]–[10].

Much of the work on proving correctness of programs is based on the inductive assertions method suggested by Floyd [7]. Manna [18] provides a good survey of this and other approaches, together with a further bibliography. The survey by Elspas et al. [5] is also a useful starting point.

Optimization transformations for programs are surveyed in the second of the two volumes by Aho and Ullman [1]. Ledgard [15] provides an introduction to "structure-improving" transformations on programs. Knuth's more informal survey [13] is also useful.

All of these topics are the focus for much current work aimed at providing solid formal mathematical foundations for improving the practice of computing. The newness of the field and the complexity and *ad hoc* nature of much of the current practice make such mathematics both a considerable intellectual challenge and an exciting endeavor in which the results may have immediate and striking impact.

REFERENCES

1. A. Aho and J. Ullman, *The Theory of Parsing, Translation, and Compiling*, Prentice-Hall, Englewood Cliffs, N.J., 1973.

2. C. Bohm and G. Jacopini, "Flow diagrams, Turing machines and languages with only two formations rules," *Comm. ACM*, **9**, no. 5 (May 1966), 366–371.
3. R. Constable and S. Muchnick, "Subrecursive program schemata," *J. Comp. and Sys. Sci.* **6**, no. 6 (December 1972), 480–537.
4. J. W. DeBakker, "Semantics of programming languages," in J. Tou, ed., *Advances in Info. Sys. Sciences*, vol. 2, Plenum Press, New York, 1969, pp. 173–227.
5. B. Elspas, K. Levitt, R. Waldinger, and A. Waksman, "An assessment of techniques for proving program correctness," *Comp. Surveys*, **4**, no. 2 (June 1972), 97–147.
6. E. Engeler, ed., *Symposium on Semantics of Algorithmic Languages*, Lecture Notes in Mathematics, vol. 188, Springer-Verlag, New York, 1971.
7. R. Floyd "Assigning meanings to programs," *Proc. Amer. Math Soc. Symp. in Appl. Math*, vol. 19, 1967, pp. 19–32.
8. C. A. R. Hoare, "An axiomatic approach to computer programming," *Comm. ACM*, **12**, no. 10 (October 1969), 576–580.
9. C. A. R. Hoare and P. Lauer, "Consistent and complementary formal theories of the semantics of programming languages," *Acta Informatica*, **3** (1974), 135–153.
10. C. A. R. Hoare and N. Wirth, "An axiomatic definition of the programming language PASCAL," *Acta Info.*, **2**, no. 4 (1973), 335–355.
11. K. Jensen and N. Wirth, *PASCAL-User Manual and Report*, Lecture Notes in Computer Science, vol. 18, Springer-Verlag, Berlin, 1974.
12. S. Kleene, *Introduction to Metamathematics*, Van Nostrand, New York, 1952.
13. D. Knuth, "Structured programming with *go to* statements," *Comp. Surveys*, **6**, no. 4 (December 1974), 261–301.
14. P. Landin, "A correspondence between Algol 60 and Church's lambda notation," *Comm. ACM*, **8**, nos. 2 and 3 (February and March 1965), 89–101 and 158–165.
15. H. Ledgard and M. Marcotty, "A genealogy of control structures," *Comm. ACM*, **18**, no. 11 (November 1975), 629–639.
16. J. Lee, *Computer Semantics*, Van Nostrand Reinhold, New York, 1972.
17. P. Lucas and K. Walk, "On the formal description of PL/I," *Ann Rev. in Auto. Prog.*, **6**, no. 3 (1969), 105–182.
18. Z. Manna, *Mathematical Theory of Computation*, McGraw-Hill, New York, 1974.
19. R. Milne and C. Strachey, *A Theory of Programming Language Semantics*, Wiley, New York, 1976.
20. T. Pratt, "Application of formal grammars and automata to programming language definition," in R. T. Yeh, ed., *Applied Computation Theory*, Prentice-Hall, Englewood Cliffs, N.J., 1976, pp. 250–273.
21. R. Rustin, ed., *Formal Semantics of Programming Languages*, Prentice-Hall, Englewood Cliffs, N. J., 1972.
22. R. Tennent, "The denotational semantics of programming languages," *Comm. ACM*, **19**, no. 8 (Aug. 1976), 437–453.
23. P. Wegner, "The Vienna Definition Language," *Comp. Surveys*, **4**, no. 1 (March 1972), 5–63.
24. ———, "Programming language semantics," in R. Rustin, ed., [21], pp. 149–248.
25. J. H. Wensley, "A class of non-analytic iterative processes," *Computer J.* , **1** (1958), 163–167.

COMPUTATIONAL COMPLEXITY*

Franco P. Preparata

1. INTRODUCTION: OBJECTIVES AND MODELS

As noted by several authors, computational complexity is one of the most rapidly developing areas of research in the theory of computation; in some enthusiast's words, it is at the heart of computer science. Naturally, the phrase computational complexity means different things to different people, and a rather sophisticated taxonomy has emerged in the literature, both according to the research objectives and according to the specific problem areas. Although the discussion of a classification scheme of the various brands and facets of computational complexity is not the main theme of this article, we shall now and then touch upon this aspect in order to place in the appropriate context the notions to be presented.

Since ancient times people have tried to devise procedures for the solution of specific problems. Once a way to solve a problem is

---

* This article was originally written for this collection in 1976. Only minimal changes have been made to the original version, which reflects the state-of-the-art at the time of writing. (Note added in proof.)

found, the difficulty of the task of obtaining a solution to an instance of the problem is assessed on the basis of the efforts it involves, such as solver's time, use of specific instruments, etc. The desire to give a precise quantification of the difficulty of a procedure, or algorithm, became particularly felt with the advent of digital computers, leading to a discipline called "computational complexity." As an incidental remark, the phrase "degrees of difficulty" was the technical precursor of the fortunate expression "computational complexity," introduced by Hartmanis and Stearns in 1965 [1].

Complexity is clearly a measure of effort, that is, of cost. Cost, on the other hand, is precisely quantified in the context of the model which is selected for carrying out computation. The choice of the computing model essentially reflects the viewpoint and the flavor of the analysis, as we shall see below. In all models, however, complexity is measured by the usage of "resources" required by a specific computation. These resources are, broadly speaking, *time* and *space*. Time is frequently expressed as the number of conventionally defined steps required by the computing device in order to complete the computation. Space may assume different connotations, such as the number of required memory locations in a somewhat idealized computer, the required length of the tape(s) of a Turing machine, or, as is the case in models of parallel computation, the number of individual processors simultaneously cooperating for the completion of a computing task.

The most abstract approach, known indeed as *abstract computational complexity* [2], [3], aims at a theory of complexity which is machine-independent. It concerns essentially the complexity of all possible computations, that is, the computation of all partial recursive functions from the integers to the integers. Its results, which hold essentially for any measure of complexity and for any computing device, are mathematically very sophisticated and sometimes surprising, but are not particularly relevant to the applied worker who is seeking a solution of a specific problem on a conventional computer.

All the other approaches constitute what is known as *concrete computational complexity*. Since we shall not deal with the abstract approach, we shall hereafter omit for brevity the attribute "concrete." In this brand of complexity, which is by far the one which has attracted the greatest research interest, specific models of computation are selected for a definition of measures of complexity. Without analyzing the various interesting shades which have been considered, the two most prominent models are the Turing machine and a suitably idealized version of the digital computer (Random access machine (RAM), Random access stored program machine (RASP), etc., discussed in [4, Chapter 1]).

The advantages of the Turing machine model are the simplicity of its instruction set, the well-definedness of its computation "step," and the reasonable assumption that all steps have identical duration; these facts greatly simplify the assessment, for example, of the amount of time taken by a Turing machine to complete a given task. On the other hand, real (or, realistically idealized) computers have a much more complex instruction set than a Turing machine, and, although simulation of any computer by a Turing machine is possible, very complicated Turing machine programs may be needed to simulate single computer instructions. Thus analyses for the Turing machine may be scarcely significant for the applied analyst, except for problems whose known solution algorithms require such an inordinate amount of time that replacing a Turing machine with an actual computer does not change the order of magnitude of the complexity; we shall reconsider this class of problems later in this section.

A great part of the current results, and particularly those which are of interest to the computer practitioner, make reference to models whose repertoire of elementary operations resemble those of conventional computers. One immediate difficulty is that, in order to be as realistic as possible, each instruction should be assigned a specific execution time. One could, of course, follow this approach and develop on paper, in a very detailed fashion, a "pedagogical computer," such as the well-known MIX used throughout Knuth's work [5]; and there are instances, as we shall shortly see, in which such detailed analysis is the only reasonable approach. In most cases, however, one contents oneself with an analysis that gives as a measure of complexity the number of times some selected key operations are performed when running an algorithm, the justification being that the time actually required by the algorithm is proportional to that number. The choice of the key operations greatly varies from application to application. For example, for sorting, merging, and pattern-matching algorithms, it is natural to use the number of two-element comparisons as a measure of complexity; in numeric or algebraic applications, one considers the numbers of arithmetic operations, such as addition, multiplication, and division; and so on. Frequently, however, since the execution of a program consists of a concatenation of two types of processing—straight-line sequences and loops of instructions—and the bulk of the time is customarily attributable to the loops, one may ignore the time taken by straight-line sequences and simply count the number of times loops are executed.

A simple example may clarify this point. Suppose we want to design an algorithm to test whether a given positive integer n is prime or not. For simplicity, assume that the word size of our computer is sufficiently large to contain the representation of n. A naive solution involves testing the divisibility of n by any integer j such that $1 < j < n$; n is a prime if and only if n is not divisible by any such j. This approach gives the following algorithm:

1. If $n = 1, 2$, then n is prime; halt.
2. Set $j \leftarrow 2$.
3. While $j < n$ do:
   - 4. If n is divisible by j, then n is composite; halt.
   - 5. Let $j \leftarrow j + 1$.
6. n is prime; halt.

Clearly if n is prime, the loop consisting of steps 4 and 5 is executed $(n - 2)$ times, whereas steps 1, 2, and 6 are executed exactly once. Notice that either step 4 or step 5 may involve several computer instructions, although a fixed sequence in both cases. Thus the total running time is proportional to the number of times the loop is executed. A less trivial approach to this problem is based on the consideration that a composite integer n is the product of two integers $j_1 \cdot j_2$, the smaller of which cannot exceed the value $\sqrt{n}$. This leads to a variant of the preceding algorithm, where step 3 is replaced by "while $j^2 < n$ do." Notice that, in this formulation, when n is prime the loop is executed approximately $\sqrt{n}$ times, although the test "$j^2 < n$" involves squaring the index j and is therefore somewhat more complicated than the test "$j < n$." We observe that, without referring to a specific computer, we are unable to quantify upper-bounds to the times taken by the two algorithms for a specific n; we can say, however, that the variant runs faster than the original procedure, and, specifically, that the running times "grow like n and $\sqrt{n}$," respectively. This determination of a "rate-of-growth" or "order of magnitude" of measures of complexity of an algorithm, whether time or space, as functions of the input, is a central feature of computational complexity and deserves some additional attention.

A complexity measure for an algorithm is normally expressed by a function $\phi(\ )$ of some significant indicators, which specify the "problem size." No general definition of problem size is possible, since the latter essentially depends upon specific characteristics of the mathematical objects constituting the problem input. For example, we have seen that, in a test for primality of n, n itself is taken as the problem size; in graph problems, one may take the number v of vertices or the number e of edges of the graph, or the pair (v, e); in matrix problems, one usually takes the order of the matrix, and so on. The only characterization of problem size which has a vague appearance of generality is the number of items in the input set (called the *input size*), since this choice is natural for a large number of problems. In some cases, the input size is given by the number of bits required to express the input set. With these precautions, we now assume that a single natural integer n specifies the problem size. Then, according to a recent proposal by Knuth [6], we say that some complexity measure $\phi(n)$ of an algorithm is $O(f(n))$ (read "of order (at most) f(n)," for some function f( )), if there exist positive constants C and n_0 such that $|\phi(n)| \leq C f(n)$ for $n \geq n_0$.* The "O" notation (also called "big-oh") refers to an upper-bound to the complexity measure; thus, when we want to refer to a lower-bound, we shall use the notation $\Omega(f(n))$ (read "of order at least f(n)") if there exist positive constants C and n_0 such that $\phi(n) \geq C(f(n))$ for $n \geq n_0$. Obviously, these definitions apply when $n \rightarrow \infty$ and are therefore called "asymptotic measures."

The consideration of the order of magnitude not only allows one to ignore the particular value of the constant of proportionality between the actual running time and the function f(n), but, more

* According to current jargon, sometimes the phrase "an O(f(n)) algorithm" is used with the meaning of "an algorithm whose complexity is O(f(n))."

important, allows the comparative evaluation of the asymptotic performances of several algorithms devised for a given problem. The asymptotic behavior of algorithms is very significant because it determines the choice of the algorithm when the input size becomes very large. For rather small values of n, however, this approach must be taken with some discretion, since in such cases it is the actual running time and not its order of magnitude which determines the choice of the algorithm.

When discussing the algorithms for testing primality, we estimated the largest value of the running time, which occurs when the integer n is prime; however, shorter times occur when n is composite. Our simple analysis of those algorithms obtained a "worst-case" measure of complexity. The performance of an algorithm can be evaluated either on a worst-case input or on an average-case input, assuming some distribution over the set of possible inputs. Both types of analyses are significant; a large majority of the known results, however, are worst-case analyses both because such analyses are normally simpler and because, but for a few cases, there is little agreement on the choice of the probabilistic model.

The preceding discussion merely sketches the main features of an important aspect of concrete computational complexity, known as *analysis of algorithms*. The performance of an algorithm which solves a given problem implicitly provides an upper bound to the performance, or complexity, of the set of algorithms which may be devised for that problem. Closely related to the analysis is the research effort known as *design of algorithms*, whose objective is the development of procedures for the solution of a given problem, so that their performance is provably superior to that of previously-known algorithms. A complementary aspect of computational complexity is the determination of lower-bounds to the performance of any possible algorithm for a given problem. Clearly, when the two bounds come close or, better, coincide, one has discovered an *optimal* algorithm, i.e., one has obtained a characterization of the inherent difficulty of a problem. One must readily add, however, that this fortunate event occurs only for a small minority of the problems considered and that researchers in this area are much more skillful in designing and analyzing algorithms for a specific problem than in proving their optimality.

Finally, to provide an appreciation of the great development of this research area, we simply list the following facts which hold at the time of this writing: four specific textbooks [4], [7], [8], [9] are available and more are announced as forthcoming, not to mention Knuth's monumental work [5], which can be appropriately considered as an encyclopedia of computational complexity; journals of theoretical computer science devote increasing attention to computational complexity, and so do several prestigious symposia. In consideration of these extensive developments, even a survey of the field would be a project largely exceeding the scope of this article. Therefore, in order to concretely illustrate some of the approaches and techniques used in this field, we shall confine ourselves to the case-study format and describe in some detail some significant results in several problem areas.

None of the case studies planned for discussion concerns the so-called NP-complete problems, mainly because of the very extensive and rapidly growing literature on this subject (see, for example, [4], [9] and the excellent textbook by M. Garey and D. S. Johnson [23]). However, due to the central importance of this topic in the theory of computational complexity, we cannot close this introductory section without mentioning some of its salient aspects. For several years computer scientists and practitioners have been confronted with very difficult problems, mainly arising in combinatorics, operations research, and graph theory, such as problems of scheduling, assignment, sequencing, etc. All these problems appear to require an inordinate amount of time for their solution, specifically a number of computational steps exponential in the input size. More recently, through the combined efforts of S. A. Cook [10] (who pioneered the topic), of R. M. Karp [11], [12], and of several other workers, it was shown that almost all the classical combinatorial problems reputed to be intractable are equivalent in the sense that, if one of them is solvable by a polynomial-time-bounded algorithm, all of them are. The equivalence is based on the fact that any problem in that class can be reformulated as any other problem in the class by means of a transformation which requires at most a polynomial time. Thus, since the transformation, albeit complex, can be carried out in polynomial time, it is ensured that the distinction between an exponential-time effort and a polynomial-time effort is preserved through the transformation. This justifies our earlier remark that the Turing machine is an appropriate computation model for this class of problems. The term "NP-complete" is an abbreviation of "nondeterministic polynomial-time complete," where *nondeterministic polynomial-time* means that there exists a nondeterministic Turing machine for solving the problem (or, equivalently, a back-track search algorithm of polynomial-bounded depth for that problem), and *complete* refers to the mentioned problem transformability. The interested reader should consult the cited references to familiarize himself with this fascinating topic.

2. A COLLECTION OF CASE STUDIES

In this section we shall examine in some detail some representative problems from different areas. Although the common objective is the design of algorithms which are either optimal or whose complexity measures improve over previous results, nonetheless in each particular problem area conventions, measures of complexity, and techniques have a distinguishing flavor. Hopefully, the common features as well as the dissimilarities will emerge from the following presentation.

2.1. Multiplication of Integers

2.1.1. *Generalities.* The problem of multiplying two integers falls in the area of arithmetic or numeric computation. The two operands are assumed to be given as two sequences of n digits each, and, without loss of generality, we shall assume the digits are binary. Since n may be arbitrarily large, the operands cannot be stored in the memory of the computing device. Therefore the natural model for this type of problem is a tape machine with a bounded memory, which, for that matter, could be a Turing machine. The computational steps to be counted are conveniently chosen as operations with operands consisting of one bit each. This choice may appear a little artificial at first sight, since the arithmetic instructions of any practical computer act on strings of several bits; however, when the operand size is much larger than the computer word size, our simplification will only change the computation time by a multiplicative constant. These elementary operations are usually referred to as "bit operations," to contrast them against "arithmetic operations" which act on word size operands.

The simplest and best-known procedure for integer multiplication is the so-called "schoolboy method," which consists of the addition of appropriately shifted partial products. The number of bit operations is clearly $O(n^2)$ for operands of length n, since there are n partial products, each of which consists of n bits. This method can be speeded up by the following modification. Let

$$A \equiv a_{n-1}a_{n-2}\cdots a_0 \qquad \text{and} \qquad B \equiv b_{n-1}b_{n-2}\cdots b_0$$

be the two operands and assume, for simplicity, that n is even. We split the sequence $a_{n-1}\cdots a_0$ into two halves $a_{n-1}\cdots a_{n/2} \equiv A_1$ and $a_{(n/2)-1}\cdots a_0 \equiv A_0$ so that $A = A_1 2^{n/2} + A_0$ and do likewise for the operand B. Then we have

$$AB = A_1B_1 2^n + (A_1B_0 + A_0B_1)2^{n/2} + A_0B_0. \tag{1}$$

This expression indicates that the multiplication of two n-bit operands is reduced to four multiplications of two (n/2)-bit operands, with no apparent computational advantage. Suppose now we compute the functions

$$C_1 = A_1B_1, \quad C_0 = A_0B_0, \quad \text{and} \quad C_2 = (A_0 + A_1)(B_0 + B_1);$$

next we note that $A_1B_0 + A_0B_1 = C_2 - C_1 - C_0$, i.e., the three terms of the expression (1) can be computed with *three* multiplications and four additions of (n/2)-bit operands. It is easily realized that addition can be done in a number of bit operations which is proportional to the operand length. It follows that, denoting by M(n) the number of operations or, briefly, the time to multiply two n-bit operands, we obtain a recurrence equation

$$M(n) = 3M(n/2) + O(n),$$

which defines M(n). The solution of this equation is obtained by standard methods and is found to be $M(n) = O(n^{\log_2 3})$, which shows a substantial improvement over the naive schoolboy method. Incidentally, we have just seen an instance of the "divide-and-conquer" technique. This technique, which is widely used in the design of algorithms, consists in reducing the original problem to a collection of simpler problems.

Before trying to push further the previous approach, let us take a critical look at what we did. If we replace the number $2^{n/2}$ by an indeterminate x in the expressions $(A_1 2^{n/2} + A_0)$ and $(B_1 2^{n/2} + B_0)$, we map the integers A and B to two polynomials $(A_1 x + A_0)$ and $(B_1 x + B_0)$, respectively. Thus the terms $A_1 B_1$, $A_1 B_0 + A_0 B_1$, and $A_0 B_0$ are the coefficients of the polynomial $C(x) = A(x)B(x)$. Each of these coefficients is an n-bit binary number, and, when $2^{n/2}$ is substituted back for x, the relative alignment of their representations is as shown in Figure 1. Thus, the product AB is obtained by adding these three numbers as shown in Figure 1, and this operation is normally referred to as "releasing the carries."

We have therefore transformed an integer multiplication operation into the following sequence of operations:

1. Multiplication of two polynomials.
2. Release of the carries.

As we shall later see, the carry release is a simple operation which can be done in time proportional to the number of bits of the result. Therefore, we shall concentrate on polynomial multiplication.

2.1.2. *Multiplication of two polynomials (evaluation and interpolation).* A straightforward method for polynomial multiplication involves distributing one polynomial into the other and collecting terms with identical powers of x. But there is a more subtle method. Suppose $A(x) = \sum_{j=0}^{p-1} A_j x^j$ and $B(x) = \sum_{j=0}^{p-1} B_j x^j$; then their product C(x) has degree $2p - 2$, i.e., it has $(2p - 1)$ coefficients. Select now $(2p - 1)$ distinct real values $x_1, \ldots, x_{2p-1}$, called "points," and *evaluate* A(x) and B(x) at each of these points, thereby obtaining two sets of values $\{A(x_i)\}$ and $\{B(x_i)\}$. Then clearly, for any i in the

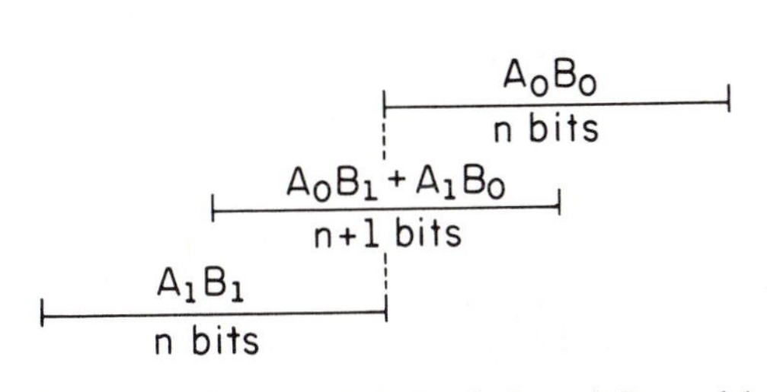

FIG. 1. Relative alignment of $A_0 B_0$, $A_0 B_1 + A_1 B_0$, and $A_1 B_1$.

range [1, 2p − 1], $C(x_i) = A(x_i) \cdot B(x_i)$. From the set of values $\{C(x_i)\}$ we can now *interpolate* the polynomial C(x) of degree (2p − 2). Thus we have the following algorithm:

Polynomial multiplication

Input: A(x) and B(x), both of degree (p − 1), and distinct real values $x_1, \ldots, x_{2p-1}$.
Output: $C(x) = A(x) \cdot B(x)$.
Step 1. Evaluate $\{A(x_i)\}$ and $\{B(x_i)\}$.
Step 2. For $i = 1, \ldots, 2p - 1$, compute $C(x_i) = A(x_i)B(x_i)$.
Step 3. Interpolate C(x) of degree (2p − 1) from $\{C(x_i)\}$.

Evaluation and interpolation are mutually inverse and can be given a very compact description. Notice in fact that

$$[C_0, \ldots, C_{2p-2}] \begin{bmatrix} 1 & 1 & \cdots & 1 \\ x_1 & x_2 & \cdots & x_{2p-1} \\ \vdots & & & \\ x_1^{2p-2} & x_2^{2p-2} & \cdots & x_{2p-1}^{2p-2} \end{bmatrix} = [C(x_1), \ldots, C(x_{2p-1})].$$

Letting $\mathbf{C} \triangleq [C_0, \ldots, C_{2p-2}]$ and $\mathbf{C(x)} \triangleq [C(x_1), \ldots, C(x_{2p-1})]$ and denoting by V the above matrix, we have

$$\mathbf{C}V = \mathbf{C(x)} \qquad \text{(evaluation)},$$

and, since V, a Vandermonde matrix, is nonsingular,

$$\mathbf{C} = \mathbf{C(x)}V^{-1} \qquad \text{(interpolation)}.$$

We see that both evaluation and interpolation are equivalent to multiplying a vector by a matrix. It is now possible to develop an integer multiplication algorithm based on the outlined polynomial multiplication method. We choose the integer p sufficiently large, select (2p − 1) integer values $x_1, \ldots, x_{2p-1}$ and construct two fixed $(2p-1) \times (2p-1)$ matrices V and V^{-1}. Next, we split the n-bit sequence representing the operand A into p segments of n/p bits each; each of these segments can be viewed as the binary representation of an integer. Next we regard these integers as coefficients of a polynomial A(x) of degree (2p − 2) (notice that the (p − 1) higher degree coefficients of A(x) are 0). We do likewise for the other operand B. We now evaluate $\{A(x_i)\}$ and $\{B(x_i)\}$. Notice that multiplication of s-bit integers by a constant with a fixed number of bits requires time proportional to s; it follows that the $p(2p-1)$ multiplications of n/p-bit integers by constants, as specified by the evaluation step, collectively require time $O(n)$, and the same can be said for interpolation. We also observe that the numbers $A(x_i)$ and $B(x_i)$ are $(n/p+k)$-bit integers, where k is a constant depending upon the values $x_1, \ldots, x_{2p-1}$. It follows that the computation of each $C(x_i)$ is a multiplication of two $(n/p+k)$-bit integers. With the usual meaning of the function $M(\ )$, we have $M(n/p+k) = M(n/p) + O(n/p)$, and we can establish the following recurrence equation which defines $M(n)$:

$$M(n) = (2p-1)M\left(\frac{n}{p}\right) + O(n); \tag{2}$$

the solution of this equation is

$$M(n) = O(n^{\log_p(2p-1)}) = O(n^{1+1/\log_2 p}).$$

Thus by choosing p sufficiently large the integer multiplication time can be bounded from above by $O(n^{1+\varepsilon})$ for any $\varepsilon > 0$. Recurrence equation (2) has a somewhat undesirable feature: its right member consists of two terms, the first of which dominates the other and determines the form of the solution. In similar cases one seeks a modification of the algorithm which tends to equalize the two terms; we recognize that the functional form of $M(n)$ is due to the fact that p is fixed, which, on the other hand, is exactly why we were able to estimate as $O(n)$ the evaluation and interpolation times. Thus possible improvements can arise by making p a function of n; this is one of the key ideas of the remarkable integer multiplication algorithm due to Schönhage and Strassen [13], based on the Discrete Fourier Transform.

2.1.3. *The Discrete Fourier Transform and the FTT.* Of course, if p is no longer fixed, the Vandermonde matrix V and its inverse V^{-1} cannot be precomputed as auxiliary devices, but are determined by the operand sizes. Therefore, one must look for a pair (V, V^{-1}) which can be easily computed in each case. Under rather general hypotheses (i.e., that we deal with a *commutative ring* S), a surprising solution is obtained by choosing $x_i = \omega^{i-1}$, for $i = 1, \ldots, q$, where ω is a primitive root of unity of order

$q \triangleq 2p - 1$ in S. With this choice V becomes

$$V = \begin{bmatrix} 1 & 1 & 1 & \dots & 1 \\ 1 & \omega & \omega^2 & \dots & \omega^{q-1} \\ \vdots & \vdots & \vdots & \dots & \vdots \\ 1 & \omega^{q-1} & \omega^{2(q-1)} & \dots & \omega^{(q-1)(q-1)} \end{bmatrix}$$

and is known as a $q \times q$ Fourier matrix. Notice that $(V)_{ij} = \omega^{(i-1)(j-1)}$. A remarkable property of this matrix is that

$$\sum_{j=0}^{q-1} \omega^{sj} = \begin{cases} q & \text{if } s \bmod q = 0, \\ 0 & \text{otherwise.} \end{cases}$$

It follows that if we take the square of this matrix, the entry $(V^2)_{ij}$ is given by

$$(V^2)_{ij} = \sum_{t=1}^{q} (V)_{it}(V)_{tj} = \sum_{t=1}^{q} \omega^{(i-1)(t-1)+(j-1)(t-1)}$$

$$= \sum_{t-1=0}^{q-1} \omega^{(t-1)(j+i-2)} = \begin{cases} q & \text{if } (i+j-2) \bmod q = 0, \\ 0 & \text{otherwise.} \end{cases}$$

We conclude that V^2 has the form

$$V^2 = \begin{bmatrix} 1 & 0 & 0 & \dots & 0 & 0 \\ 0 & 0 & 0 & \dots & 0 & 1 \\ 0 & 0 & 0 & \dots & 1 & 0 \\ \vdots & \vdots & \vdots & \dots & \vdots & \vdots \\ 0 & 0 & 1 & \dots & 0 & 0 \\ 0 & 1 & 0 & \dots & 0 & 0 \end{bmatrix}$$

i.e., V is its own inverse except for a row and column permutation. But a most attractive property of the Fourier matrix is the ease with which it can be multiplied by a $(2p - 1)$-dimensional vector, by resorting to a technique known as the Fast-Fourier-Transform (FFT), due to Cooley and Tukey [14], which we shall now outline.

Suppose, for simplicity, that $q = 2r$ and $\mathbf{a} = [a_0, \dots, a_{2r-1}]$. We must compute $\mathbf{a}V$, the *Discrete Fourier Transform* (DFT) of $\mathbf{a}$. Let us compare the components $(\mathbf{a}V)_{j+1}$ and $(\mathbf{a}V)_{j+1+r}$. The former is given by the expression:

$$(\mathbf{a}V)_{j+1} = \sum_{i=0}^{2r-1} a_i \omega^{ij} = \sum_{i=0}^{r-1} a_{2i} \omega^{2ij} + \sum_{i=0}^{r-1} a_{2i+1} \omega^{2ij+j}$$

$$= \sum_{i=0}^{r-1} a_{2i} (\omega^2)^{ij} + \omega^j \sum_{i=0}^{r-1} a_{2i+1} (\omega^2)^{ij}. \tag{3}$$

Similarly, the component $(aV)_{j+1+r}$ is given by the expression:

$$(\mathbf{a}V)_{j+1+r} = \sum_{i=0}^{2r-1} a_i \omega^{(j+r)i} = \sum_{i=0}^{r-1} a_{2i}(\omega^2)^{ij} \cdot (\omega^{2r})^i$$

$$+ \omega^j \omega^r \sum_{i=0}^{r-1} a_{2i+1}(\omega^2)^{ij}(\omega^{2r})^i$$

$$= \sum_{i=0}^{r-1} a_{2i}(\omega^2)^{ij} - \omega^j \Sigma\, a_{2i+1}(\omega^2)^{ij} \tag{4}$$

since $\omega^{2r} = +1$ and $\omega^r = -1$ (ω is the primitive root of unity of order 2r). Notice now that $\Sigma a_{2i}(\omega^2)^{ij}$ is the $(j + 1)$-st component of the discrete Fourier transform of the r-component vector formed by the even-indexed coefficients of **a** (this Fourier transform obviously uses the root of unity ω^2, which is of order r). Similarly, $\Sigma a_{2i+1}(\omega^2)^{ij}$ is the $(j + 1)$-st component of the discrete Fourier transform of the odd-indexed coefficient sequence. Thus, we find that $(\mathbf{a}V)_{j+1}$ and $(\mathbf{a}V)_{j+1+r}$ are computable by the arrangement shown in Figure 2. We see that the DFT of a 2r-component vector is obtained with r multiplications, r additions, and r subtractions from the DFT's of two r-component vectors. Assuming that q is a power of 2, the same analysis can be carried out for each of the two FFT-computers for r inputs, and we reach the conclusion that the FFT calculation requires O(q log q) arithmetic operations. A similar result holds when q is a highly composite number, although we are explicitly interested in the power of 2 case.

The use of the Fourier matrix certainly simplifies the operations of interpolation and evaluation; however, there is still the difficulty

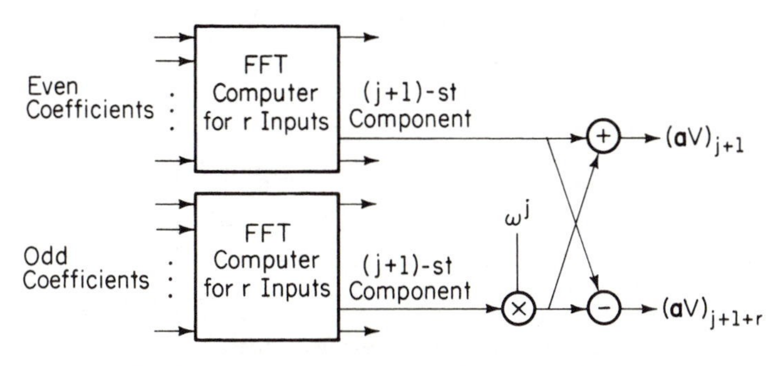

FIG. 2. A scheme for computing $(\mathbf{a}V)_{j+1}$ and $(\mathbf{a}V)_{j+1+r}$.

that it requires O(q log q) complex multiplications if ω is the qth primitive root of unity in the complex field. Schönhage and Strassen avoided this difficulty by the following extremely brilliant solution. Suppose that the two operands A and B are integers representable with at most $n = 2^{m-1}$ bits. Then, the product $AB < 2^{2^m} < 2^{2^m} + 1$. The integer $2^{2^m} + 1$, denoted by F_m, is called the Fermat number of order m and the set $\mathbb{Z}_m$ of integers modulo F_m is known as the *Fermat ring* of order m. Notice that as long as $A \cdot B \in \mathbb{Z}_m$, we can assume that arithmetic be done in $\mathbb{Z}_m$. A crucial property of $\mathbb{Z}_m$ is that 2 is a primitive root of unity of order 2^{m+1}; in fact $2^{2^m} = -1$ in $\mathbb{Z}_m$. It follows that multiplications of an integer by powers of the root reduce to shifts of the representation of this integer, provided we use 2^{m+1} bits for representing the operands. We now have developed all the preliminaries to the description of the integer multiplication algorithm.

2.1.4. *The Schönhage-Strassen integer multiplication algorithm.* In Figure 3 we sketch the relative length of the nonzero portions of the representations of the operands and of the product. Letting $m = 2s - 1$ (the case of even m is analogously handled), we chop the bit sequence of each operand into 2^{s+1} segments of 2^{s-1} bits each and let $A_{i-1}(B_{i-1})$ be the integer represented by the ith segment in A (in B), for $i = 1, \ldots, 2^{s+1}$. Thus we have

$$A = \sum_{i=0}^{2^{s+1}-1} A_i 2^{i2^{s-1}}, \qquad B = \sum_{j=0}^{2^{s+1}-1} B_j 2^{j2^{s-1}}.$$

Next, we compute

$$AB = C = \sum_{k=0}^{2^{s+1}-1} \left(\sum_{\substack{i+j=k \\ \text{mod } 2^{s+1}}} A_i B_j \right) 2^{k2^{s-1}} \triangleq \sum_{k=0}^{2^{s+1}-1} C_k 2^{k2^{s-1}}. \quad (5)$$

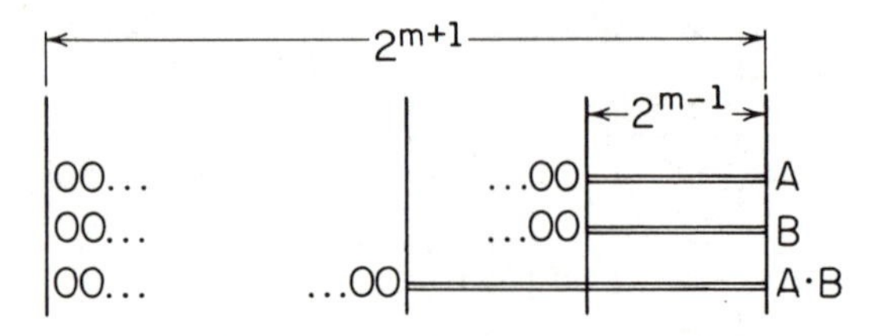

FIG. 3. Alignment of operands and result.

It is convenient to split the last summation into two parts, i.e.,

$$C = \sum_{k=0}^{2^s-1} C_k 2^{k2^{s-1}} + \sum_{k=2^s}^{2^{s+1}-1} C_k 2^{k2^{s-1}}$$

$$= \sum_{k=0}^{2^s-1} C_k 2^{k2^{s-1}} + \sum_{k=0}^{2^s-1} C_{k+2^s} 2^{k2^{s-1}} \cdot 2^{2^m}.$$

Recalling that $2^{2^m} = -1$ we have

$$C = \sum_{k=0}^{2^s-1} (C_k - C_{k+2^s}) 2^{k2^{s-1}}. \tag{6}$$

Notice now that $(C_k - C_{k+2^s})$ could be a negative integer. However, $C_j < 2^{s+1} \cdot 2^{2^s}$ for $j = 0, 1, \ldots, 2^{s+1} - 1$ (since, by (5), it is the sum of 2^{s+1} products whose moduli are bounded by 2^{2^s}), whence $(C_k - C_{k+2^s}) > -2^{s+1+2^s}$; since it will be convenient to deal with positive coefficients, if we add and subtract the same quantity from the right member of (6), we obtain

$$C = \sum_{k=0}^{2^s-1} (C_k - C_{k+2^s} + 2^{s+1+2^s}) 2^{k2^{s-1}}$$

$$+ \sum_{k=2^s}^{2^{s+1}-1} 2^{s+1+2^s} 2^{k \cdot 2^{s-1}} \triangleq \sum_{k=0}^{2^{s+1}-1} z_k 2^{k2^{s-1}}$$

where each z_k satisfies the inequalities

$$0 < z_k < 2^{s+2+2^s} \leq 2^{s+2}(2^{2^s} + 1) = 2^{s+2} F_s.$$

(*Notice that* 2^{s+2} *and* F_s *are relatively prime.*) At this point, the original multiplication is reduced to the problem of releasing the carries when adding the appropriately shifted numbers $z_0, z_1, \ldots,$ which are shown pictorially in Figure 4. We must still find an

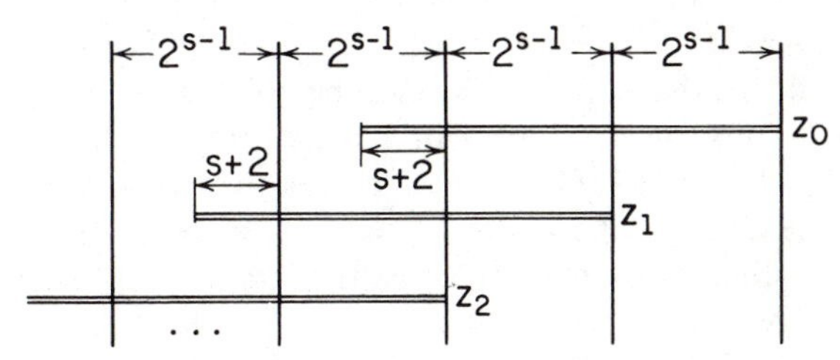

FIG. 4. Alignment of $z_0, z_1, z_2, \ldots$.

efficient way for computing the coefficients z_k. A most unexpected method for effecting this computation is provided by the following algorithm:

Computation of z_k

1. Compute $z'_k \triangleq z_k \bmod 2^{s+2}$;
2. Compute $z''_k \triangleq z_k \bmod F_s$;
3. Reconstruct z_k from z'_k and z''_k.

The correctness of this procedure is supplied by the Chinese Remainder theorem, since z'_k and z''_k are the remainders modulo two relatively prime integers. We shall now discuss how steps 1, 2, and 3 can be implemented and analyze their complexities.

Letting $\alpha_i = A_i \bmod 2^{s+2}$ and $\beta_j = B_j \bmod 2^{s+2}$ (for $i, j = 0, \ldots, 2^{s+1} - 1$), we obtain

$$z'_k = \left[\sum_{i+j=k} (\alpha_i \beta_j) + \sum_{i+j=k+2^{s+1}} (\alpha_i \beta_j)\right] - \left[\sum_{i+j=k+2^s} (\alpha_i \beta_j) + \sum_{i+j=k+3\cdot 2^s} (\alpha_i \beta_j)\right] \pmod{2^{s+2}};$$

The terms of the form $\Sigma_{i+j=k} (\alpha_i \beta_j) \bmod 2^{s+2}$ are easily computed by forming two integers A′ and B′ whose representation are shown below.

$$A' \equiv 0 \ldots 0\, \alpha_{2^{s+1}-1}\, 0 \ldots \quad \ldots \quad 0 \ldots 0\, \alpha_1\, 0 \ldots 0\, \alpha_0$$

$$B' \equiv 0 \ldots 0\, \beta_{2^{s+1}-1}\, 0 \ldots \quad \ldots \quad 0 \ldots 0\, \beta_1\, 0 \ldots 0\, \beta_0$$

where the number of the zeros separating α_i and α_{i+1} is chosen as the minimum required for avoiding any propagation between two consecutive columns when performing the multiplication $A' \cdot B'$. Since each column is the sum of at most 2^{s+1} terms upper-bounded by the values $(2^{s+2})^2$ we see that $(3s + 5)$ bits suffice to hold such sum. It follows that A′ and B′ can be formed as integers with $2^{s+1}(3s + 5)$ bits each, which can be multiplied in time at most $O([2^{s+1}(3s + 5)]^{\log_2 3}) < O(2^{2^s})$, i.e., in time less than linear in the length of the original operands.

Next, we shall consider step 3, i.e., the reconstruction of z_k from z'_k and z''_k. Notice at first that $z''_k = z_k - jF_s$, for some unknown integer $j < 2^{s+2}$. If we can determine j from z''_k and z'_k, we obtain

z_k. Now we have:

$$z_k'' = z_k - jF_s = z_k - j2^{2^s} - j,$$

whence $z_k'' \bmod 2^{s+2} = z_k \bmod 2^{s+2} - j$, that is,

$$z_k'' \bmod 2^{s+2} = z_k' - j.$$

We conclude that $j = z_k' - z_k'' \bmod 2^{s+2}$. Thus the computation of j requires time $O(s + 2)$ and the computation of z_k from z_k'' and j requires time $O(2^{s+1})$. We conclude that the reconstruction of all z_k's from the corresponding pairs (z_k', z_k'') requires time which is linear in the length of the original operands.

Finally, we consider the computation of z_k'', which turns out to dominate with its complexity the operations leading to the computation of z_k. Since $z_k'' \in \mathbb{Z}_s$, it is natural to carry out the operations in $\mathbb{Z}_s$, i.e., we must compute

$$z_k'' = [(C_k - C_{k+2^s}) \bmod F_s + 2^{s+1+2^s}] \bmod F_s.$$

The term $(C_k - C_{k+2^s}) \bmod F_s$ is expediently computed by using the FFT algorithm. Specifically we have

1. Compute the forward FFT in $\mathbb{Z}_s$ of the sequences $(A_0, A_1, \ldots, A_{2^{s+1}-1})$, $(B_0, B_1, \ldots, B_{2^{s+1}-1})$;
2. Multiply corresponding terms of the transforms (this is done by recursive calls of the multiplication algorithm in $\mathbb{Z}_s$).
3. Compute the inverse FFT.

To analyze the complexity of the overall multiplication algorithm, let $N(m)$ be the number of operations required to multiply two integers of length $2^m = 2^{2s-1}$. Steps 1 and 3 have global complexity 3 × (arithmetic operations for FFT of 2^s terms) × (length of the operands) $= \gamma_0[2^{s+1}(s + 1)]2^s$ for some constant γ_0; Step 2 involves 2^{s+1} multiplications in $\mathbb{Z}_s$, i.e., it has complexity $2^{s+1}N(s)$. Thus we obtain

$$N(m) \leq \gamma_0 2^{2s+1}(s + 1) + 2^{s+1}N(s) + O(2^{2s}).$$

If we solved this equation, we would obtain the result $N(m) = O(2^m m^2)$; however, there is a further simplification that, almost magically, allows another reduction in complexity. In fact, we need not compute the 2^{s+1} terms $C_k \bmod F_s$, but only 2^s differences of the form $(C_k - C_{k+2^s}) \bmod F_s$. It is easily shown from the properties of the FFT that these differences are completely determined from the odd-indexed terms of the transform product; hence step 2, in this revised form, requires only 2^s multiplications in $\mathbb{Z}_s$. It follows that

$$N(m) \leq \gamma_0 2^{2s+1}(s+1) + 2^s N(s) + O(2^{2s}).$$

If we assume now for simplicity that $m = 2s - 1 = 2(2s_1 - 1) - 1 = \ldots$, we obtain after one step

$$\begin{aligned} N(m) &\leq \gamma_0 2^{2s+1}(s+1) + 2^s[\gamma_0 2^{2s_1+1}(s_1+1) \\ &\qquad + 2^{s_1} N(s_1) + O(2^{2s_1})] + O(2^{2s}) + o(2^{2s}) \\ &= \gamma_0 2^{2s+1}(s+1) + \gamma_0 2^{2s+2} \frac{s+3}{2} \\ &\qquad + 2^{s+s_1} N(s_1) + O(2^{2s}) + o(2^{2s}). \end{aligned}$$

Thus, each time that we approximately halve the number of bits of the operands, we add a term of the form $O(2^m m)$; this process will be repeated $O(\log m) = O(\log \log n)$ times, whence we obtain the upper-bound expressed by the following theorem:

THEOREM. Two n digit operands can be multiplied with at most $O(n \log n \log \log n)$ bit operations.

2.2. Sorting by Merging. The problem of sorting n elements of a totally ordered set A (typically, n numbers) is one of the most celebrated and thoroughly studied examples in the area, commonly called *combinatorial* or *nonnumerical computation.* The fact that the elements dealt with are numbers is basically accidental, since the key operation used is "comparison," i.e., a test by which we can detect the ordering relationship between two elements of the set A, and the algorithms specify the strategy of execution of the comparisons to obtain the desired sorting.

There is a voluminous literature on sorting (see, e.g., [4], [9], [15]), concerning both practical algorithms and some deep theoretical questions, some of which are only partially answered; the interested reader is encouraged to refer to it. Our objective in this section is to discuss a specific and very important sorting technique, called sorting by merging (or, briefly, merge-sort), in different processing environments: the single conventional processor, the network of comparators, and the parallel processing system.

Let us first consider the basic idea of merge-sort. Assume that the elements to be sorted are placed in a unidimensional array A and let A[i] denote the ith element of this array; also, let A[i: j] denote the segment A[i], A[i + 1], ..., A[j]. Suppose now we have an algorithm called MERGE (A_1, A_2), which accepts two ordered sequences A_1 and A_2 and combines them into a single ordered sequence. We then have the following algorithm:

Algorithm SORT (A[1:n])

Step 1. Set $A[1:\lceil n/2 \rceil] \leftarrow \text{SORT}(A[1:\lceil n/2 \rceil])$ and $A[\lceil n/2 \rceil + 1:n] \leftarrow \text{SORT}(A[\lceil n/2 \rceil + 1:n])$.
Step 2. Set $A[1:n] \leftarrow \text{MERGE}(A[1:\lceil n/2 \rceil], A[\lceil n/2 \rceil + 1:n])$ and halt.

Notice, incidentally, that this is a recursive algorithm which contains among its steps "calls" to itself for operating on inputs of increasingly smaller size (in this case, geometrically decreasing). In words, we split the original array into two segments of approximately the same size, sort them separately, and finally use the MERGE operation to combine them. Thus, we can have as many distinct algorithms which comply with the description given above as we have ways of specifying the MERGE operation.

A simple merge algorithm constructs the combined sequence term by term. The first term is the smaller of the respective smallest terms of the two sequences to be merged; this term is removed from the input sequence to which it belongs and transmitted to the output sequence, and exactly this process is repeated until the input sequences are exhausted. Less descriptively, let A[1:n] and B[1:m] be two input sequences with $A[1] \leq \cdots \leq A[n]$ and $B[1] \leq \cdots \leq B[m]$, and $C[1:n+m] = \text{MERGE}(A[1:n], B[1:m])$. The sequence $C[1:n+m]$ is constructed as follows:

Algorithm MERGE (A[1:n], B[1:m])

Step 1. Set $i \leftarrow 1, j \leftarrow 1, k \leftarrow 1$.
Step 2. If $i > n$, set $C[k:m+n] \leftarrow B[j:m]$ and halt.
Step 3. If $j > m$, set $C[k:m+n] \leftarrow A[i:n]$ and halt.
Step 4. If $A[i] \leq B[j]$, set $C[k] \leftarrow A[i]$ and $i \leftarrow i + 1$; else set $C[k] \leftarrow B[j]$, and $j \leftarrow j + 1$.
Step 5. Set $k \leftarrow k + 1$ and go to step 2.

We now analyze the time complexity of the given algorithms. Since, in our scheme, sorting depends critically on merging, we begin from the latter. Notice that each comparison between A[i] and B[j] is accompanied by a fixed amount of work in the loop consisting of steps 2, 3, 4, and 5. Since this loop is executed at most $(n + m - 1)$ times, we conclude that the running time $M(n, m)$ required to merge two sequences of respective lengths n and m is at most $O(n + m)$. With this result, denoting by $S(n)$ the time required to sort n numbers, a straightforward analysis of our sorting algorithm shows that step 1 runs in time $2S(n/2)$ and step 2 runs in time $M(n/2, n/2)$. Thus we have the recurrence equation

$$S(n) = 2S(n/2) + M(n/2, n/2) = 2S(n/2) + O(n)$$

which is solved by standard methods as $S(n) = O(n \log n)$. It can be shown that this order of complexity is optimal (see Section 2.3); moreover, if one exclusively counts comparisons, merge-sort requires a number of comparisons which differs from the optimal only by additive terms which are asymptotically negligible with respect to $n \log n$.

The merge algorithm we have just described is of the sequential type, i.e., the operations are executed in sequence. Such is the processing environment of a conventional computer, also referred to as the *sequential processor*. We now want to investigate whether sorting by merging lends itself to implementation by a computing system in which several operations can be performed concurrently, or, equivalently, in which several sequential processors can simultaneously operate on the same data set. Any such system is normally referred to as a parallel system.

The advantage that one expects in going from a sequential algorithm to a parallel algorithm is a speed-up of the computation; in other words, one trades time with equipment. More specifically, assuming that all processors considered are constructed with the same technology (i.e., have the same operational speed) and assuming that T_1 is the running time of the best-known algorithm to solve a given problem on a single processor, the most that one can hope for in employing k processors for the same problem is to achieve a running time $T_k = T_1/k$; normally, however, except for some very particular problems, $T_k > T_1/k$, that is, there is some loss with respect to the optimum speed-up. It is also appropriate to mention that the case in which k is fixed is referred to as *bounded parallelism*, whereas *unbounded parallelism* denotes the case in which as many processors are available as one sees fit.

In connection with the sorting by merging problem, we shall consider two instances of parallel computing: the network of comparators and the fully parallel system (shared-memory-machine).

The network of comparators is an interconnection of modules called comparators. A comparator is a two-input, two-output device which receives two numbers on its input lines and places the larger number on a specified output line and the smaller on the other. A network of comparators has an arbitrary interconnection, except for the constraints that there is no feedback and that each comparator output line is either connected to exactly one comparator input line or is a network output line; moreover, each network input line is connected to exactly one comparator input line (fan-out restriction). It is easy to realize that a network of comparators is a parallel system. In fact it may be convenient to think of it as a cascade of stages, which form a partition of the network modules, with the property that the modules of a stage can operate in parallel. A partition of a network into stages can be obtained very simply; think of the network as a directed graph, whose vertices are the comparators and the network input lines and whose arcs are the connections, directed toward comparator input lines. With each comparator we associate an integer i, called the level, which is the length of the longest path from the input lines to that comparator. Defining as a time unit the time required for a comparator to operate, it is clear that a comparator at level i will not have its operands available before the $(i - 1)$-st time unit; therefore, if all the comparators at level i are placed in the ith stage, it is clear that they can operate in parallel. Notice also that, since at any given level each operand can be at most on one comparator input line, each stage contains at most $\lfloor n/2 \rfloor$ comparators. This indicates that a comparator network with n input lines and with the stated fan-out restriction is equivalent to a system of $\lfloor n/2 \rfloor$ processors operating in parallel and accessing an n-cell memory. At each time unit, or step, each processor reads two operands from memory, compares them, and stores into memory the two results. The read-store scheme at the ith time unit is completely specified by the wiring of the ith stage of the equivalent network; it follows that this scheme is fixed and is not influenced by the outcomes of previous comparisons—that is, the algorithm embodied by the network is *nonadaptive*.

We shall now analyze the number of stages, i.e., the time required by a network that sorts n numbers by repeated merging. For simplicity we shall assume that n is a power of 2, that is, $n = 2^k$. A 2^k-number sorting network can be constructed as shown in Figure 5, i.e., it consists of two 2^{k-1}-number sorting networks operating in parallel followed by a 2^k-number merging network. Clearly the structure of this network entirely reflects the organization of the algorithm SORT previously illustrated, and our problem is reduced to the analysis of the merging network. The latter can be constructed as follows. (This very interesting construction is due to K. E. Batcher [16].) Let $A = (a_0, \ldots, a_{2^{k-1}-1})$ and $B = (b_0, \ldots, b_{2^{k-1}-1})$ be the two sorted sequences to be merged with $a_0 \leq a_1 \leq \cdots \leq a_{2^{k-1}-1}$, and $b_0 \leq b_1 \leq \cdots \leq b_{2^{k-1}-1}$. The merging scheme is also of a recursive type (Figure 6); that is, we separately merge the even-indexed terms and the odd-indexed terms by means of two 2^{k-1}-number merging networks operating in parallel, and combine the outputs of the latter. This combination is very simple to implement. In fact, referring to Figure 6, assume inductively that the elements on the output lines of the merging

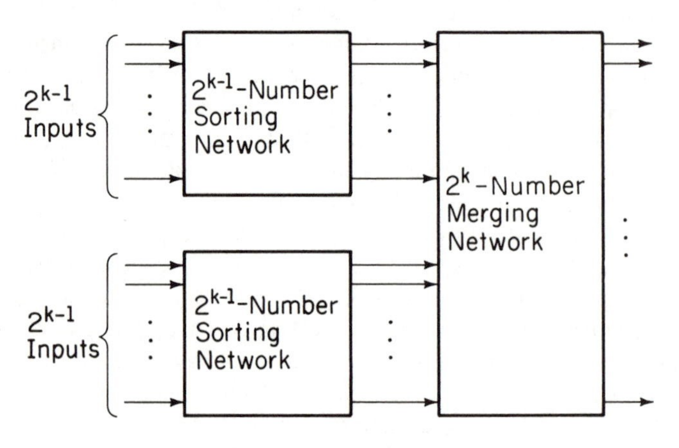

FIG. 5. A merge-sort sorting network.

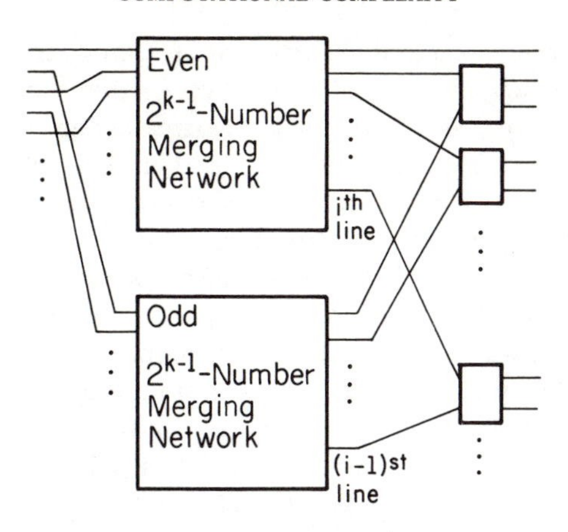

FIG. 6. Batcher's merging network.

networks appear in increasing order from top to bottom. Consider the element c_i which appears on the ith output line of the EVEN merging network and suppose, without loss of generality, that it coincides with b_{2s} (for some $0 \leq s \leq 2^{k-2} - 1$). Clearly on the first $(i - 1)$ lines of this EVEN network there are s elements of sequence B and $(i - 1 - s)$ elements of sequence A; this implies that the first $(i - s - 2)$ odd-indexed terms of sequence A are no larger than $c_i = b_{2s}$. It follows that the elements appearing on the $(i - s - 2) + s = (i - 2)$ top lines of the ODD merging network are known to be no larger than c_i. By a similar argument one can show that the elements appearing on the bottom $(2^{k-1} - i + 1)$ lines of the ODD merging network are known to be not smaller than c_i, so that we conclude that c_i must be compared only with the element on the $(i - 1)$-st line of the ODD merging network. Thus the merging process is completed by a single stage consisting of $(2^{k-1} - 1)$ comparators following the EVEN and ODD merging networks; hence a 2^k-number merging network can be constructed which consists of $\log_2 2^k = \log_2 n$ stages of comparators. This result can be used to evaluate the number of stages of the sorting network, which is

$$\log_2 2^k + \log_2 2^{k-1} + \cdots + \log_2 2 = \tfrac{1}{2} \log_2 2^k(\log_2 2^k + 1)$$
$$= \tfrac{1}{2} \log_2 n(\log_2 n + 1).$$

Thus we see that this parallel sorting algorithm does not achieve an optimal speed-up with respect to the best-known sequential sorting algorithm, the loss being a factor approximately equal to $\frac{1}{4}\log_2 n$.

A most surprising fact, however, is that the order of time required by merging with a network of comparators cannot be improved upon, as the following argument, due to Floyd (see [15, vol. 3, p. 230]) shows. Let $C(2t, 2t)$ be the minimum number of comparators of a network that sorts an input sequence $a_1, a_2, \ldots, a_{4t}$, consisting of two interleaved sorted sequences $a_1 \leq a_3 \leq \cdots \leq a_{4t-1}$ and $a_2 \leq a_4 \leq \cdots \leq a_{4t}$. Each comparator is characterized by a pair of indices (i, j) if it compares a_i and a_j. Divide the comparators into three classes: Class 1: $i \leq 2t$, $j \leq 2t$; Class 2: $i \geq 2t + 1$, $j \geq 2t + 1$; Class 3: $i \leq 2t$, $j \geq 2t + 1$. Clearly Class 1 must form a merging network for 2t inputs, since $a_{2t+1}, \ldots, a_{4t}$ may already be in their final arrangement; so does Class 2. Finally, the input sequence for which $a_{2s} > a_{2r-1}$ for $s, r = 1, \ldots, 2t$ requires at least t exchanges between the first half and the second half of the input sequence, so that Class 3 must contain at least t comparators. We conclude that

$$C(2t, 2t) \geq 2C(t, t) + t$$

i.e., $C(2t, 2t) \geq t \log_2 t$. Since each stage can contain at most 2t comparators, the number of stages is at least $\frac{1}{2}\log_2 t$, thus proving the original claim.

It is now natural to ask the question whether additional speed-ups can be obtained by employing $O(n)$ processors, without the restrictions embodied by the network constraints. In other words, we assume that a given element can be simultaneously compared with more than another element, and we let the algorithm be adaptive. No better answer was available until very recently, when Valiant [16] proposed the following interesting parallel merging algorithm.

Let $A[1:n]$ and $B[1:m]$ be two sequences sorted in ascending order, with $m \geq n \geq 4$, and assume that $\lfloor\sqrt{nm}\rfloor$ processors are available. We partition sequence A into segments, the first elements of which are $A[i\lceil\sqrt{n}\rceil + 1]$ for $i = 0, 1, \ldots, \lfloor\sqrt{n}\rfloor - 1$; similarly, B is partitioned into segments whose first elements are $B[i\lceil\sqrt{m}\rceil + 1]$. Let $A[i\lceil\sqrt{n}\rceil + 1] \triangleq A'[i]$ and $B[j\lceil\sqrt{m}\rceil + 1] \triangleq B'[j]$ for

$i, j = 0, 1, \ldots$. We now perform the following operations:

Algorithm PARALLEL MERGE

Step 1. Compare in parallel each element of A' with each element of B'.

Comment: The total number of processors required is $\lfloor\sqrt{n}\rfloor \cdot \lfloor\sqrt{m}\rfloor \le \lfloor\sqrt{nm}\rfloor$, i.e., there are enough processors to carry out this step in one unit of time.

Step 2. For each $A'[i]$ $(i = 0, 1, \ldots, \lfloor\sqrt{n}\rfloor - 1)$ decide the segment of B into which it must be inserted.

Comment: Let P_{ij} be the processor assigned to compare $A'[i]$ and $B'[j]$ for $i, j = 0, 1, \ldots$. Since $B'[0] \le \cdots \le B'[\lfloor\sqrt{m}\rfloor - 1]$ the *sequence* consists of two not simultaneously void segments $P_{i0}, \ldots, P_{ik}$ and $P_{i,k+1}, \ldots, P_{i,\lfloor\sqrt{m}\rfloor - 1}$, such that for $j \le k$ we have $A'[i] \ge B'[j]$ and for $j > k$ we have $A'[i] < B'[j]$. Thus k is determined in fixed time by letting each P_{ij} compare the outcome of its comparison with those of $P_{i,j-1}$ and $P_{i,j+1}$ (ignoring for simplicity the end-effect).

Step 3. Insert each $A'[i]$ into the segment of B determined in Step 2, by comparing it simultaneously with each element of the latter except the first.

Comment: There are $(\lceil\sqrt{m}\rceil - 1)$ comparisons for each $A'[i]$; thus $\lfloor\sqrt{n}\rfloor(\lceil\sqrt{m}\rceil - 1) < \lfloor\sqrt{nm}\rfloor$ processors are sufficient. The insertion can be done in unit time, by an argument analogous to the one in the Comment to Step 2. This insertion induces a new segmentation of B into $B[k_i + 1 : k_{i+1}]$, where $B[k_i] \le A'[i] < B[k_i + 1]$ for $i = 1, 2, \ldots, \lfloor\sqrt{n}\rfloor$.

Step 4. For $i = 0, \ldots, \lfloor\sqrt{n}\rfloor - 1$ simultaneously merge in parallel $A[i\lceil\sqrt{n}\rceil + 1 : (i + 1)\lceil\sqrt{n}\rceil]$ and $B[k_i + 1 : k_{i+1}]$.

Comment: This step specifies the simultaneous execution of $\lfloor\sqrt{n}\rfloor$ parallel merges of sequence pairs. The ith merge operation involves two sequences of respective lengths x_i and y_i, where $x_i \le \lceil\sqrt{n}\rceil - 1$ and y_i is the number of elements in $B[k_i+1 : k_{i+1}]$. We now assume inductively that this merge can be done in parallel with $\lfloor\sqrt{x_i y_i}\rfloor$ processors. The total number of required processors is given by $\Sigma\lfloor\sqrt{x_i y_i}\rfloor$. By Cauchy inequality we have

$$\Sigma\sqrt{x_i y_i} \leq \sqrt{(\Sigma x_i)(\Sigma y_i)};$$

recalling that $\Sigma x_i = n - \lfloor\sqrt{n}\rfloor$ and $\Sigma y_i = m$ we obtain:

$$\Sigma\lfloor\sqrt{x_i y_i}\rfloor \leq \Sigma\sqrt{x_i y_i} \leq \sqrt{(\Sigma x_i)(\Sigma y_i)} = \sqrt{(n - \lfloor\sqrt{n}\rfloor)m}$$
$$= \sqrt{nm} \cdot \sqrt{1 - \frac{\lfloor\sqrt{n}\rfloor}{n}} \leq \lfloor\sqrt{nm}\rfloor,$$

where the latter inequality holds for $m \geq n \geq 4$. Thus there is a sufficient number of processors to be distributed to execute the simultaneous merge operations.

We can now evaluate the running time of the described algorithm. Steps 1, 2, and 3 each require a fixed amount of time; the recursive call represented by Step 4 involves sequence pairs whose smaller member has size at most $\lceil\sqrt{n}\rceil - 1$. Thus the problem size has been reduced according to the square root. It follows that the merge is completed with a recursion of depth at most $\log_2 \log_2 n$, i.e., the running time of the parallel-merging algorithm is $O(\log \log n)$.

This merging algorithm can now be used to obtain a fast parallel algorithm for sorting n numbers with n/2 processors. Assume for simplicity that $n = 2^k$. We begin by forming sequences of length 2 and at each subsequent step we merge sequence pairs whose common length doubles at each step. At the ith step we merge 2^{k-i} sequenced pairs of common length 2^{i-1}: by the preceding discussion this can be done by the described merging algorithm with $2^{k-i} \cdot (\lfloor\sqrt{2^{i-1} \cdot 2^{i-1}}\rfloor) = 2^{k-1}$ processors, which are available by hypothesis. Thus in $\log_2 n$ merging steps sorting is completed; since each merging step runs in time at most $O(\log \log n)$, sorting can be done in parallel with n/2 processors in time at most $O(\log n \cdot \log \log n)$.

Thus we see that the speed-up loss with respect to the best known sequential algorithm has been reduced to a factor $O(\log \log n)$, a substantial improvement brought about by removing the fan-out restriction and by devising an adaptive scheme. Notice, however, that the preceding analysis of Valiant's algorithm is restricted to the so-called "data dependence" of sorting and ignores the complexity of another important facet of the algorithm, called "data movements." In fact, the adopted computation model consists of a set of identical processors, each capable of random accessing a common memory; of course, an additional device—called an *interconnection network*, or, briefly, a *switch*—is needed for aligning each processor with the memory cell accessed by it. The work done by this network is referred to as "data movements."

Finally, we mention that more recent results [17], [18] exhibit, for the same computation model, enumeration-sorting parallel algorithms, which have better time performance than the known merge-sorting parallel algorithms described above.

2.3. Convex Hulls of Finite Sets of Points. Increasing attention is being currently devoted to the computational solution of problems of a geometric nature, for they occur in a number of fields, such as operations research, pattern recognition, design automation, and statistics. This interest has coalesced into a new branch of computational complexity, adeptly called *computational geometry* by Shamos [19]. The objective of computational geometry is to recast geometric problems, some of which are classical, into a framework which makes them amenable to efficient computer solution. The techniques used in connection with geometric problems are essentially those employed in other areas of concrete computational complexity, with the additional feature that one must discover properties of the geometric objects involved which will simplify the computational task.

An interesting problem in this area is the determination of the convex hull of a finite set of points in Euclidean space. As is well known, the convex hull of a finite set of points S is the intersection of all convex sets which contain S. Clearly, this definition is totally useless from a computational standpoint. In this respect, a more useful definition regards the convex hull H(S) of S as a subset of S with the property that any point of S is obtainable as a convex linear combination of the points of H(S).* In general, the convex

* A linear combination is said to be convex if its coefficients are nonnegative and add to 1.

hull H(S) of a d-dimensional set S is a convex polytope in d-space, which becomes a convex polyhedron and a convex polygon in three and two dimensions, respectively. A d-dimensional polytope K is bounded by $(d-1)$-dimensional polytopes, called *facets* or *faces*, each of which lies in a hyperplane: in two dimensions the faces are line segments and in three dimensions they are convex polygons.

With this nomenclature, it is relatively easy to understand a general convex hull algorithm, due to Chand and Kapur [20]. This algorithm is based on an idea, which is commonly referred to as the "gift-wrapping principle." Let K be a d-dimensional convex polytope and assume that a face f of K is given. This face is delimited by $(d-2)$-dimensional polytopes, called *hyperedges* (line segments or points in three and two dimensions, respectively), and let e be a hyperedge of f. The edge e and every vertex of K determine a hyperplane. For any such hyperplane we compute the inner product of the orthogonal unit vector with the unit vector orthogonal to the hyperplane containing f. The hyperplane for which this inner product has largest absolute value contains a new face of K. Thus, by going from face to face, one can identify all the faces of the convex polytopes, and the choice of the phrase "gift-wrapping principle" is entirely justified in its intuitive three-dimensional interpretation. The preceding informal discussion also shows that, given a set of n points in 3-space, application of the gift-wrapping principle will identify the vertices of the convex hull. To analyze the computational effort, notice that the determination of each new vertex of the hull involves work $O(n)$; since all n points could belong to the hull, the total resulting work is $O(n^2)$ in the worst case.

The question has been raised whether faster algorithms can be designed in cases of low dimensionality. The answer is affirmative both for two and for three dimensions, where algorithms are known with running time at most $O(n \log n)$ for sets of n points. Since the three-dimensional algorithm is quite involved [21], it will not be discussed here; rather, we shall discuss two two-dimensional procedures involving quite different techniques and yet achieving the same order of time complexity. In this manner, it is possible to gain considerable insight into the techniques of computational geometry without the burden of complicated details.

Before discussing the specific algorithms, which obtain the so-called *ordered convex hull*, i.e., the *sequence* of the vertices of the convex hull polygon, it is worth pointing out that the order of their time performance is optimal. To show that, we must first specify the computation model. We shall adopt a random access machine (RAM) with the variant that integer arithmetic is replaced by real number arithmetic. Of course this does not entail infinite length operands; it only means that finite length operands are approximations of real numbers, as it normally happens in floating-point arithmetic.

A common method to establish a lower bound for the computation time of a problem P_1 is a suitable *reduction to* P_1 of some problem P_2 for which a lower bound is known. By "suitable reduction" we mean that P_2 may be reformulated as P_1 with an effort whose complexity is not greater than the lower bound. In our case, the following argument—due to Stan Eisenstat [19]—shows that the problem of sorting n numbers can be reduced to finding the convex hull of n points in the plane. This transformation is doable in time O(n), as follows. Let $U = \{x_1, \ldots, x_n\}$ be a set of n numbers; corresponding to each x_i construct—with a single multiplication—the point (x_i, x_i^2). The set of points $S = \{(x_i, x_i^2) | x_i \in U\}$ lie on the parabola $y = x^2$ (a convex curve), so that the ordered convex hull of S gives the sorting of S. Since sorting is known to require $\Omega(n \log n)$ operations, the argument is complete. We shall now illustrate in some detail two convex hull algorithms for the plane.

The first procedure we shall consider, which was also one of the first to appear, is due to R. L. Graham [22]. Let S be a set of n points in the plane. The points of S are assumed to be expressed in polar coordinates (ρ, θ) with respect to some point P internal to H(S) and some arbitrary reference half-line. (Should the points be expressed in a different coordinate system, the conversion can be effected in time O(n): in fact, P can be found as the centroid of S in time O(n) and the coordinate transformation requires a fixed amount of work per point.) The points (ρ_k, θ_k) are then sorted in order of increasing θ and let $(r_1, \phi_1), (r_2, \phi_2), \ldots, (r_n, \phi_n)$ be the resulting sequence with $0 \leq \phi_1 \leq \cdots \leq \phi_n < 2\pi$. This sorting operation requires work O(n log n). Next the algorithm scans the sequence, each step consisting of the application of the following rule to the three-point configuration illustrated in Figure 7:

If $\alpha + \beta \geq \pi$, delete (r_{k+1}, ϕ_{k+1}) and substitute the index triplet $(k - 1, k, k + 2)$ for the triplet $(k, k + 1, k + 2)$; else, set $k \leftarrow k + 1$ and proceed.

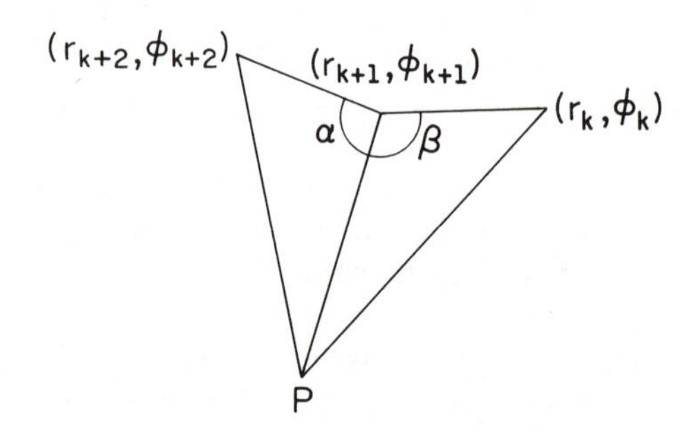

FIG. 7. Illustration of key feature of Graham's algorithm.

Clearly, each application of this rule either eliminates one previously examined point (r_{k+1}, ϕ_{k+1}) or it examines a new point (r_{k+3}, ϕ_{k+3}). It follows that the rule can be applied at most twice per point, for a total of 2n applications. Thus, the initial sorting pass determines the complexity of the algorithm.

The following convex hull algorithm is due to Shamos [19] and involves an entirely different technique. It is based on finding the *hull of the union* of two convex polygons. It can be shown that the latter can be found in time at most proportional to the total number of vertices of the two polygons. To avoid inessential details, we assume for simplicity that n be a multiple of 3. The points of S are then arbitrarily partitioned into n/3 subsets, each containing 3 points. Each such subset determines a triangle, i.e., a convex polygon (Figure 8(a)). We consider disjoint pairs of triangles and replace them with the hull of their union, a convex polygon with at most 6 edges; this can be achieved with work at most $O(k_1 n/3)$ for some constant k_1 (Figure 8(b)). Next we pair these polygons and again find the hulls of their unions: this stage also requires at most $O(2k_1 n/6) = O(k_1 n/3)$ operations. Therefore at each stage we use $O(k_1 n/3)$ operations to halve the number of polygons: clearly, after

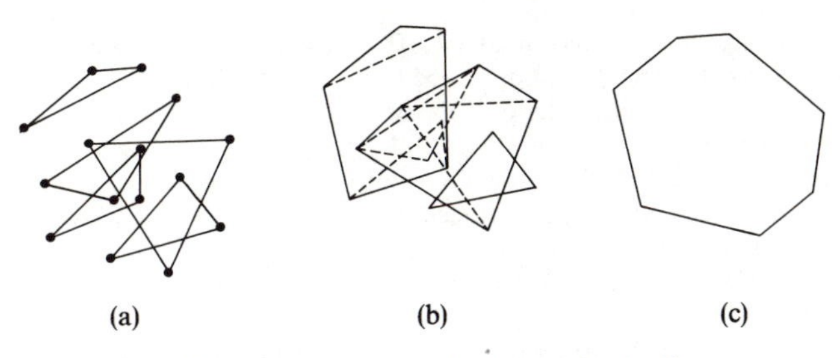

FIG. 8. Illustration of a convex hull algorithm by Shamos.

$\lceil \log_2 n \rceil$ stages, i.e., with O(n log n) operations, H(S) has been computed (Figure 8(c)).

Notice that both algorithms have optimal order of time complexity. Only the former uses sorting explicitly, whereas the latter mimics the general merge-sort technique (see Section 2.2) in the construction of convex polygons with increasing numbers of vertices.

Other optimal convex-hull algorithms for planar sets are known, but their presentation exceeds the scope of this chapter; the interested reader is referred to [19] for further details on this topic and for further exposure to a variety of problems of computational geometry.

REFERENCES

1. J. Hartmanis and R. E. Stearns, "On the computational complexity of algorithms," *Trans. Amer. Math. Soc.*, **117** (1965), 285–306.
2. M. Blum, "A machine independent theory of the complexity of recursive functions," *Journal of the ACM*, **14** (1967), 322–336.
3. J. Hartmanis and J. E. Hopcroft, "An overview of the theory of computational complexity," *Journal of the ACM*, **18** (1971), 444–475.
4. A. V. Aho, J. E. Hopcroft, and J. D. Ullman, *The Analysis and Design of Computer Algorithms*, Addison-Wesley, Reading, Mass., 1974.
5. D. E. Knuth, *Fundamental Algorithms*, The Art of Computer Programming, vol. 1, Addison-Wesley, Reading, Mass., 1969.
6. ———, "Big omicron and big omega and big theta," *ACM SIGACT News*, **8** (April–June 1976), 18–23.
7. A. Borodin and I. Munro, *The Computational Complexity of Algebraic and Numeric Problems*, American Elsevier, New York, 1975.
8. J. E. Savage, *The Complexity of Computing*, Wiley-Interscience, New York, 1976.

9. E. M. Reingold, J. Nievergelt, and N. Deo, *Combinatorial Algorithms: Theory and Practice*, Prentice-Hall, Englewood Cliffs, N.J., 1977.
10. S. A. Cook, "The complexity of theorem-proving procedures," *Proc. 3rd. ACM Symposium on Theory of Computing*, 1971, pp. 151–158.
11. R. M. Karp, "Reducibility among combinatorial problems," in *Complexity of Computer Computations*, R. E. Miller and J. W. Thatcher, eds., Plenum Press, New York, 1972, pp. 85–104.
12. ———, "On the computational complexity of combinatorial problems," *Networks*, **5** (1975), 45–68.
13. A. Schönhage and V. Strassen, "Schnelle Multiplikation grosser Zahlen," *Computing*, **7** (1971), 281–292.
14. J. W. Cooley and J. Tukey, "An algorithm for the machine calculation of complex Fourier series," *Math. Comp.*, **19** (1965), 297–301.
15. D. E. Knuth, *Sorting and Searching*, The Art of Computer Programming, vol. 3, Addison-Wesley, Reading, Mass., 1973.
16. L. G. Valiant, "Parallelism in comparison problems," *SIAM J. Comput.*, **4** (September 1975), 348–355.
17. D. S. Hirschberg, "Fast parallel sorting algorithms," *Comm. ACM*, **21**, 8 (August 1978), 657–661.
18. F. P. Preparata, "New parallel sorting schemes," *IEEE Trans. Comput.*, **27**, 7 (July 1978), 669–673.
19. M. I. Shamos, *Computational Geometry*, Dept. of Comp. Sci., Yale University, New Haven, Conn., (1977).
20. D. R. Chand and S. S. Kapur, "An algorithm for convex polytypes," *Journal of the ACM*, **17** (January 1970), 78–86.
21. F. P. Preparata and S. J. Hong, "Convex hulls of finite sets in two and three dimensions," *Comm. ACM*, **20** (February 1977), 87–90.
22. R. L. Graham, "An efficient algorithm for determining the convex hull of a finite planar set," *Information Processing Letters*, **1** (1972), 132–133.
23. M. Garey and D. S. Johnson, *Computers and Intractability*, W. H. Freeman, San Francisco, 1979.

COMPUTER SCIENCE AND ARTIFICIAL INTELLIGENCE

James R. Slagle

This article examines the role of computers as intelligent machines. A pivotal point in this discussion is that these machines are performing activities often called intelligent when performed by humans. Included herein are machines that play games, prove theorems, solve calculus problems, aid in the manipulation of mathematical expressions, discern and differentiate among chemical structures, and direct the activities of physical robots. These are not hypothetical machines; virtually all of these projects have been reduced to practice on properly operating digital computing systems. Moreover, each machine (or, at least, its crucial component) appears extendable, so that its operating characteristics can be applied to more difficult problems.

1. CHARACTERISTICS OF ARTIFICIAL INTELLIGENCE

A fundamental motivation in this field has been to devise information processing systems (i.e., computing equipment executing appropriate algorithms) whose behavior is considered to be intelligent. That is, we are willing to ascribe that term to the same activity when observed in a human.

Because the term itself is so highly evocative, it will be helpful to state some dictionary definitions of intelligence so that there is a common reference point. *Webster's New Collegiate Dictionary* (1956 edition) defines intelligence as:

> A. The power of meeting any situation, especially a novel situation, successfully by proper behavior adjustments. B. The ability to apprehend interrelationships of presented facts in such a way as to guide action toward a desired goal.*

While we may resist the idea emotionally, it is not difficult to apply these definitions to the behavior of a machine as well as to that of a human. Intelligence is multipurpose in nature and involves the ability to learn. This characteristic (or, at least, an excellent imitation of it) is fundamental to heuristic algorithms and programs, and we shall have more to say about the desirability of incorporating it into a wide spectrum of computing activities.

Almost all of the machines to be discussed in this article are high-speed general-purpose stored-program electronic digital computers. In their most important aspects, they fall into the general category of Von Neumann machines outlined in the first article of this study. Accordingly, for purposes of this article, we are concentrating on activities that are implemented as sequential processes.

1.1. Approaches to Artificial Intelligence. Researchers characterize artificial intelligence in one of three ways:

1. artificial networks
2. artificial evolution
3. heuristic programming.

Since it is currently impossible to say anything conclusive about the preferability of one approach over the others as being the model that parallels (natural) intelligent activities most closely, research toward and implementation of intelligent machines contin-

---

* By permission from *Webster's New Collegiate Dictionary*, copyright 1916, 1925, 1931, 1936, 1941, 1949, 1951, 1953, and 1956, G. and C. Merriam Company, Springfield, Massachusetts.

ues to reflect the approaches most congruent with the perceptions of the respective investigators. Accordingly, this article will emphasize the heuristic approach, with the others receiving only brief mention.

A network consists of an arbitrary number of simple elements along with the interconnections among them. An artificial network may be a physical reality or it may be simulated on a computer. Very often it is helpful to perceive of each element as an artificial neuron. One advantage of this approach is that the network usually is adaptive; that is, it can "learn" from experience. Researchers who take the artificial-network approach tend to perceive natural intelligence as being based on (natural) neural networks alone. At present, artificial networks have "learned" to recognize simple visual and aural patterns, a level of performance considered to be short of intelligent behavior. One difficulty with this approach is that there is little prospect of producing an artificial network that approximates the size and complexity of a brain (i.e., in the order of magnitude of 10^{10} neurons). Another deterring factor is our incomplete understanding regarding the operating characteristics of and interconnections among neurons.

In the artificial-evolution approach to artificial intelligence, computer-simulated systems are designed to evolve by "mutation" and "selection." Using predefined criteria for determining such selections, systems have been devised which evolved into vehicles for solving simple equations. Proponents of this approach point out that many people think that human intelligence evolved through a process in which mutations and natural selection played crucial roles. Here, again, natural evolution is not understood sufficiently to enable close parallels to be drawn. Moreover, the analogous process in computing systems must proceed at an enormously accelerated rate (in comparison to natural evolution) in order for its results to have substantial practical impact.

The approach addressed here revolves around the use of heuristics. These are rules of thumb, strategies, methods, or tricks aimed at improving the efficiency of a system which tries to discover solutions to complex problems. Another trip to the same dictionary produces the following information for "heuristic": "Serving to discover." It is related to the word "eureka" ("I have found it," from the Greek *heuriskein*, to discover, to find). Heuristics, then, form the centerpieces for many artificial intelligence systems. Some of the heuristic programs to be discussed can play checkers and chess; others can deduce answers to questions from a store of given facts; still others solve calculus problems or prove theorems in mathematical logic and geometry. For example, there are heuristic geometry programs that can prove the following rather difficult theorem:

> If the segment joining the midpoints of the diagonals of a trapezoid is extended to intersect the side of the trapezoid, it dissects that side.

Some other heuristic programs can "learn" from their experience. A few are multipurpose in the sense that they are useful in solving several kinds of problems, i.e., they are applicable to several domains. For example, a heuristic for "working backward" (in a sense, "undoing" an activity that was found to be futile or unproductive) is useful in many theorem-proving and pattern-matching domains. Other heuristics are very specific, limited to one problem-solving domain (such as theorem-proving in geometry).

1.2. Purposes of Heuristic Programming. When used in the context of artificial intelligence, a heuristic program reflects the following motivations:

1. an attempt to gain additional understanding of natural intelligence;
2. development and use of machine intelligence to acquire knowledge and solve intellectually difficult problems.

A researcher concerned with the first aspect, for example, may be a psychologist interested in exploring some facet of human behavior regarded as being intelligent. Based on personal observations, available experimental data, and reports in the literature, the investigator will define a logical structure intended to model the behavioral aspect of interest. When implemented as a heuristic program, that model can be run and its results compared against those observed experimentally. This validation process is repeated, with each cycle serving to identify (and remove) discrepancies in the model so that it becomes a progressively improving reflection of the perceived reality. (Further discussion of this validation process, in a more general context, is found elsewhere in this study, in Mark Franklin's article on computer simulation.) Deficiencies in the model revealed by this iterative process may prompt further experimentation, and the process continues until the modeled results are in stable agreement with those obtained over the observable domain. Once that has been achieved to the investigator's satisfaction, the behavioral aspect under scrutiny then can be "observed" simply by running the program on the computer.

A researcher motivated by the second purpose is interested in producing intelligent behavior, with less concern as to whether or not the underlying process duplicates or parallels that employed by humans. His hope is that the computer, driven by a heuristic program, eventually will solve important complex problems in the physical, biological, and social sciences.

In this article we shall be concerned primarily with the second of these two purposes. While routine use of complex problem-solving computing systems still lies in the future, there are intelligent systems in current use whose behavior attests to the fact that there has been impressive progress toward that end. Some of these will be examined and discussed in subsequent sections.

2. PROGRAMS THAT PLAY GAMES

An important conceptual operation in many heuristic programs involves the selection of logical possiblilites whose respective consequences are more desirable than others. The collection of these possibilities can be specified quite effectively when represented in the form of a tree. (As Figure 1 shows, this data structure is characterized more accurately as an upside-down tree, since the "root" is at the top.) While tree-structured data are useful in a wide variety of information-processing contexts, we shall focus on two types that are of particular interest in heuristic programs: game trees and goal trees.

The branches in a game tree represent moves, replies, and counterreplies. In a goal tree, some ultimate goal is shown to be achievable if certain subgoals are achievable. A subgoal, in turn, may be shown to be achievable if certain of its subgoals are achievable, and so on, to some arbitrary level. (As we shall see later on, it is possible to establish equivalence between certain types of game trees

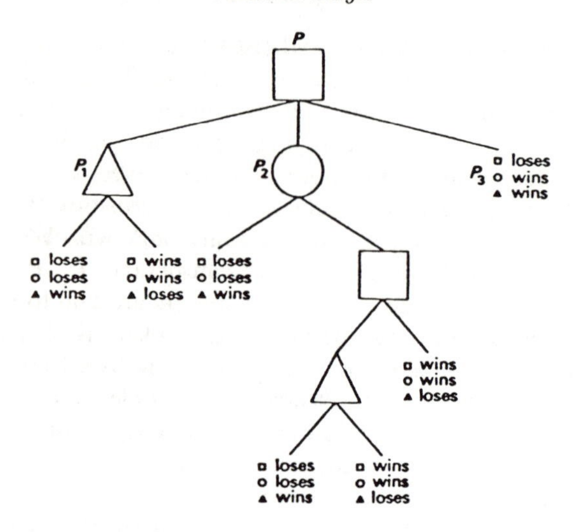

FIG. 1. An explicit game tree.

and goal trees.) Since trees have a tendency to become very large, even when representing modest collections of logical possibilities, a crucial component in many heuristic programs is a procedure for searching these trees effectively. Consequently, a substantial amount of associated work is concerned with finding ways to minimize the search for relevant parts of the tree by "anticipating" (and, therefore, avoiding) ultimately fruitless searches.

Since our immediate concentration will be on game-playing programs, we shall examine game trees more closely.

2.1. Characterization of Game Trees. A game tree is either *explicit* or *implicit*. As the name implies, an explicit game tree is shown in its full structure (Figure 1). Each move is depicted, along with its consequences. An implicit game tree, on the other hand, is described only in terms of an initial position and a set of rules for generating the tree. The game of checkers is an appropriate example: By knowing the characteristics of the board, the starting positions of the 24 pieces, and the rules governing their legal movements, we can generate a tree that depicts completely the consequences of each possible move under each possible set of circumstances.

Referring to the explicit tree in Figure 1, we see that the nodes (squares, circles, and triangles) represent game positions, with the top node representing the starting point. The connecting line segments, then, represent moves. A node's shape defines the player whose turn it is to move. Thus, the square player makes the first move from position P. If the square moves to position P_3, then the square loses while the circle and triangle both win. We shall say that position P is at level 0 and its *successors* (i.e., P_1, P_2, and P_3) are at level 1. The successors of P's successors are said to be at level 2, and so on. In Figure 1, assuming a good play on the part of the other players, the square player can force a win by selecting a move to position P_2: the only good move for the circle player, then, is to the square-position from which the square (expectedly) will cause himself (and the circle) to win. Note that an actual move transforms a game tree into another game tree. For example, the move to position P_2 transforms the tree in Figure 1 into the tree in Figure 2.

An implicit tree consists of a top position together with rules which can be used to generate successors of many positions. These rules include termination criteria so that no successors can be generated for a position meeting these criteria. A procedure which starts at a particular position and follows the game's rules is termed a generation procedure and, in effect, such a procedure converts an implicit tree into an explicit one.

Generation procedures vary in the order in which they generate positions of the tree. A rough but helpful categorization contrasts

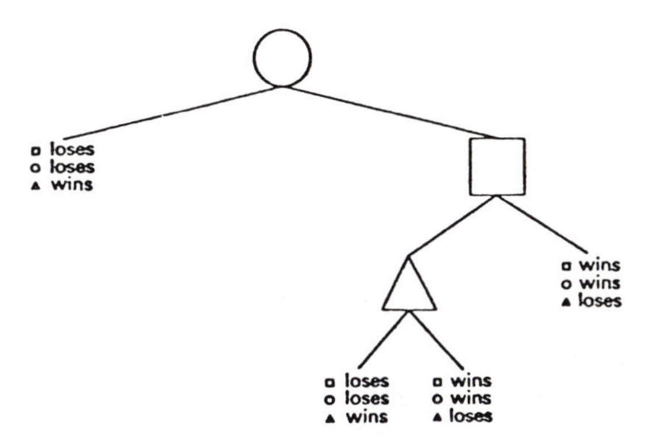

FIG. 2. Transformation of Figure 1 resulting from movement to position P_2.

breadth-first procedures with depth-first procedures. Generally speaking, the former type generates from the top. That is, it produces all the positions at level 1, then all the positions at level 2, and so on. For example, suppose that the top position of an implicit tree is the square-position P shown in Figure 3 and the generating rules are as follows:

1. A square-position generates two positions at the next lower level: a circle-position to the right and a triangle-position to the left.
2. A circle-position generates two lower-level positions: a triangle-position to the right and a square-position to the left.
3. A triangle-position generates two positions at the next lower level: a circle-position to the right and a square-position to the left.

Application of these rules in a breadth-first procedure, then, generates the explicit trees in the order shown in Figure 3.

A somewhat more complicated parallel can be cited using the familiar game of checkers. The rules of the game, applied in a

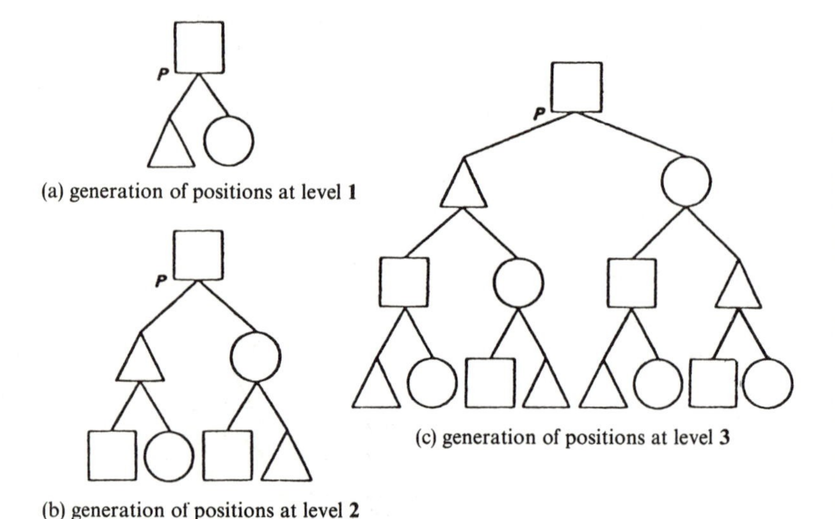

FIG. 3. Breadth-first generation procedure.

breadth-first procedure to the front row (the only one that can move when the system is at the top position), generate seven successor positions. Completed application of the procedure to the successors generates seven level 2 successors for each one, so that we produce an explicit tree at level 2 with 49 positions.

A depth-first procedure generates a tree from left to right in the following sense: The procedure starts by generating the top position's first successor and then, in turn, its first successor at the next level, and so on. To illustrate, suppose that we have an implicit tree with the same top position and generating rules as the one in Figure 3. In addition, we shall define level 3 as the maximum depth, thereby imposing a termination criterion. The square-position at level 0 is labeled A in Figure 4 for convenience. Now, if we apply the generating rules in a depth-first procedure, the positions will be generated in the order A, B, C, D, and E. Since the termination criteria have been met for that branch of the tree, the procedure would then pick up the alternative branch, starting from position F and working to complete that branch to the maximum depth. This progression is shown in Figure 4(a) and 4(b). Implicit in this progression is the fact that a termination criterion always must be given for a depth-first procedure.

When a depth-first procedure is applied to checkers, the 49 positions comprising the top two levels of the checker tree would be

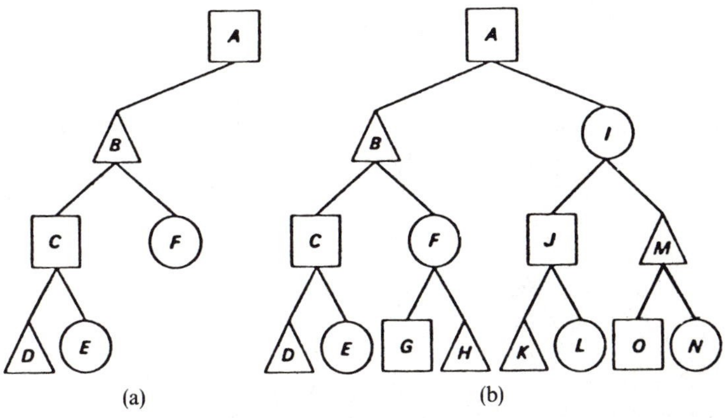

FIG. 4. Depth-first generation procedure. The positions are generated in alphabetical order.

generated as before. However, the sequence in which they are generated is different: After generating the first successor position, its seven reply positions will be generated. Only then will the depth-first procedure return to level 0 to generate the second successor position, and so on until each group of 7 reply positions has been generated from its respective level 1 position.

2.2. Purposes for Programming Game-Playing Computers. Aside from the obvious recreational interests in producing game-playing computing systems, these games are useful vehicles because, despite their relative simplicity, they resemble many important real problems. In many cases, the complexity of the real problems makes them highly resistant to direct attack, so that the researcher hopes that the methods developed to handle simple problems can be extended systematically to encompass a wider range of complexity. The resemblance between games and real problems certainly is not farfetched: The intelligent participant chooses actions based on his search of the tree of future possiblilites, on his rough evaluation of possible future situations, and on expectations about what others will do. Business games, medical games, and war games are intended to serve as models for real problems.

Games tend to be relatively simple because their rules are well defined and relatively straightforward. The selection of games as arenas for developing and improving heuristic programs reflects a conscious decision on the part of researchers to the effect that the advantages of simplicity outweighed those associated with closer approximations to reality. (The implication is that a working heuristic for a well-defined system can ultimately be extended to accommodate a more realistic situation that is less well-defined.) Moreover, there are systems that are both realistic and well-defined on which heuristics can be made to work quite nicely. Theorem proving and assembly-line balancing are two such examples.

2.3. General Description of Game-Playing Programs. Most of the games implemented by the programs described in these sections can be characterized as two-person strictly competitive games. The first of these characterizations is self-explanatory; "strictly competitive" (or zero sum) refers to the fact that whatever one player wins the other player may be considered to lose. Thus, the outcome of such a game may be defined by describing what happens to just one of the players. (By implication, then, cooperation in a zero sum game is never worthwhile.) Very often such programs play against other programs but, for interest, we shall frame our discussion around a machine player and a human opponent.

Clearly, the ability to follow rules and generate positions is only a small part of the overall game-playing structure. Another important ingredient is the ability to evaluate how good a position is for the machine (i.e., how bad it is for the opponent). The program evaluates well to the extent that it assigns high (positive) values to good positions and low (negative) values to bad ones. Since the game is strictly competitive, the negation of a particular value serves as an estimate of how good that position is for the human opponent. Using the perspective established by the foregoing discussion, we can turn to a general description of the consecutive steps followed by game-playing programs. Exceptions will be noted as necessary. (For purposes of calibration, a "step" in this context corresponds to a sequence of program instructions numbering in the hundreds.)

A. The human submits an arbitrary position in the game. Proceed to step B or step E according to whether it is the human's or the computer's turn to move.
B. The human submits his move.
C. Generate the new position.
D. If the termination criteria are met (i.e., the game is over), print the result and stop; otherwise go to step E.
E. Generate some or all successors to the (top) computer position.
F. Evaluate each successor. (Note that this step may be arbitrarily complex, involving elaborate searching of future possibilities and their respective consequences.)
G. Move to the successor with the highest value.
H. Print the computer's move.
I. If the game is over, print the result and stop; otherwise, go to step B.

The success of these heuristics, even with surprisingly complicated games, is manifested in the proliferation of amusing, challenging (and even ego-debilitating) computer-based competitive games.

Before looking more closely at checkers and chess (the traditional paradigms in artificial intelligence for examining heuristics potentially useful in other problems), it will be helpful to redefine the major ingredients in a heuristic search procedure:

1. A generation procedure.
2. A (static) evaluation function.
3. A backing-up procedure.

The first of these has been discussed before and needs no further elaboration for our current purposes. The property of an evaluation function that makes it static is the fact that it assigns a value to a given position without generating any of its successors. In contrast, a backing-up procedure assigns a value to a position based on the values of that position's successors.

2.4. Evaluating a Position in Checkers and Chess. The most commonly used backing-up procedure is the minimax procedure: Since the program's static evaluation function assigns a numerical value to each game position such that the greater the value of the function, the better the position tends to be, we can think of the machine as being the maximizing player (or, simply, Max). Correspondingly, its opponent can be considered the minimizing player (or Min). A position in which it is Max's turn to move is called a Max-position. For example, in Figure 5, square (i.e., the computer

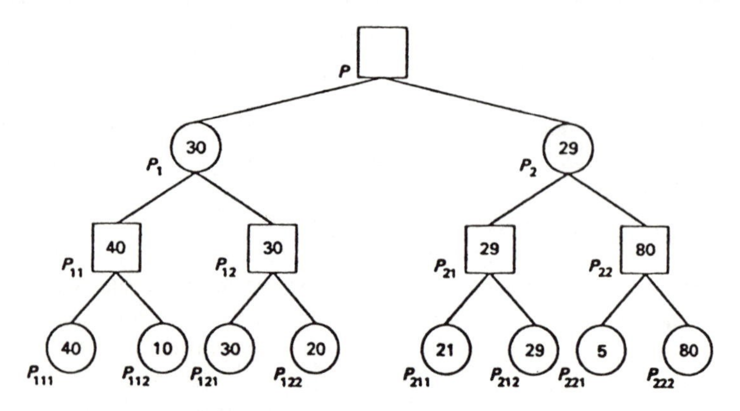

FIG. 5. The minimax backing-up procedure.

or maximizing player) uses its given termination criteria to determine not to generate any positions below level 3. Having established these terminal nodes, let us suppose that the static evaluation function assigns the value $v\!:_{111} = 40$ for position P_{111}. Similarly, the value $v\!:_{112} = 10$ is assigned to position P_{112}, $v\!:_{121} = 30$ for position P_{121}, and so on, for all 8 positions as shown in the figure. We shall not dwell here on how these static evaluations are assigned. It should be said, however, that much of the exploratory work in an artificial intelligence system centers around the determination of a meaningful evaluation function. Having obtained these values, the minimax procedure will back up the value of the best successor of the Max position, i.e., the one with the maximum value. In the case of Figure 5, the maximum successors for each of the four level 2 Max-positions are P_{111} (value = 40), P_{121} (value = 30), P_{212} (value = 29) and P_{222} (value = 80). Since the positions at level 2 are Max-positions (i.e., successors to their respective Min-positions), procedure will back up the value of the best successor of the Min-position. By definition, this is the one with a minimum value. Hence the procedure backs up the value of 30 (at position P_{12}) to position P_1, and the value of 29 (at postion P_{21}) to position P_2. As a result, then, the backing-up procedure would lead square to select a move to position P_1 over P_2. This minimizes the "damage" that the opponent can do by making his best moves. If square moves to P_1, circle's best move is to position P_{12}, in reply to which square's best move is to P_{121}, with a terminal value of 30. The alternative (i.e., an initial move to position P_2) prompts a reply on circle's part to position P_{21} from whence square's best move (to P_{212}) produces a terminal value of only 29.

2.5. Elaborations on Search Procedures. Having established the roles of a static evaluation function, a backing-up procedure, and a generation procedure in an overall search system, we shall look at some possible improvements in these processes.

One version of the basic search procedure combines a static evaluation function and a minimax backing-up procedure with a depth-first generation procedure. This approach has been employed successfully in a number of game-playing programs and will serve as a helpful precursor to the more effective alpha-beta search. To illustrate its nature, let us suppose that the tree of Figure 5 had been given implicitly and we start again at the top level, i.e., position P in Figure 5. This has been relabeled A in Figure 6(a) in order to emphasize a sense of sequence. As Figure 6(a) indicates, the depth-first procedure generates positions B, C, and D. Since the termination criteria are met at that level, the procedure uses its static evaluation function on position D, producing a value, say, of 40. It then generates position E and computes a value for that position (namely, 10). The better of these two values, namely, 40, is backed up to position C. The procedure then generates F, the alternative Max-position at level 2, from which it generates and

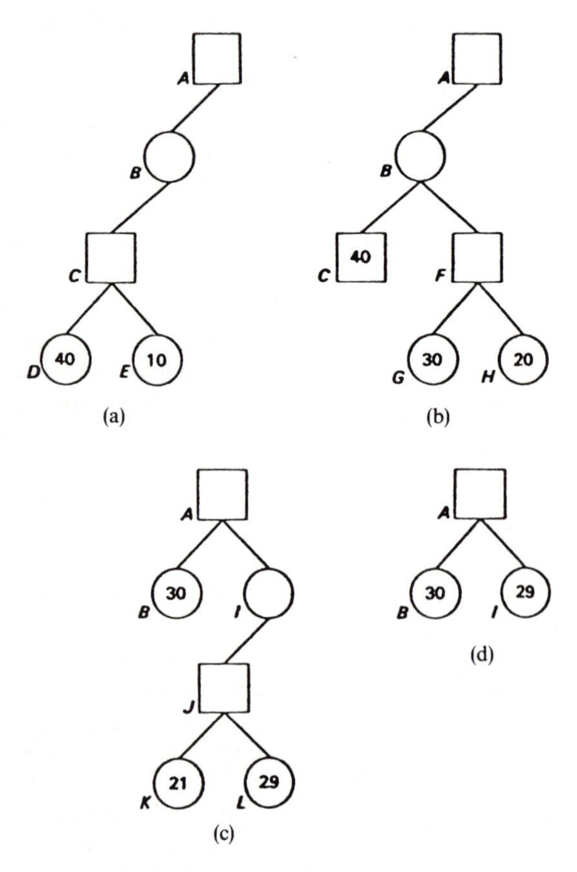

FIG. 6. Depth-first minimax procedure.

evaluates positions G and H. This is shown diagrammatically in Figure 6(b) where the two terminal nodes are shown to have assigned values of 30 and 20, respectively. The better of these two, namely, 30, is backed up to F, at which point it can select the better value between C and F and back it up to B. (Recall that "better" in this instance is the minimum value, i.e, 30.) Now, the procedure can generate I and J, followed by generation and evaluation of K and L (Figure 6(c)). The backing-up and generation process continues until the procedure obtains the result shown in Figure 6(d). Consequently, the choice will be a move to position B, since it has a higher value than the alternative position, i.e., I.

From the foregoing discussion, it is seen that the generation of successor positions is separate from their evaluation. That is, a series of successors is prepared before any evaluation is undertaken. Although this procedure works (limited, of course, by the efficacy and reality of the evaluation function), it is inherently inefficient since it generates a number of ultimately inferior positions. The "alpha-beta" procedure overcomes this deficiency to a considerable degree. It is equivalent to the depth-first minimax procedure in that it will always choose the same move as the other, given the same top position, termination criteria, and evaluation function. However, the primary difference lies in the fact that generation and evaluation are interleaved rather than sequenced. Thus the alpha-beta procedure almost always chooses its move after generating only a very small fraction of the tree produced by the equivalent depth-first minimax procedure. The strategy pivots around the computation of two limiting values at a given Max-position: Alpha is a backed-up value for a position that is computed after a depth-first procedure followed the generating rules until termination criteria were met. This is not necessarily a final value; rather, it is an interim evaluation, serving as a minimum limit. In this capacity, it regulates the generation of other parts of the search tree. If the procedure finds that the value of another Max-position at that level does not exceed alpha, then there is no reason to continue generating successors from that position. Similarly, a maximum limit, beta, is established for a Min-position.

Additional insights into the alpha-beta procedure can be gained by considering these simple exercises: In Figure 7, we see part of a tree in which an interim backed-up value $v{:}_1 = 3$ already has been

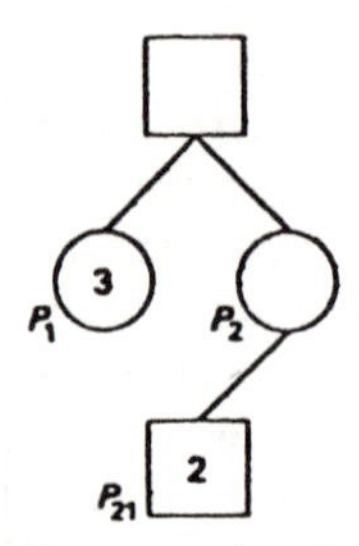

FIG. 7. The alpha-beta procedure finds an alpha cutoff.

established for position P_1. Consequently, it uses that as a limit (alpha) to be imposed on position P_2. Now, suppose that the procedure, through evaluation, assigns a value of 2 to position P_{21}. If this value were to be backed up to its predecessor (position P_2), it would fall below the limiting value (alpha) established for that position. Consequently, the procedure reaches what is termed an alpha cutoff: We see that there is no point in generating any other successors to P_2 (i.e., P_{22}, P_{23}, and all of their successors). The procedure can "conclude" that Max will not choose the move to position P_2 because it is already better off (at least so far) by moving to P_1. Having eliminated this move from contention, there is no need to generate its successors, and the alpha-beta procedure, instead, can go on to generate another position at P_1's level, i.e., P_3.

A similar example is seen in Figure 8: after obtaining $v{:}_{11} = 4$ from position P_{11}, the alpha-beta procedure sets a limit of beta = 4 at P_{12}. Now, suppose that the evaluator assigns $v{:}_{121} = 8$ to position P_{121}. If this value were to be backed up to position P_{12}, it would exceed the provisional limit of 4 established at that position. This, then, would be a beta cutoff. From Max's point of view, a move from position P_{12} is less desirable in comparison to a move from position P_{11}. Therefore, there is no need to generate any other successors of P_{12} (such as P_{122}, P_{123}, and so on). Instead, the alpha-beta procedure can drop back to P_1 and generate another successor (i.e., P_{13}) from there and repeat the process of comparison against beta.

It is evident, then, that the savings realized with the alpha-beta procedure increased dramatically with the number of levels that have to be generated before the termination criteria are met. This forms the basis for further improvements in the search procedure:

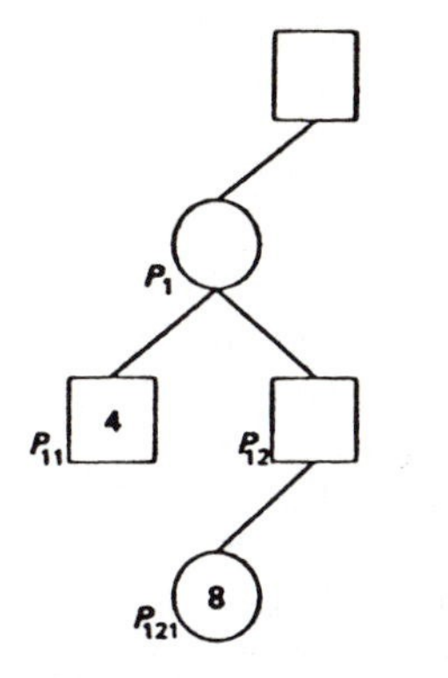

FIG. 8. The alpha-beta procedure finds a beta cutoff.

Clearly, there are further gains to be realized if the number of alpha and beta cutoffs can be increased. Moreover, those gains would be intensified if such cutoffs could be made as early (i.e., as high in the tree) as possible. Accordingly, additional variations can be introduced to reduce the somewhat arbitrary nature of the depth-first alpha-beta procedures. Inevitably, these enhancements complicate the overall search procedure, so that there always is a necessity to consider the tradeoff between the increased efficiency and attendant complexity.

Many of these enhancements are centered around the idea of shallow searching. That is, instead of following a depth-first search down to its terminating nodes, a depth-first minimax procedure or an alpha-beta procedure may be interjected across several positions at a given (relatively high) level. Intermediate results obtained can then be used to prejudice subsequent behavior of the deeper searches. One way of exploiting the shallow search's results is to order subsequent searches so that the first one in a series is likely to be the best one (i.e., the one producing high alpha/low beta values), thereby resulting in early cutoffs for subsequent searches. This approach is called plausibility ordering.

In another, related approach, called forward pruning, not all successors of a given position are searched. The time saved by avoiding (presumably) unpromising branches may be used in searching more fruitful ones to a deeper level. This advantage must be balanced against the risk of failing to search relevant branches. One method of forward pruning, called n-best forward pruning, allows the search to proceed below only the small n seemingly best successors of a position.

2.6. The Evaluation Function. It was mentioned earlier that the determination of the static evaluation function for a given game-playing system is itself the subject of considerable inquiry. For the sake of simplicity, many evaluation functions often are chosen to be linear. That is of the form $c_1 y_1 + c_2 y_2 + \cdots + c_n y_n$. This may be represented as the scalar product of 2 vectors, i.e., $C \cdot Y$. Each y_j is some real-valued function called a feature of the position. For example, such a feature might be a piece advantage, relative mobility, etc. Each coefficient c_j is the weight of the corresponding y_j. A miniature evaluation function in checkers, for instance, is $6k + 4m + u$, where k is the king advantage, small m is the (plain) man advantage, and u is the undenied mobility advantage. The coefficients for these features are 6, 4, and 1, respectively. Suppose, then, in a position to be evaluated, the computer has two more kings, two fewer men, and one more unit of undenied mobility than does its opponent. The evaluation function would assign a value of $6(2) + 4(-2) + 1$ or 5 to that position.

2.7. A Program That Plays Chess. There are numerous chess-playing programs at various levels of proficiency that embody the types of procedures outlined earlier. An example is a program written by Richard Greenblatt, Donald Eastlake III, and Stephen Crocker (1967) at the Massachusetts Institute of Technology. This program used the alpha-beta procedure combined with plausibility ordering and n-best forward pruning of moves. In addition, the program is equipped with book openings and other items of "chess knowledge." Typically, the program makes its move (in a game) in about a minute. A good player (not to mention an expert or master) usually beats this or any other chess program, but the task is becoming increasingly difficult as these program designs improve. This particular program is an honorary member of the United States Chess Federation and the Massachusetts Chess Association under the the name of Mac Hack Six. The program plays in tournaments and is operated via telephone lines from a teletype at the tournament sight. In an April 1967 tournament, the program won the Class D trophy. Over the past few years computer-based chess tournaments have become a staple feature of most national

TABLE 1(a)
A Tournament Game Lost by the Mac Hack Six Program.
Black is Mac Hack Six; White is a human rated 2190.

Move No.	White	Black	Move No.	White	Black
1	P-KN3	P-K4	29	R-Q3	R-K7
2	N-KB3	P-K5	30	R-Q2	RxR
3	N-Q4	B-B4	31	QxR	N-K4
4	N-N3	B-N3	32	R-Q1	Q-QB2
5	B-N2	N-KB3	33	B-Q5	K-N3
6	P-QB4	P-Q3	34	P-QN4	B-N3
7	N-B3	B-K3	35	Q-B2	N-B3
8	P-Q3	PxP	36	B-K6	N-Q5
9	BxP	QN-Q2	37	RxN	BxR
10	PxP	R-QN1	38	QxPch	K-N2
11	B-N2	0-0	39	Q-N4ch	K-R3
12	0-0	B-N5	40	QxB	Q-K2
13	Q-B2	R-K1	41	Q-R4ch	K-N3
14	P-Q4	P-B4	42	B-B5ch	K-N2
15	B-K3	PxP	43	QxRPch	K-B1
16	NxP	N-K4	44	Q-R8ch	K-B2
17	P-KR3	B-Q2	45	Q-R8	Q-B2
18	P-N3	B-QB4	46	Q-Q5ch	K-N2
19	QR-Q1	Q-B1	47	K-N2	Q-K2
20	K-R2	N-N3	48	P-KR4	K-R3
21	B-N5	R-K4	49	P-N4	K-N2
22	BxN	PxB	50	P-R5	Q-K7
23	N-K4	P-B4	51	P-R6ch	K-B1
24	N-KB6ch	K-N2	52	P-R7	QxKBPch
25	NxB	QxN	53	KxQ	K-K2
26	N-B6	QR-K1	54	P-R8Qun.	P-R3
27	NxR	RxN	55	Q-K6mate	
28	Q-B3	P-B3			

BxP: Bishop takes Pawn
Castling: 0-0 = king side;
0-0-0 = queen side.

computer conferences, with programs playing other programs. As a matter of general interest, Tables 1(a) and 1(b) reproduce two tournament games played by Mac Hack Six. These are of particular interest because the former (a loss) was the first tournament game ever played by a computer (January 21, 1967), and the latter was the first tournament game ever won by a computer (a rating of 2190 represents that of an expert, almost a master).*

2.8. Conclusions to Be Drawn from Checkers and Chess-Playing Programs. The programs described in this section typically require

---

* In 1979, Hans Berliner's backgammon program defeated a world champion in a tournament.

in the order of one minute per move, whereas tournament rules in chess and checkers allow four or five minutes per move. Checker programs play an excellent game (by human standards), whereas chess programs are not quite so adept. In checkers, the successful player is able to perform fast and accurate searches on rather large trees. In chess, however, success is associated with the ability to invent and use strategies. The successful chess player mixes abstract thinking with move-by-move analysis. This selectivity enables him to restrict his searches to relatively small trees. Accordingly, researchers are working to mechanize these types of heuristics.

In a sense, the heuristics embodied in these game-playing programs emulate a kind of learning. The checker program, for example, uses a type of generalization in developing evaluation functions that are more effective than their predecessors. This has led to practical procedures for good (but not optimal) evaluation functions in a variety of contexts. Many researchers in artificial intelligence feel that these techniques will provide valuable insights into human learning processes. Such accomplishments, however, still lie in the future; it is safe to say that, at present, no program learns better than the checker program.

TABLE 1(b)
A Tournament Game Won by the Mac Hack Six Program.
White is Mac Hack Six; Black is a human rated 1510.

Move No.	White	Black
1	P-K4	P-QB4
2	P-Q4	PxP
3	QxP	N-QB3
4	Q-Q3	N-B3
5	N-QB3	P-KN3
6	N-B3	P-Q3
7	B-B4	P-K4
8	B-N3	P-QR3
9	0-0-0	P-QN4
10	P-QR4	B-R3ch
11	K-N1	P-N5
12	QxQP	B-Q2
13	B-R4	B-N2
14	N-Q5	NxP
15	N-B7ch	QxN
16	QxQ	N-B4
17	Q-Q6	B-KB1
18	Q-Q5	R-B1
19	NxP	B-K3
20	QxNch	RxQ
21	R-Q8mate	

3. PROGRAMS THAT SOLVE PROBLEMS IN CHESS, GEOMETRY, AND CALCULUS

A wide variety of intellectually difficult problems, including many problems in chess, geometry, and indefinite integration, share a certain common structure which can be represented in terms of implicit trees. Heuristic programs have been written to search such trees, and experiments with working computer programs have produced successful solutions to some fairly difficult problems. For example, Baylor and Simon (1966) performed such work with chess problems; Gelernter and his co-workers (1959, 1960) did related work with geometry problems, and this author (1963) performed such experiments with problems in integration. The speed of these programs compares favorably with that of a skilled human problem-solver. Accordingly, research continues toward machine solutions for increasingly difficult and important problems.

3.1. Representation of a Problem as an Implicit Tree. Geometry and chess problems are good examples of a fairly general kind of problem that can be represented by two kinds of implicit trees. These two representations ultimately will be shown to be equivalent. A chess problem, for instance, may be represented as an implicit, two-person, strictly competitive game tree. Recall that the trees discussed in the previous section were conceptually similar, with the exception that they were explicit. The tree in Figure 9

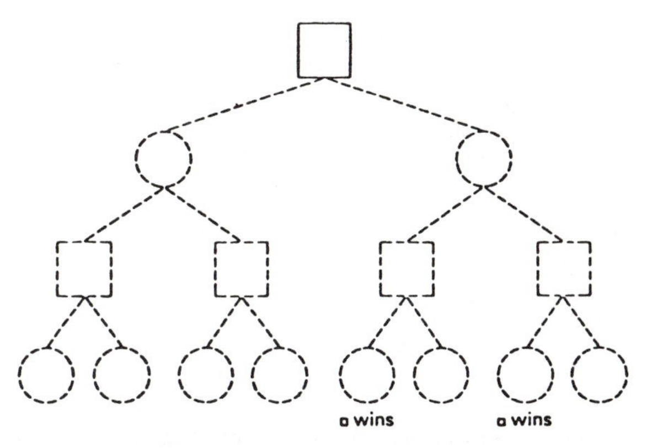

FIG. 9. Top three levels of an implicit game tree.

illustrates this structure. We see that there are terminating nodes which, if reached, guarantee a win for the computer (i.e., white, or square). The problem, then, is to search (i.e., make explicit) enough of a game tree to prove that the computer can force a win. Figure 10 shows the appropriate solution: The heavy solid lines represent a proof that the computer indeed can force a win by selecting the right-hand move from the given starting position.

A geometry problem may be represented as an implicit and/or goal tree. The top three levels of such a tree are illustrated in Figure 11. In this context, the problem is to prove some geometric conclusion, e.g., that two angles are equal, given certain hypotheses. Looking for a proof corresponds to the search of an implicit goal tree whose top goal (the node labeled G in Figure 11) is "to prove the conclusion given originally." As implied in Figure 11, the top goal G is achievable if the disjunction (i.e., the inclusive OR) of goals G_1 and G_2 is achievable. This disjunction is represented by the square shape of G. If G, say, is to prove two angles equal, the goal G_1 might be to prove that the angles are corresponding parts of congruent triangles, and the goal G_2 might be to prove that the angles are alternate interior angles of parallel lines. Referring again to Figure 11, we see that the goal G_2 is achievable if the conjunction (i.e., the logical AND) of G_{21} and G_{22} is achievable. This conjunction is represented by the circular shape of goal G_2. Figure

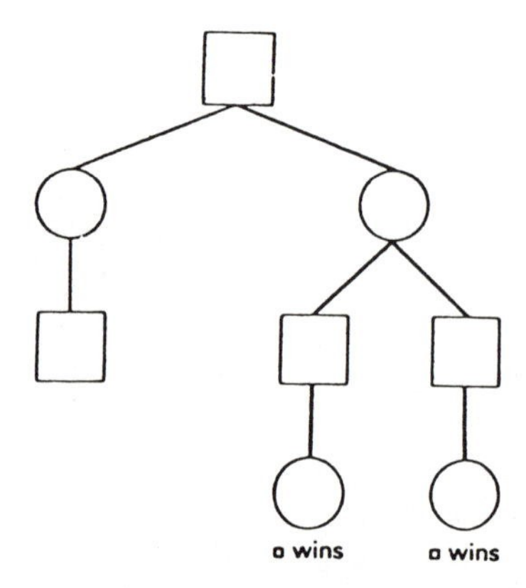

FIG. 10. Explicit game tree and proof. Heavy solid lines represent the proof.

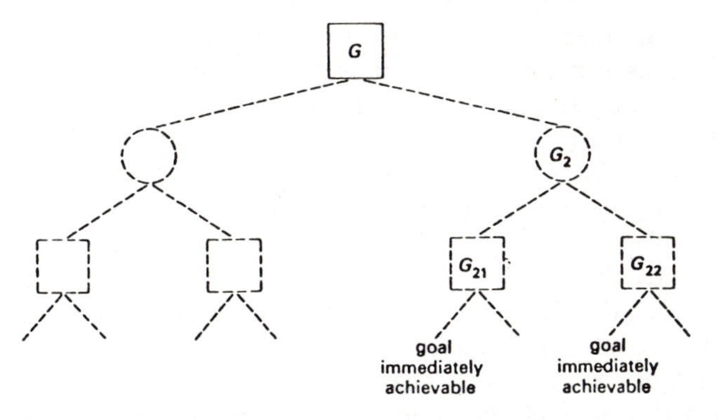

FIG. 11. Top three levels of an implicit goal tree.

12 shows the solution by making the appropriate parts of the tree explicit. Here again (as in Figure 10), the heavy solid lines represent the proof.

Comparison of Figure 9 with Figure 11, and Figure 10 with Figure 12, shows that the two representations are equivalent. Specifically, the chess problem could have been represented just as well by an implicit, and/or goal tree, and the geometry problem could have been represented as an implicit, two-person, strictly competitive game tree. We shall see that a problem in indefinite integration also can be represented in either way.

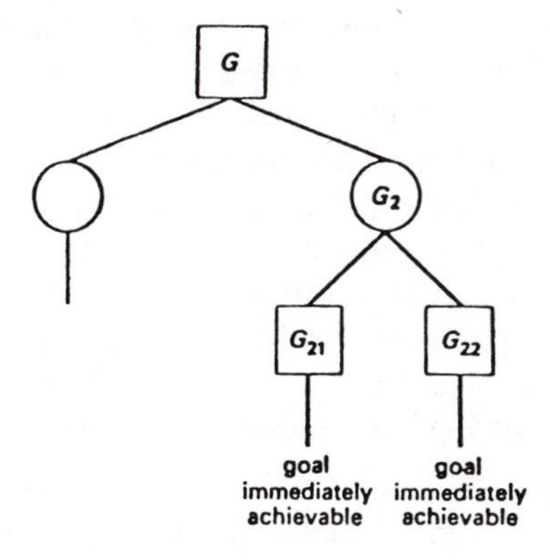

FIG. 12. Explicit goal tree and proof. Heavy solid lines represent the proof.

3.2. Objectives in Developing Problem-Solving Programs. The motivations that prompt the study of programs for solving problems in chess, geometry, and indefinite integration are similar to those underlying the development of heuristics for game-playing programs. These problems, in effect, are simplified versions of real, potentially important ones. Moreover, since many people are familiar with these kinds of problems, they are in an excellent position to evaluate the solutions given by a program, comparing its performance with that of humans. Thus these problem types serve as vehicles for the development of procedures to handle a more general range of problems, i.e., those problems representable by an implicit, two-person, strictly competitive game tree or, equivalently, by an implicit, and/or goal tree. Such procedures have the potential of yielding valuable insights into the problem-solving process.

As for purposes specific to each program, the geometry program studies the use of models and the integration program is potentially useful in itself. (The latter, for example, was extended to a calculus program capable of handling definite and multiple integration.) The geometry program, in using a diagram as a model, rejects subgoals which do not conform to this model. This is particularly interesting, since humans use models to great advantage in successful problem-solving.

3.3. General Description of the Procedure. While a detailed examination of these problem solutions is well beyond the scope of this introductory overview, we can characterize the general approach in terms of a sequence of steps designed to navigate through the goal tree to a satisfactory arrival at the originally stated goal or to a definite conclusion that the goal cannot be reached. At the start, the procedure "knows" what the original goal is and has at its disposal a spectrum of defined resources: allowable transformations that it can perform when in a given state, and knowledge (e.g., axioms, other already proved theorems) that it can use to reduce the problem to a combination of subproblems that are handled more easily. Within this general framework, we can enumerate the following basic sequence of steps:

A. If the program succeeds in its attempt for an immediate solution with the original goal, print the answer and stop.
B. If the program cannot proceed because it has exhausted its available resources, print this fact and stop.

C. If no untried goals remain, print this fact and stop.
D. Select an untried goal as a basis for generating additional parts of the tree.
E. If no more (sub)goals can be generated from the selected goal, go to step B.
F. Generate the next untried goal from the selected goal.
G. If the program fails in its attempt for an immediate solution with the newly generated goal, go to step E. On the other hand, if the try succeeds, prune the goal tree with respect to this goal. The pruning will have one of the following three results:
   1. If the original goal is met, print the answer and stop.
   2. If the original goal is not met but the selected one is met, go to step B.
   3. If neither the original goal nor the currently selected one is met, go to step E.

Research in the improvement of these problem-solving programs continues, fueled by a growing conviction that the resulting methodology constitutes the beginning of a general problem-solving theory.

4. AUTOMATIC THEOREM-PROVING USING THE RESOLUTION PRINCIPLE

An alternative to proof-finding as an approach to automatic theorem-proving is the idea of consequence-finding. Programs using this latter approach, such as the one written by R. C. T. Lee in 1967, start with a collection of axioms and try to deduce consequences from these axioms and select those that are "interesting." It turns out that both approaches use the resolution principle, a natural and powerful rule of inference.

It is this use of deduction that provides a substantial part of the motivation for studying automatic theorem-proving. Of course the basic activity itself is intrinsically interesting, since proving a nontrivial theorem is an intellectually difficult problem. Interest is heightened, moreover, by the application of first-order predicate calculus to the design of such programs. In mathematical logic, one can express fairly conveniently almost all kinds of deductive arguments. Thus writing a theorem-proving program that uses predicate calculus allows the researcher to study deduction in its purest form. This ability (on the part of a program) to make deductions from given facts has been characterized by Professor John McCarthy (1959) as common sense. When exercised in humans, it is considered an important part of natural intelligence. The obvious extrapolation, then, is irresistible: A program that uses mathematical logic to find proofs can be extended to deduce answers to questions. Such extensions have beem implemented successfully using frameworks of knowledge ranging from a set of major league baseball statistics to the text of a children's encyclopedia.

Clearly, the extension also can be taken in another direction: a future program that proves new and interesting theorems would be useful in itself. It would be a tremendous achievement, for instance, if some undefined but imaginable program of the future proved or disproved the famous Fermat or Goldbach conjecture. Providing additional motivation is the fact that mathematical logic is well suited to computers. Since logicians have labored for decades to make their rules of inference "mechanical," it is an attractive idea to develop computerized algorithms based on mathematical logic, since this is a well-formulated, well-studied branch of mathematics. Thus, writing and using theorem-proving programs is a way to study mathematical logic.

P. C. Gilmore (1960), H. Wang (1965), and M. Davis and H. Putnam (1960) are among the early investigators who used first-order predicate calculus in the design of theorem-proving programs. Using formal inference rules, the programs progressively substituted constant terms for variables in well-formed formulas, checking at each step to see whether the theorem had been proved. In 1965, J. A. Robinson developed an inference rule called the resolution principle, which served to unify many of the relatively fragmentary theorem-proving algorithms in use at that time. The resolution principle seeks to draw the most general possible conclusion from two given statements, serving to delay (or even eliminate) the need for progressive substitution of literals in place of variables. The resolution principle is more natural, more intuitive, and easier to use than are the formal inference rules it replaces.

4.1. Characteristics of the Resolution Principle. We shall examine the basic use of the resolution principle by charting its course through some simple examples. This will provide background from which we can establish some generalizations about it.

EXAMPLE 1. We are going to seek proof of a simple theorem through use of the resolution principle. The statements given below are understood to hold for all values of their variables. For instance, the statement P1 holds for all x, for all v, and for all y. This is true in the same sense in which the identity $x^2 - y^2 = (x + y)(x - y)$ holds for all x and y.

THEOREM 1. Suppose

P1: If x is part of v, and if v is part of y, then x is part of y.
P2: A finger is part of a hand.
P3: A hand is part of an arm.
P4: An arm is part of a man.

From P1 through P4, we may conclude that a finger is part of a man.

Proof of Theorem 1. A procedure that tries to find proofs using the resolution principle first takes the denial of the conclusion and then tries to deduce a contradiction. In our current example, this denial can be expressed as follows:

P5: A finger is not part of a man.

To reach this denial, and its accompanying contradiction, the procedure produces the following consequence:

P6: If hand is part of y, then finger is part of y.

This is called a resolvent of P1 and P2. (More about this later.) P6 is obtained by first "matching" (making identical) the clause P2 and the first portion of clause P1 by letting x be "finger" and y be "hand." This substitution in P1 gives the following intermediate result, a logical consequence of P1:

P1′: If a finger is part of a hand and a hand is part of y, then a finger is part of y.

This is indeed a logical consequence of P1, since P1 is asserted to be true for all x, v, and y, and therefore it must be true in the special case when x is "finger" and v is "hand." The clause P6 is an immediate consequence of P1′ and P2. Usually, the resolvent P6 would be given directly without expressing the intermediate result P1′.

Each of the three portions of the clause P1 is called an atom. For example, the second atom in P1 is "v is part of y." Thus, each of the clauses P2, P3, and P4 consists of one atom; clause P6 consists of two atoms. Returning to the proof, we match the clause P3 and the first atom in P6 by letting y be "arm." This yields the intermediate result

P6′: If a hand is part of an arm, then a finger is part of an arm.

This, together with P3, yields the resolvent

P7: A finger is part of an arm.

Similarly, the resolvent obtained by matching P7 and the first atom of P1 is:

P8: If an arm is part of y, then a finger is part of y.

The resolvent of P4 and the first atom of P8 is:

P9: A finger is part of a man.

Matching this with P5 gives a contradiction of the denial, thereby completing the proof of Theorem T1. This proof is outlined in the first and second columns of Table 2. In this table the corresponding

TABLE 2
Proof of Theorem 1.

Clause name	*Proof in words*	*Proof in symbols*	
		Clause	*Reason*
P1	If x is part of v and if v is part of y, then x is part of y.	Part (x,v) & Part $(v,y) \rightarrow$ Part (x,y)	Given
P2	A finger is part of a hand.	Part (finger, hand)	Given
P3	A hand is part of an arm.	Part (hand, arm)	Given
P4	An arm is part of a man.	Part (arm, man)	Given
P5	A finger is not part of a man.	−Part (finger, man)	Denial of conclusion
P6	If a hand is part of y, then a finger is part of y.	Part (hand, y)$\rightarrow$ Part (finger, y)	r[P1a,P2]
P7	A finger is part of an arm.	Part (finger, arm)	r[P3,P6a]
P8	If an arm is part of y, then a finger is part of y.	Part (arm, y)$\rightarrow$ Part (finger, y)	r[P1a,P7]
P9	A finger is part of a man.	Part (finger, man)	r[P4,P8a]
P10	Contradiction.	Contradiction	r[P5,P9]

proof in symbols is self-explanatory, except for the following: The expression r[P3, P6a] denotes the resolvent obtained by matching the clause P3 with the first atom in P6. The symbol & means "and," the symbol → means "if ..., then ..." or "implies," and the symbol – means "not."

EXAMPLE 2. The resolution principle consists of only factoring and resolution. This example illustrates factoring, disjunctive notation, and the way in which the theorems are formulated for application of the resolution principle.

THEOREM 2. In any associative system which has left and right solutions s and t for all equations, $s \cdot x = y$ and $x \cdot t = y$, there is a right identity element.

Proof. Application of the resolution principle to produce a proof for this theorem is outlined in Table 3. The notes given below elaborate on that proof:

A1: There exists a left solution. This means that for all x and y there is an s such that $s \cdot x = y$. In other words, there exists a function $g(x, y) = s$ such that $g(x, y) \cdot x = y$.

A2: There exists a right solution. This means that for all x and y there exists t such that $x \cdot t = y$. That is, there is a function $h(x, y) = t$ such that $x \cdot h(x, y) = y$.

TABLE 3
Proof of Theorem 2.

Clause name	*Clause*	*Reason*
A1	$g(x,y)\cdot x = y$	Given (existence of a left solution)
A2	$x\cdot h(x,y) = y$	Given (existence of a right solution)
A3	$(x\cdot y = u)$ & $(y\cdot z = v)$ & $(x\cdot v = w)\rightarrow(u\cdot z = w)$	Given (part of associativity)
A4	$k(x)\cdot x \neq k(x)$	Denial of conclusion
A5	$(x\cdot y = u)$ & $(y\cdot z = y)\rightarrow(u\cdot z = u)$	f[A3,a,c]
A6	$(y\cdot z = y)\rightarrow(u\cdot z = u)$	r[A1,A5a]
A7	$y\cdot z\neq y$	r[A4,A6b]
A8	Contradiction	r[A2,A7]

A3: Establishing associativity. In this instance, $(x \cdot y) \cdot z = x \cdot (y \cdot z)$. Actually, we need and use only the following:

$$x \cdot (y \cdot z) = w \rightarrow (x \cdot y) \cdot z = w.$$

Hence,

$$(x \cdot y = u) \,\&\, (y \cdot z = v) \,\&\, (x \cdot v = w) \rightarrow (u \cdot z = w).$$

A4: Denial of conclusion. For this theorem, the denial is that there is no right identity element. For any proposed right identity element x, there exists u such that $u \cdot x$ is not equal to u. In other words, there exists a function $k(x) = u$ such that $k(x) \cdot x$ is not equal to $k(x)$.

In the proof given in Table 3, clause A5 is obtained from clause A3 by "factoring" A3 with respect to its first and third atoms. The clause A3 implies its special case A5, and, accordingly, A5 is called a factor of A3. This factor is obtained by making identical the first and third atoms of A3, namely, $x \cdot y = u$ and $x \cdot v = w$: y is substituted for v, and u replaces w throughout A3; the third atom (which has become identical to the first and therefore redundant) is canceled. From these operations, we see that, in general, factoring a clause with respect to two or more of its atoms is performed as follows:

1. Find the minimal substitution (if any) which makes the atoms identical. ("Minimal" is clarified a little later.)

2. Make the substitution throughout the clause and cancel all but one of the redundant atoms.

Example 2 also is used to illustrate implication and disjunction. Representing $x \cdot y = z$ by P(x, y, z) transforms the mathematical notation of Table 3 into the notation of symbolic logic (implication) of Table 4. In the latter table, the "disjunctive" notation is logically equivalent to the implication notation and is the computer-based formalism recommended by Robinson and used in programs embodying his resolution principle. (The symbol $\vee$ indicates disjunction.)

To illustrate the idea of minimal substitution, consider the process of factoring the following clause with respect to its first two atoms:

$$P(f(w), w) \vee P(y, g(x)) \vee P(w, f(y)).$$

TABLE 4
Symbolic logic (predicate calculus) proofs of Theorem 2.

Clause name	*Implication notation*	*Disjunctive notation*	*Reason*
A1	$P(g(x,y),x,y)$	$P(g(x,y),x,y)$	Given (existence of a left solution)
A2	$P(x,h(x,y),y)$	$P(x,h(x,y),y)$	Given (existence of a right solution)
A3	$P(x,y,u)$ & $P(y,z,v)$ & $P(x,v,w) \rightarrow P(u,z,w)$	$-P(x,y,u) \vee -P(y,z,v) \vee -P(x,v,w) \vee P(u,z,w)$	Given (part of associativity)
A4	$-P(k(x),x,k(x))$	$-P(k(x),x,k(x))$	Denial of conclusion
A5	$P(x,y,u)$ & $P(y,z,y) \rightarrow P(u,z,u)$	$-P(x,y,u) \vee -P(y,z,y) \vee P(u,z,u)$	f[A3,a,c]
A6	$P(y,z,y) \rightarrow P(u,z,u)$	$-P(y,z,y) \vee P(u,z,u)$	r[A1,A5a]
A7	$-P(y,z,y)$	$-P(y,z,y)$	r[A4,A6b]
A8	Contradiction	Contradiction	r[A2,A7]

We try simply to match the two atoms from left to right. The first required substitution is f(w) for y. (Of course, the reverse, i.e., substitution of y for f(w), would not be valid.) Substituting f(w) for y throughout the clause yields

$$P(f(w), w) \vee P(f(w), g(x)) \vee P(w, f(f(w))).$$

The next required substitution is one in which g(x) replaces w. Note that the nonminimal substitution of h(z) for x and g(h(z)) for w also leads to a match, but the "factor" thus obtained would be a special case of (and, therefore, worse than) the factor that we shall obtain this way. Substituting g(x) for w yields

$$P(f(g(x)), g(x)) \vee P(f(g(x)), g(x)) \vee P(g(x), f(f(g(x)))).$$

Thus, the substitution of g(x) for w and small f(g(x)) for y in the original clause is minimal. The factor, obtained by deleting one of the two redundant atoms, is

$$P(f(g(x)), g(x)) \vee P(g(x), f(f(g(x)))).$$

If the above-mentioned nonminimal substitution were used, the "factor" would be as indicated above, except that h(z) would be substituted for x throughout the factor. It happens that there are no other factors of the original clause, since the third atom cannot be matched with either the first or second atoms.

EXAMPLE 3. This example, whose proof is given in Table 5, illustrates the minimal-substitution aspect of resolution. The list of arguments for B1 and B2 can be made to match with each other by implementing minimal substitutions. Accordingly, each list of arguments becomes:

$$\begin{array}{l} s, \\ g(s), \\ m(g(s)), \\ h(s, m(g(s))), \\ n(g(s), h(s, m(g(s)))), \\ k(s, m(g(s)), n(g(s), h(s, m(g(s))))). \end{array}$$

4.2. Definition of the Resolution Principle. A more precise definition of resolution and factoring now can be considered. Using disjunctive notation, the resolution principle combines the following two basic ideas:

1. The syllogism principle of propositional calculus, i.e., from

$$a \vee b$$

and

$$-a \vee c$$

one may infer

$$b \vee c.$$

2. The instantiation principle of predicate calculus: from the formula

$$F(v_1, v_2, \ldots, v_n)$$

TABLE 5
Illustration of the minimal-substitution aspect of resolution.

Clause name	*Clause*	*Reason*
B1	$P(s,g(s),t,h(s,t),u,k(s,t,u))$	Given
B2	$-P(v,w,m(w),x,n(w,x),y)$	Denial of conclusion
B3	Contradiction	r[B1,B2]

which is understood to hold for all values $v_1, v_2, \ldots, v_n$, one may infer the formula

$$F(t_1, t_2, \ldots, t_n)$$

obtained by substituting the "terms" $t_1, t_2, \ldots, t_n$ for the variables $v_1, v_2, \ldots, v_n$, respectively. By definition, a term is either:

(a) an individual constant, for example, "finger"
(b) an individual variable, e.g., x
(c) a function of other terms, e.g., g(x, y) and h(s, m(g, (s))).

General resolution is part of the overall resolution principle: A disjunctive clause consists of the disjunction of literals. By definition, a literal is an atom of the negation (denial) of an atom. The resolvent, if any, of a literal in one clause and a literal in another clause is implied by the two clauses taken together. Such a resolvent is obtained as follows:

A. Rename variables so that all individual variables in one clause are distinct from the individual variables in the other clause.
B. Find the minimal substitution, if any, which makes the literals identical, but opposite in "sign."
C. Perform the indicated substitution throughout both clauses.
D. If exactly the same literal occurs more than once in a clause after the substitution, cancel all but one copy of that literal in that clause.
E. Delete the two literals which were made identical but opposite in sign.
F. The resolvent, then, is the disjunction of the literals remaining in the first and second clauses.

General factoring is another part of the resolution principle. The factor, if any, of two or more literals in a clause is implied by that clause and is obtained as follows:

A. Find the minimal substitution, if any, which makes the literals identical (and the same in sign).
B. Perform the indicated substitution throughout the clause.
C. Cancel all but one copy of the literals which were made identical as a result of the substitution.

D. The factor, then, is the disjunction of the remaining literals.

J. A. Robinson (1965) proved that the resolution principle is effective, sound, and complete for proof-finding. Subsequently, R. Lee (1967) showed that resolution is complete for consequence finding as well. Its effectiveness means that one can write a computer program, which, in a finite number of steps, will find the factors of any clause and the resolvents of any two clauses. The principle's soundness means that a clause logically implies each of its factors and that two clauses, taken together, logically imply each of their resolvents.

THEOREM. If a finite set of clauses is unsatisfiable, then a finite number of applications of the resolution principle will find a contradiction.

THEOREM. If a clause C is a consequence of a finite nonempty set of clauses, then a finite number of applications of the resolution principle will find a clause T such that C is an immediate consequence of T alone.

Several researchers have strengthened these theorems by showing that certain restricted forms of the resolution principle are still complete. This is of practical importance to automatic theorem-proving because research results have indicated that restricted (yet still complete) resolution tends to be more efficient than unrestricted resolution.

5. OTHER PROGRAMS FOR MATHEMATICS

In addition to integration and theorem-proving in calculus and geometry, researchers have written a variety of heuristic programs for mathematical processing. These include a successful regression analysis program and a program for manipulating arbitrarily complicated symbolic mathematical expressions. These are excellent vehicles for trying out artificial intelligence algorithms and techniques since the problems are well formulated. In addition, many researchers are interested in eventually obtaining a practical program suitable for developing desired solutions to realistic mathematical problems. The additional examples outlined in this section are motivated by both of these major forces.

5.1. A Programmed Aid for Manipulating Mathematical Expressions. Several researchers have written programs to help people manipulate mathematical expressions. William Martin (1967) and Joel Moses (1967) have written one such program for the time-shared computer system at MIT. Among other things, the program can simplify, differentiate, and integrate mathematical expressions, and it can solve some simple differential equations. In a typical episode the human user submits some initial mathematical expressions, along with directions for their manipulation. In response, the program produces intermediate expressions resulting from the classified operations. Then the user describes the way in which he wants those intermediate expressions manipulated. This interactive dialogue may go through an arbitrary number of cycles until the user obtains a set of output expressions that he considers to be in final form. In this way, for example, the program can be used to simplify and differentiate very large expressions quickly and accurately. Its operating capabilities remove it from the class of being merely a replacement for work that had been done manually; rather, it can handle routinely expressions whose size and complexity place them beyond contemplation for manual processing.

As such programs are augmented with improved algorithms resulting from more recent research, they become more intelligent assistants, thereby enhancing their usefulness. For example, suppose someone wants to write a heuristic program for performing some specified task. Instead of trying to do this directly, it would be possible to describe the task formally to a "task-generating assistant," which, in response to these instructions, would produce the required task directions. While such capabilities still are limited, they certainly are not farfetched: programs that generate exact sequences of detailed activities to be performed by industrial robots already are in use, and increasingly natural-like languages are being developed to instruct these task-building programs.

5.2. A Heuristic Regression Analysis Program. Floyd Miller (1967) wrote a practical heuristic regression analysis program that learned. Input to the program consists of a set of (n + 1)-tuples (i.e.,

$x_1, x_2, \ldots, x_n, y$), where n is no greater than 40. The independent variables x_j are the predictors, and the dependent variable y is the response. The other input item is k, which is either 3, 4, or 5 and subject to the constraint that it does not exceed $n - 2$. The program computes the model associated with each of the 2n distinct sets of k predictors. By definition, the model associated with the k predictors, (i.e., $x_{i_1}, x_{i_2}, \ldots, x_{i_k}$) is the function having the form

$$y = a_0 + a_1 x_{i_1} + a_2 x_{i_2} + \cdots + a_k x_{i_k},$$

which provides the "best fit" for the sets of $(n + 1)$-tuples. The a_j, of course, represents the multiple regression coefficients. Output generated by the program consists of the three best models thus produced. The user often is interested primarily in the three good sets of k predictors which he obtains as part of the models. Once a set of k predictors is chosen, the program applies standard least-squares procedures to compute the multiple regression coefficients a_j. This latter aspect is of secondary interest since it does not depart seriously from traditional methodology. Of greater relevance in our present context, however, is the procedure whereby the heuristic regression system learns to select good predictors:

A. Initialization of the run includes input of the observed data (i.e., the set of $(n + 1)$-tuples). Assuming no prior knowledge about the relative usefulness of each of the n predictors, the initialization process assigns equal probabilities of usage. Thus, each $P_j = 1/n$, where P_j is the probability of choosing predictor x_j in step C.
B. Initialize the trial processing. During each trial, a set of k predictors is chosen.
C. Choose a predictor in accordance with the relative probabilities P_j.
D. If the selected predictor already has been chosen during this trial, discard this choice and go to step C.
E. If a set of k predictors has not yet been chosen, go to step C.
F. If the current set of k predictors is the same as the set used in some previous trial, discard the set and go to C.
G. Use regression analysis to find the model associated with the set of k predictors.
H. Save this model if it is one of the three best models obtained so far during the run.

I. Reward and punish each of the predictors. That is, adjust the probability P_j of choosing predictor x_j. (Space prohibits detailed discussion of these techniques.)

J. If fewer than two n models have been generated, go to step B; otherwise, print the three best models and stop.

The performance of this type of program is excellent. Since its inception in 1964, it has solved well over a thousand different practical problems at various industrial companies. (Examples are described in articles by Blackmore et al. (1966) and Drattell (1968).) Applications have ranged from production planning in the paint industry to the analysis of traffic fatalities in Florida.

5.3. Implications of Mathematical Problem-Solvers. Progress in this area of artificial intelligence is sufficiently striking to place the efficacy of mathematical problem-solvers well beyond doubt. Since these systems produce useful solutions to problems considered intellectually nontrivial in human terms, there are grounds for a strong argument that such programs meet most traditional criteria for intelligent behavior.

6. A PROGRAM THAT FINDS CHEMICAL STRUCTURES

Heuristic Dendral is a program designed by Edward Feigenbaum (1968) and others at Stanford University. Input consists of the empirical formula and the mass spectrum of some acyclic organic molecule. In response, the program produces a list of structural formulas (i.e., molecular graphs) that explain the given input in the light of the program's model of the mass spectrometry and stability of organic molecules. The list is ordered starting with the most satisfactory explanation. Comparisons always are interesting: In this instance, for certain classes of organic molecules, the program's performance (i.e., its speed and accuracy) approaches or exceeds that of postdoctoral laboratory workers in mass spectrometry.

Heuristic Dendral consists of four basic processes: the preliminary inference maker, the hypothesis generator, the predictor, and the evaluator. The first of these components embodies a set of pattern recognition heuristic rules. It makes a preliminary interpretation of the data in terms of the presence of key functional groups, absence of other indicator groups, weights of radicals attached to key functional groups, and so on.

These activities pave the way of the hypothesis generator, a component that "knows" the valences of atoms and is capable of generating all of the topologically possible isomers for the empirical formula submitted as input. In a very real sense, then, the hypothesis generator and the empirical formula determine an implicit tree. At the top node, we have all the atoms but no structure. At the terminal nodes, then, there are complete structures but no unallocated atoms. The search within this tree is guided by various heuristic rules and chemical models such as the output from the preliminary inference maker, the a priori model, and the zero-order theory of mass spectrometry. The a priori model is a model of the chemical stability of organic molecules based on the presence of certain denied and preferred subgraphs of the chemical graphs. Zero-order theory is a relatively crude but rather efficient explanation of the behavior of molecules in mass spectrometry. It screens out entire classes of structures because they are not valid with respect to the data, even within the latitude tolerated by a crude approximation. Output from the preliminary inference maker and hypothesis generator consists of a list of molecular structures that our candidate hypothesizes for explaining the empirical spectrum. This generation process embodies a rather complex theory of mass spectrometry. The evaluator is a heuristic algorithm which matches the predicted spectrum for each candidate with the empirical spectrum submitted previously. After discarding some candidates, it displays the remainder as mentioned earlier. More recent enhancements to this program include nuclear magnetic resonance data as part of the criteria for identifying candidates and evaluating their suitability.

7. DATABASE MANAGEMENT SYSTEMS AND ARTIFICIAL INTELLIGENCE

A typical database management system consists of one or more large computer programs that can make additions, deletions, retrievals, and modifications in a large collection of data. Such a system can answer a question submitted to it by displaying facts extracted from the data collection. More sophisticated systems can deduce answers from retrieved facts. In a corporate setting, a database management system might be used to maintain current information about the firm (and even about its competitors). In a military situation, similar systems help keep track of equipment and personnel. Other collections ranging from medical diagnoses to stock market transactions also are processed by such systems.

Such collections of data usually are either formatted or text-based. A typical example of a former organization is a record-based collection consisting of large tables of numbers and letters in accordance with an underlying categorization. Computerized card catalog systems and census data are common instances. A query submitted to such a database generates search requests aimed at isolating subsets of the collection based on the presence or absence of categorized criteria. In contrast, databases consisting of bodies of information like the *New York Times* or the *Encyclopaedia Britannica* are instances of text-based systems. Typically, these are searched using as criteria the occurrence of specified key words or phrases. With either type of system the searches, while operationally quite complex, are conceptually straightforward. If we were to anthropomorphize, we could liken such searches to the efforts of a rather dull-witted, lazy, and completely literal-minded reference librarian.

One of the most active areas in computer science concerns itself with the development of concept and methodologies that will improve the quality and responsiveness of searches on database systems. A considerable part of that activity is relevant here because it seeks to apply artificial intelligence techniques to the specification, analysis, and processing of search requests. Anthropomorphizing again, the objective is to make the system more like an industrious librarian who exercises some initiative to help the user. One aspect of this work seeks to replace the highly structured category sensitive search requests with a more natural syntax. For example, William Woods (1978) at Bolt, Beranek and Newman wrote a program called the Lunar System, a natural language system for retrieving information from a database of chemical analyses of moon rock samples. The interface between the user and the system is equipped to handle inquiries such as, "How many samples contain more than 10 percent iron?" Improvement in such systems often is manifested in their ability to accept less and less precise inquiries. David

Waltz's system (1975), developed at the University of Illinois, successfully handles vague questions such as, "Are there any common features among the planes that crashed last month?" Of course, the designers must be careful to avoid the production of trivial or irrelevant answers ("All the planes were metallic," or "All the planes were made in America").

Other experimental systems are beginning to make more and more complicated deductions in producing responses to search requests. An increasing number of such systems are moving out of the research arena into the production environment. For example, Microdata's Reality is but one of several systems whose users can formulate their queries in natural language.

8. COMPUTERS THAT "UNDERSTAND" NATURAL LANGUAGE

The database management system is only one instance in a wide range of contexts where it would be desirable to communicate with a computer system using natural language. While there are many programming languages, for example, that attempt to approach natural language (often by providing a syntax that imitates a very small subset of a natural language), the vast majority of man-machine interaction still requires extensive compromise on the part of the human communicant. Of course, the big problem continues to be natural languages' inherent vagueness and ambiguity. Any time a person makes an utterance he brings with it an implied wealth of world knowledge and experience shared by other people. Since, by and large, this is not available to computers, the difficulty in duplicating (or even approximating) such interpretive capabilities is understandable.

In this section, accordingly, we shall make brief mention of some systems that represent interesting instances of efforts to improve computers' ability to communicate with man on his own terms.

8.1. Speech Understanding Systems. The idea of being able to speak to a computer is tantalizing. Speech is man's most natural communication system. People can speak faster than they can type or write; they can speak when they are moving around and when their hands are busy. Consequently, a microphone or telephone would make a tremendously convenient terminal to a computer that understands speech.

Under the sponsorship of the Advanced Research Projects Agency (ARPA), Alan Newell et al. (1973) formulated a five-year plan whose objective was to produce a prototype system capable of understanding speech. Within the context of a limited domain of discourse, the system would be able to understand an American uttering an ordinary (though somewhat simple) sentence constructed from a one-thousand-word vocabulary. The speaker would enunciate in a natural manner, with no extreme regional dialect. Two of the five original participants (Carnegie-Mellon University and Bolt, Beranek and Newman) still are in active pursuit of these objectives.

These prototype systems base their processing on acoustical clues, syntactic context, and semantic context. A microphone transforms the spoken input sentence into a waveform of amplitude versus time. This, in turn, is converted into a spectrum of frequencies versus time and the amount of energy in a given frequency band ultimately is represented by a gray level. Embedded in this spectrum are bands called formants that can be tracked through the signal. These formants correspond to vocal track resonances whose trajectories represent movements of the vocal track. Certain vowels and diphthongs often can be detected by means of the formants. Unfortunately, a formant's shape also is subject to the influence of vowel context, i.e., the sounds that precede and follow the vowel. Consequently, the determination of spoken phonemes (elemental speech components) is highly context-sensitive. Because of this, these speech understanding systems take advantage of the small size of their vocabularies and use context heavily in identifying phonemes.

In fact, this is not fundamentally different from the type of processing performed in human speech recognition. On isolated phonemes, experienced speech scientists choose the correct ones less than 75 percent of the time. However, when given entire sentences, their accuracy exceeds 99 percent.

8.2. Natural Language Processing Systems. Speech recognizing systems are paralleled by a much larger effort aimed at developing processes that enable a computer to understand natural language in written form. Roger Schank et al. (1975) constructed a system called MARGIE (Memory Analysis Response Generation and Inference in English). The system consists of a set of three programs: (1) A conceptual analyzer accepts sentences in restricted English and maps them into a conceptual representation of their meaning. (2) The memory and inference program stores new input data, establishes reference, and makes inferences about the input just received. (3) Finally, the generator program encodes conceptual representations into syntactic structures which are then mapped into English sentences.

MARGIE operates in two modes: paraphrase and inference. In paraphrase mode, sentences are analyzed into a representation of their meanings, and then those meanings are written in as many ways as are available to the generator for expressing them in English. Since this internal representation of meaning uses constructs that are not part of natural English, the problem, basically, is one of coding ideas into language. In inference mode, the input sentences are analyzed into their meaning representation and then processed to find out what other information might be true or might arise as a consequence of the input. Inferences are used to motivate further inferences. These, in turn, are transformed into English and printed for display.

The processing is rather massive, so that the response is slow in human terms. For example, when the system executes on a million-instruction-per-second machine at Stanford's Artificial Intelligence Laboratory, the response time in paraphrase mode is a few seconds; when operating in inference mode, the response is measured in minutes.

We can develop an inkling of the system's operating level by looking at a processing example in each mode. In these illustrations, the human communicant was familiar with the contents and extent of the program's vocabulary.

Inference Mode

Input: John told Mary that Bill wants a book.

Output 1: A book about what?
Output 2: Mary knows that Bill wants a book.
Output 3: Bill wants to come to have a book.
Output 4: Bill wants someone to cease to have a book.
Output 5: Bill wants to read a book.

Input: John gave Mary an aspirin.

Output 1: John believes that Mary wants an aspirin.
Output 2: Mary is sick.
Output 3: Mary wants to feel better.
Output 4: Mary will ingest the aspirin.

Paraphrase Mode

Input: John advised Mary to drink the wine.

Output: John told Mary that drinking the wine would benefit her.

Input: John prevented Bill from giving a banana to Mary by selling a banana to Rita.

Output 1: Bill was unable to give a banana to Mary because Rita traded John some money for a banana.
Output 2: Because Rita bought a banana from John, Mary could not get a banana from Bill.

MARGIE is of particular significance because it represents a fundamental departure from earlier natural language processing systems which were based on a stored inventory of words and grammatical rules. The basis of this system is conceptual dependency theory. In essence, this theory contends that natural language has an underlying meaning structure which should be used for all pertinent processing. The claim is that people think by using this meaning structure which is independent of a language's specific words. Schank et al. have proposed such a structure and used it in MARGIE for representing the output of a meaning analyzer. It also serves as the basis of inference and memory programs. The structure requires that any two sentences having the same meaning should have only one representation. An important component of such a structure is a set of primitive semantic elements into which words with complicated meanings can be mapped. At this fundamental level of meaning, concepts combine into conceptualizations. A conceptualization is a statement about an actor performing an act; an actor, in this context, is either an animate object or a natural force. Correspondingly, a physical act is performed on a physical object while a mental act is performed on a mental object, i.e., a conceptualization. In this context, for example, "hurt" is not considered an act. Rather, it makes reference to an unknown act that results in a "hurt" state for the object acted upon. By the same reasoning, "prevent" is not an act either. Instead, it is relational between two acts.

This formulation of the nature of meaning structures allows the creation of a set of basic primitive acts, each of which is defined by the inferences that are true when it is present. A given verb is represented by a combination of primitive acts. The current system uses eleven such acts (e.g., propel, move, grasp, expel). Another primitive act is "atrans" ("give" is an instance of atrans). This primitive act requires an actor, an object, and a recipient consisting of a source and a goal. The source and goal must be animant and the object must be physical.

Even from this superficial overview of the basic ideas, it is clear that MARGIE is an attempt to simulate the processes involved in understanding natural language in a nonartificial (i.e., concept-based) environment. Here again, as we have seen in the case of game-playing and theorem-proving programs, the vehicle itself provides a stimulus for further study about the way we interpret and process ideas.

9. CONCLUDING REMARKS

Based on earlier discussions of a number of heuristic programs, we now are in a position to examine such programs more abstractly and ask general questions about them: Into what categories or aspects can heuristic programming be divided? What are the future applications of heuristic programming? What are the philosophical and social implications of the advent of intelligent machines?

9.1. Aspects of the Heuristic Programming Problem. The basic issues of heuristic programming can be categorized conveniently in terms of the following six aspects:

(1) generality
(2) searching
(3) functions that make evaluations and recognize patterns

(4) the matching of data structures to determine appropriate substitutions for variables
(5) learning
(6) planning

A heuristic program is said to be multipurpose (i.e., general or broad) if it can solve a wide variety of problems or answer a wide variety of questions. Since the intelligence of a human is multipurpose, there is motivation in the artificial intelligence community toward getting heuristic programs to be multipurpose, even if this means sacrificing the solution of some difficult problems. The hope is that once multipurpose programs can be written to solve relatively simple problems, these programs will be extendable to more difficult ones. There is serious speculation that such programs ultimately will be equipped to solve their own problems. For example, a program that needs to perform a search will include an operational component that will define how the search will be conducted.

The idea of heuristic searching is applicable to problems that cannot be solved directly. Under such circumstances, there are instances (Slagle, 1970) in which programs have been written to search for a solution. If the number of possibilities to be searched is sufficiently small, the problem is trivial since the program can consider all possibilities. For an intellectually difficult problem, however, the number of possibilities is sufficiently large so that an exhaustive procedure is not feasible. In most kinds of theorem proving, including those of predicate calculus, the number of possibilities to be searched is potentially infinite. Consequently, it is much better to consider alternative ways of defining and modifying the search. In fact it often is desirable (and even necessary) to replace a search procedure guaranteed to work in principle with an alternative procedure that is not guaranteed but is good in practice. This happens, for example, when a search procedure examines only the top two levels of a game tree rather than the complete one. After being modified in this way, a search can sometimes be replaced by a more efficient equivalent search. Thus, a depth-first minimax search may be replaced by an alpha-beta search. Wherever they can be identified, it is desirable to search the most promising possibilities first, thereby allowing alpha-beta cutoffs to occur.

In general, then, it is safe to predict that improvements in hardware performance, dramatic as they may be, will not bring exhaustive searches and other brute-force approaches into the realm of practicality. Consequently, the six aspects of heuristic programming listed above will play significant roles in the formulation and improvement of methods for reducing the number of possibilities to be examined in the pursuit of a desirable solution.

9.2. Future Applications of Heuristic Programming. The impact of heuristic programming is growing rapidly. Pertinent algorithms and techniques have been refined to a sufficient extent so that there already is a trend away from game playing and the solution of toy problems toward the solution of problems having real economic and social value. Many solution methods which were technically feasible but not cost effective in the past now find everyday use in business, industry, and government because of the rapid decline in computing costs.

Robots already are a reality, and many of them are driven by heuristic programs. Originally designed for use in hostile environments (e.g., no oxygen, high radioactivity, or inhospitable temperatures and pressures), improved heuristics have helped make them profitable in a wider variety of contexts. Consequently, such robots are performing complex assemblies and installations in the aircraft and automotive industries.

Continuing miniaturization of computers has made it possible to sever the physical connection between robot and computer, so that prospects of a mobile robot (with the computer as an intrinsic component) are realistic. For example, John McCarthy and his colleagues at Stanford University are doing research on a self-guided cart designed (ultimately) to travel unaided on existing roads. The Beast, designed by George Carlton, John G. Chubbuck, and others at the Applied Physics Laboratory of Johns Hopkins University, is equipped with hardware logic and steering, and it has tactile, sonar, and optical apparatus as well. This combination of facilities, motivated by heuristic programs, enables the Beast to find its way down the center of a hall. When its battery becomes sufficiently run-down, it "looks" for (i.e., optically locates) an electric outlet and plugs itself in to recharge its battery. The heuristics for locating such outlets have been developed to a sufficient extent so that the system was able to "survive" in a building of halls and offices for periods exceeding 40 hours before it "starved" as the result of its inability to find another outlet.

An assembly robot developed by Ambler et al. (1975) offers some interesting insights regarding future developments in robotics. This particular system, using connected overhead television cameras, is able to assemble objects such as toy cars and boats using part descriptions developed from the input signals supplied by the cameras. Production robots start with such predefined descriptions as part of their "knowledge." With part descriptions thus developed, in conjunction with predefined assembly instructions, this robot is able to select and assemble the appropriate parts from a larger heap containing a mixture of required and unnecessary parts.

An area of artificial intelligence in which research results have been disappointing is that of automatic language translation. Thus far, the idiomatic nature of natural languages and their great sensitivity to geographical, historical, and other cultural contexts have been serious obstacles to the design and implementation of successful (accurate and natural-sounding) general translators. However, this work has produced innumerable valuable insights with regard to the underlying structure of natural language, and these discoveries have been instrumental in the development of highly improved programming languages and language processors. Many of the heuristics that drive today's natural language database query systems stem from the work in automatic language translation. While the opinion is not universally shared, many computer scientists working in artificial intelligence are convinced that continuing progress in heuristic programming and in our understanding of meaning will lead eventually to an effective natural language translating system.

Earlier in this article, mention was made of the power of heuristic programs as evaluative vehicles. Instead of merely embodying the operational characteristics of a particular model, such heuristic systems include a description of the model itself. Consequently, it is possible to simulate a particular system much more realistically; moreover, much more realistic systems can be simulated. Accordingly, it will be much more helpful to use such systems as aids to defining plans and strategies in a widening variety of endeavors. In addition, these systems already are proving to be effective teaching vehicles. One of the reasons for this effectiveness lies in the fact that the student can participate in the model as player or problem solver. By learning the content of the model, the student gains a deeper understanding of the complex system under study.

9.3. Implications of Intelligent Machines. The prospect of really intelligent machines already has tremendous philosophical and social implications. In philosophy, the presence of such machines will shed light on mechanism, the perennial "mind-body problem," and perhaps even the role of man in the universe. In itself, the existence of intelligent machines would bolster the claims of mechanists that man is nothing but a machine and that the answer to the mind-body problem is that there is only a body and nothing that can be called "mind." However, in the process of developing intelligent machines, certain intrinsic differences between man and machines may or may not reveal themselves, and this will constitute evidence for or against Mechanism. Many people claim that such differences already are apparent. They argue, for example, that a computer can do only what it is told to do and that people can do more. A Mechanist would counter this contention by saying that people can do only what they are told to do in the same sense that they, too, are operating under a set of imposed restrictions. That is, man's heredity "tells" him what to do, including how to learn from his environment. Arguments that "show" that a machine cannot in principle be as intelligent as a man are equally debatable. The question is awaiting final proof or refutation. Presence of intelligent machines will compel man to face the idea that he is not the only intelligent creature. The effect on man's image of himself will be even greater than the effect of the realization that man inhabits a minor planet revolving around a minor star in a minor galaxy. Perhaps man's most cherished claim to his uniqueness is that his intelligence cannot be matched by a mere machine.

What are the social implications of intelligent machines? Will even our most skilled workers be displaced by the new automation? What will be the impact of intelligent machines on the right to privacy? What does this author believe lies in the distant future?

The computer will be our slave and, in a sense, our brother. It is up to us to guarantee this by taking proper precautions. Without such precautions there is some danger that intelligent machines eventually will "take over." Nature builds into each human a set of primary goals or desires. If the analogy holds, then we are in a position to build such goals into a highly intelligent machine. Hence we must be careful that these objectives are congruent with the welfare of humanity. Similarly, we must be sure that no private individual (or group) can "subvert" intelligent machines toward purposes that are to the detriment of society as a whole. It will be relatively easy to take these precautions if enlightened people learn the capabilities and limitations of highly intelligent machines. A computer will be our brother in the sense that human and machine will work together to solve problems.

The computer-motivated threat to privacy is well known and well documented, requiring no further discussion here. However, it is worthwhile to point out that intelligent machines using powerful heuristics have intensified that threat: instead of merely extracting, summarizing, and displaying data about an individual, such systems are perfectly capable of deducing "new" facts from existing ones. This danger, once recognized, can readily be met by taking precautions similar to those already mentioned.

As is usual with technological change, the development of intelligent machines will lead to a mixture of benefits and dislocations. So far, automation has taken away jobs at the unskilled and semiskilled levels, causing particular problems for certain segments of the labor force. With the development of very intelligent machines, even highly skilled workers will be displaced. There will be a need, then, to reexamine the "Protestant ethic" that hard work is good in itself. Many people will be able to transfer their energies to social service work; others will need to adjust their lives to a much wider spectrum of leisure activities.

Of course, no individual can predict the future. However, I am motivated to make the following predictions: Before the end of this century, computer-based solutions of intellectually difficult problems will play a dominant role in bringing enormous material prosperity to the world. In less than a century, computer systems will be making substantial progress on the solution of social problems, including the overriding problem of war and peace. Then, at last, the world may be able to live in peace and prosperity.

REFERENCES

Ambler, A. P., H. G. Barrow, C. M. Brown, R. M. Burstall, and R. Popplestone (Edinburgh), "A versatile system for computer controlled assembly," *Artificial Intelligence* (1975) 129–156.

Baylor, G., and Herbert A. Simon, "A chess mating combinations program," *Proceedings of the AFIPS Annual Spring Joint Computer Conference*, 1966, pp. 431–447.

Blackmore, W. R., G. Cavadies, D. Lach, F. A. Miller, and Twery, "Annual CORS Survey for 1965," *Canadian Operations Research Society Bulletin* (Fall 1966).

Carbonell, J. R., and A. M. Collins, "Natural semantics in artificial intelligence," **3** *IJCAI*.

Davis, M., and H. Putnam, "A computing procedure for quantification theory," *Journal of the ACM*, **7** (July 1960), 201–215.

Drattell, A., "Management training ground at Glidden," *Business Automation*, **15**, no. 4 (April 1968).

Ernst, Heinrich, "MH-1: A computer-operated mechanical hand," *Proceedings of the AFIPS Annual Spring Joint Computer Conference*, 1962, pp. 39–51.

Feigenbaum, Edward A., "Artificial intelligence: Themes in the second decade," *Proceedings of the IFIP Congress*, 1968.

Fikes, R., and S. Weyl, personal communication, Stanford Research Institute, Menlo Park, Calif.

Gelernter, Herbert, "Realization of a geometry theorem proving machine," reprinted from *Proceedings of the International Conference on Information Processing*, 1959.

Gelernter, Herbert, J. R. Hansen, and D. W. Loveland, "Empirical explorations of the geometry theorem machine," reprinted from *Proceedings of the Western Joint Computer Conference*, 1960, pp. 143–147.

Gilmore, P. C., "A proof method for quantification theory: Its justification and realization," IBM Journal of Research and Development (January 1960), 28–35.

Greenblatt, Richard, D. Eastlake III, and Stephen Crocker, "The Greenblatt chess program," *Proceedings of the AFIPS Annual Fall Joint Computer Conference*, 1967, pp. 801–810.

Kleene, Stephen, *Mathematical Logic*, Wiley, New York, 1967.

Lee, R. C. T., "A completeness theorem and a computer program for finding theorems derivable from given axioms," doctoral dissertation, Department of Electrical Engineering and Computer Science, University of California, Berkeley, 1967.

Levy, D. N. L., "Computer chess: A case study," in B. Meltzer and D. Michie, eds., *Machine Intelligence 6*, Edinburgh University Press, 1970, pp. 151–163.

Martin, William Arthur, "Symbolic mathematical laboratory," doctoral dissertation, Electrical Engineering Department, Massachusetts Institute of Technology, Cambridge, Mass., January 1967.

McCarthy, John, "Programs with common sense," *Proceedings of the Symposium on the Mechanization of Thought Processes*, Her Majesty's Stationery Office, London, 1959, pp. 75–84.

McDermott, D. V., "Assimilation of new information by a natural language-understanding system," *MIT AI Lab AI-TR-291*, 1975.

Moses, Joel, "Symbolic integration," *MAC-TR-47, Project MAC*, Massachusetts Institute of Technology, Cambridge, Mass., December 1967.

Nevins, Arthur J., "Plane geometry theorem-proving using forward chaining," *Artificial Intelligence* (Spring 1975), 1–23.

Newell, A., J. Barnett, J. Forgie, C. Green, D. Klatt, J. C. R. Licklider, J. Munson, R. Reddy, and W. Woods, *1971 Speech Understanding Systems: Final Report of a Study Group*, North-Holland, New York, 1973.

Newell, A., and George Ernst, "The search for generality," *Proceedings of the IFIP Congress*, vol. 1, Spartan Books, Washington, D.C., 1965, pp. 17–24.

Robinson, J. A., "A machine oriented logic based on the resolution principle," Journal of the ACM, **12** (January 1965), 23–41.

Samuel, Arthur, "Some studies in machine learning using the game of checkers," *IBM Journal of Research and Development*, **3**, no. 3 (July 1959), 210–229; reprinted in Edward Feigenbaum and Julian Feldman, eds., *Computers and Thought*, McGraw-Hill, New York, 1963.

Schank, Roger C., Neil M. Goldman, C. J. Rieger III, and C. K. Riesbeck, "Interference and paraphrase by computer," *Journal of the ACM* (July 1975), 309–328.

Slagle, James R., "A heuristic program that solves symbolic integration problems in freshman calculus," reprinted from *Journal of the ACM*, **10** (1963), 507–520.

———, "Automatic theorem proving with renamable and semantic resolution," *Journal of the ACM*, **14** (October 1967), 687–697.

———, "Heuristic search programs," in Mihajlo D. Mesarovic and Ranan B. Banerji, eds., *Theoretical Approaches to Non-numerical Problem Solving*, Springer-Verlag, 1970, pp. 246–273.

———, *Artificial Intelligence: The Heuristic Programming Approach*, McGraw-Hill, New York, 1971.

Waltz, David, "Application of artificial intelligence to data base problems," *Coordinated Science Laboratory Report*, University of Illinois, Urbana, Ill., 1975.

Wang, Hao, "Formalization and axiomatic theorem proving," *Proceedings of the IFIP Congress*, vol. 1, Spartan Books, Washington, D.C., 1965, pp. 51–58.

Wilson, Gerald A., "A description and analysis of the PAR technique: An approach to parallel inference and parallel search in problem solving systems," doctoral dissertation, University of Maryland, 1976.

Woods, William, "Semantics and Quantification in Natural Language Question Answering," in Marshal C. Yovits, ed., *Advances in Computers*, Academic Press, 1978, pp. 2–87.

THE IMPACT OF COMPUTERS ON NUMERICAL ANALYSIS

E. R. Buley and R. H. Pennington

1. INTRODUCTION

Numerical Analysis is concerned with determining specific numerical values for the variety of mathematical entities which arise as solutions of real physical problems. It began with techniques directed toward hand computation and has assumed ever increasing importance with the vastly increased computational capability offered by modern computers.

Classical Numerical Analysis is often subdivided into the problem areas associated with interpolation and curve fitting; solution of equations, with simultaneous linear equations and matrix operations forming a separate category; numerical differentiation; numerical integration or quadrature, and the solution of ordinary and partial differential equations. Overlaying most of these areas are the techniques associated with the evaluation of functions.

Depending upon one's point of view the computer has had little effect on the field of numerical analysis or it has had an overwhelming impact. One can argue that the effect has been minor in that many of the numerical methods in use today originated several hundred years ago. These methods still bear the names of the great men of mathematics—Newton, Euler, Legendre, Gauss. The methods were devised for pencil-and-paper application, and the actual steps involved were addition, subtraction, multiplication, and division. These are precisely the same operations that form the basic arithmetic instruction repertoire of today's digital computer. Thus interpolation by Newton's method or Gaussian integration can be performed on the digital computer today in a manner accurately paralleling the hand computation that might have been performed a hundred years ago. In this age of computers, the core of numerical analysis is still the set of methods devised by the great minds of an earlier era. The computer has not changed the fundamental basis of numerical analysis.

On the other hand, there are several respects in which the digital computer has had a profound influence on the field of numerical analysis. The most conspicuous of these is that numerical methods are now being used on a scale unthinkable in the days of hand computation. For example, the numerical methods for solving a set of twenty linear equations in twenty unknowns, or for finding the eigenvalues of a ten by ten matrix, have been known for centuries. One shudders at the thought of actually setting out to solve a problem of this size using hand computation, and it is probable that such massive computations were seldom, if ever, attempted. With today's computer, however, problems of this magnitude and of much larger magnitude are solved routinely.

So widespread is the use of numerical methods in computers that in many cases the person using the computer calls upon some sophisticated numerical method without even being aware of doing so. A programming language such as FORTRAN allows the user to ask for sin(x) or exp(x) by name, and causes the computer to use a Chebychev expansion or Pade approximation to compute the value when needed. All that is required of the user is a blind faith that when sin(x) is requested, the computer will produce the correct result.

One consequence of the massive application of numerical methods has been the development of an enormous library of variants on the traditional methods of numerical analysis. If one can identify any special attributes of the numbers involved in a particular problem, he can frequently tailor the numerical solution method to combine some steps to improve efficiency or to improve accuracy.

Because computer time is a commodity of value, the discovery of a more efficient computational approach can have a positive payoff. Thus a considerable amount of useful work in the field of numerical analysis has gone into the discovery of special tricks that are effective only for limited subclasses of problems. These tricks might well have fallen into the category of tricks not worth knowing in the days of hand computation, mainly because the problem solver might never find the time to address such problems anyhow. Now that massive numerical problems are being solved routinely, such special tricks or limited application methods have a useful role.

The attribute of the computer that has had the greatest impact on numerical analysis is its speed of computation. The computers of today allow more computation to be done in a single day than could have been accomplished in the days of Gauss by the entire human race computing day and night for a generation. It is this speed, of course, that has allowed the massive application of numerical methods, with the attendant growth in variety of methods mentioned above. This speed of computation has also had some other effects on the field of numerical analysis. One, which may be temporary and which may change as computer hardware changes with time, is a shift in relative emphasis on methods of numerical analysis. In the days of hand computation, the use of tables to obtain values for functions was an important activity, and consequently methods of interpolation were heavily used. With current-generation computers it is generally cheaper and quicker to compute function values from a series approximation than to store a table and look it up. Thus the relative emphasis between interpolation and series approximations has tended to shift in favor of series approximations.

It is quite possible, however, that future technological advances in the area of high-speed, high-density memories would reverse this trend. It may once more become cheaper to store tabulated values and utilize interpolation more extensively.

There is a more pervasive, and probably more enduring, by-product of the speed of computation, one which is of central importance in the application of computers to numerical analysis. From its earliest days, the field of numerical analysis has concerned itself with questions of error propagation. So long as computations were being done by hand, however, not enough computational steps could be performed in most practical problems to allow error propagation to become a serious consideration. With computer speed, however, it is not at all difficult or time-consuming to combine thousands of numbers, or to take an iterative computation through thousands of steps. In such situations the presence of even small errors with small growth rates can be quite destructive. For this reason the considerations associated with error propagation must receive close attention in computerized numerical analysis. In order to address this, it is necessary to consider in more detail the computer representation of numbers and the ways in which error accumulation occurs.

2. COMPUTER REPRESENTATION OF NUMBERS

A number is usually stored in computer memory in the form of electrical or magnetic representation of binary bits. The memory is subdivided into cells, or words, each cell being able to store all the bits of a single number. A fixed number of bit positions is used for each word of storage. About 32 is the usual number, although some machines use as few as 12 or as many as 64. Because of the ease of representing binary numbers by electrical or magnetic state, it is natural and efficient to represent numbers in terms of some radix that is a power of 2, such as 8 (octal) or 16 (hexadecimal). Arithmetic circuitry for any particular computer is constructed to match the number representation used in that computer. Ordinarily the arithmetic circuitry will allow for two different types of number representation, integer and floating point. In representing integers as 32 bit binary numbers, it is only possible to represent numbers having absolute value from 0 to 2 to the 31st power, or about 2 billion. Therefore, floating-point representation is usually used for general computation. In floating-point representation, the number is represented by a mantissa, f, and an exponent e. The number is interpreted by the arithmetic circuitry of the machine as

$$f \times \gamma^{e}$$

where γ is the radix used in the representation. In order to stay on familiar ground in discussing accuracy problems, let us assume a radix of 10, although, as already mentioned, a radix of 8 or 16 is more common in practice. For a radix of 10, the standard floating-point number is of the form

$$f \times 10^e.$$

The number of digits used to represent f is fixed by the word length of the computer. Let t stand for the number of digit positions allowed. Then every floating-point number stored in the machine will contain t digits. The number t is termed the "precision" of the floating point number.

Precision is closely related to the relative error achievable in representing a number within the computer. Customarily the computer will normalize floating-point numbers, so that there are no leading zeros. This is done by shifting the mantissa and adjusting the exponent accordingly. Thus, on a machine having 6-digit precision, a number whose computed value was

$$.003154724 \times 10^4$$

would be represented as*

$$315472 \qquad 2$$

rather than

$$003155 \qquad 4$$

if the machine rounds while storing, or

$$003154 \qquad 4$$

if the machine does not round.

For a normalized floating-point number, on a machine with precision t, the best that can be assumed generally about relative error is that it is 5 or less in the t + 1st digit position. (We will talk in terms of a machine that rounds rather than truncates to get rid of

---

* In this representation, the 315472 is the mantissa, assumed to have a decimal point in front of the 3, and the 2 is the exponent, the power of 10 by which the mantissa is to be multiplied. A computer will ordinarily carry these internally as a single number, with an offset added to the exponent so it will not be negative. If an offset of 50 were used, allowing a number range from 10^{-50} to 10^{50}, the number would appear internally as 52315472, the first two digits representing the offset exponent, and the remainder the mantissa.

extra digits.) This implies a relative error on the order of

$$.5/(.5 \times 10^{t}), \quad \text{or} \quad 10^{-t}.$$

This assumes, of course, that all digits in the stored number are significant. If the number represents some measured or estimated physical error, or if it is the result of a series of computations using stored numbers, it may have less accuracy than the precision of the machine would indicate.

For example, if the relative error were actually 1 percent, only two of the stored digits would be significant, and the remainder would be garbage. Unfortunately, the computer itself provides no warning or indication of this fact; so it is up to the user to protect himself from accepting inaccurate data.

3. ERROR ACCUMULATION

In the standard arithmetic operations of addition, subtraction, multiplication, and division, people generally assume that they start with two known operands and that the operation produces a number which is the desired result. In fact, they usually start with two operands which represent only approximately the true, but unknown, numbers that are of interest to them. Thus they deal with "approximate" numbers as representation of "true" numbers, and the goodness of the approximation is changed by arithmetic operations.

To illustrate how this happens, consider the results of performing additions on two such approximate numbers u_1 and u_2. Assume the errors in the two numbers are bounded in absolute value by Δu_1 and Δu_2, respectively. It is easy to see that the error in the sum due to the error in the operands is no larger than $(\Delta u_1 + \Delta u_2)$. Round-off introduces an additional contribution as described previously, so that the total error from the operation of addition (or subtraction) can be as large as

$$\Delta u_1 + \Delta u_2 + (1 \times 10)^{-t} |u_1 + u_2|.$$

A more meaningful description of the impact of the error is given by the "relative error," which gives the size of the error as a fraction of the magnitude of the error-free result. This is particularly revealing in the case where we subtract the two positive numbers u_1

and u_2 where the relative error is given by

$$\frac{\Delta u_1 + \Delta u_2}{|u_1 - u_2|} + (1.0 \times 10^{-t}) = \frac{u_1 + u_2}{|u_1 - u_2|}$$

$$\cdot \left\{ \frac{u_1}{(u_1 + u_2)} r_1 + \frac{u_2}{(u_1 + u_2)} r_2 \right\} + 1 \times 10^{-t}$$

with $r_1 = \Delta u_1/u_1$ and $r_2 = \Delta u_2/u_2$ being the relative errors of the two arguments. The expression in the brackets is no larger than the maximum of the incoming relative errors r_1 and r_2. However, the expression in front of the brackets is greater than one and can become very large if u_1 and u_2 are nearly equal, greatly magnifying the original errors.

It is an unfortunate fact of life that subtraction of two similar numbers in the fixed word length computer is the greatest single cause of loss of significance in typical calculations. Multiplication and division are much more benign, producing relative errors that are bounded by the sum of the relative errors in the operands. The existence of floating-point arithmetic does nothing to alleviate the problem. It is incumbent on the modern numerical analyst to be aware of the problem and to attempt to minimize it.

Fortunately, there are often steps which can be taken in the formulation and implementation of algorithms which can significantly reduce error growth. In many cases a careful rearrangement of the order in which individual operations are performed will reduce the potential error. For example, if a and b are approximately the same, it is better to compute $a^2 - b^2$ as $(a + b) \cdot (a - b)$ as opposed to $(a^2) - (b^2)$. Tracing the error caused by round-off through more complex operations involving many operands reveals that it is generally better to begin evaluations with the smallest term and evaluate the arithmetic operations with the largest operands last. For example, tracing the evaluation of

$$(a + (b + (c + d)))$$

with no errors assumed initially in a, b, c, or d gives a relative error due to round-off alone of

$$\frac{(a + 2b + 3c + 3d)}{(a + b + c + d)} \times (1 \times 10^{-t}).$$

Therefore, the error is smaller if c and d are the smallest numbers of the original set.

In most cases it is difficult to trace the error growth through a complex computer algorithm. In these circumstances the computer itself can be used to help estimate the error growth by carrying out the estimates for each individual arithmetic operation, just as was done above. This can be done with competing algorithms, for example, in order to select the one having the best error characteristics.

The error estimates that have been described so far are usually over-pessimistic. This is because they do not account for the fact that errors are often of opposite sign. This is particularly true of the errors introduced by round-off which can be expected to be random in direction and magnitude. If this is taken into account, it can be shown that the cumulative error due to round-off in N operations is less than $\sqrt{12N} \times 10^{-t}$ with very high probability as contrasted with the estimate of $N \times 10^{-t}$ which would result from the techniques described earlier. A similar approach can be taken with respect to the propagation of errors through a sequence of arithmetic operations, but the assumption of randomness becomes much harder to justify and it is safer to rely on the more conservative approach in determining error bounds.

The errors described so far are fundamental to the basic arithmetic operations of the computer. In addition to these, there are the errors introduced by the technique chosen to evaluate a particular mathematical entity on the computer. These are the errors that would be introduced even if there were no error in the operands and if the computer had infinite precision. As such, they are the type of errors which have been the object of study under classical numerical analysis. These, too, are introduced and propagated through the ongoing stream of calculations typical of the large-scale problems solved on modern computers. Since such errors are peculiar to the function being performed, we will point them out in the context of summarizing the major areas of Numerical Analysis after a brief discussion of function evaluation techniques.

As we briefly describe each of the classical analysis areas, we will limit ourselves to examples in one dimension. Extensions to multiple dimensions are often straightforward, although with an attendant penalty in terms of computer resources needed for implementation. With few exceptions, there are a variety of algorithms available to solve any numerical analysis problem on the modern computer. For normal applications it may be difficult to justify selection of one over another. The best advice to prospective problem-solvers is to identify and at least try techniques that are already available on their computers before resorting to the development of new ones.

4. EVALUATION OF FUNCTIONS

Although not considered a major research area of classical Numerical Analysis, the evaluation of functions is inherent in every application of the computer. These range from the common trigonometric and exponential functions to the less frequently used Bessel and Elliptic functions. Often there are several evaluation techniques to be considered for implementation on the computer. For many functions, tables of values are available which could be read into the computer memory and used just as an individual would look up and interpolate to derive the value at the point of interest. However, entering the tables is often a nontrivial task and the speed of modern computers generally makes direct evaluation of the function a more efficient approach. In most cases the function is to be evaluated for values of the argument in some interval. Direct evaluation of functions in the computer is usually based on approximation by a polynomial or a rational function. The prototype of the polynomial expression is given by the standard Taylor series expression for a function

$$f(x) = f(x_0) + f'(x_0)(x - x_0) + \frac{f''(x_1)(x - x_0)^2}{2} + \cdots$$

where x_0 is some point in the interval. The error introduced by this approximation depends on the number of terms retained in the series and can be estimated by the remainder term derived in any Advanced Calculus text. In the Taylor series this "truncation error" is small near the point x_0 and takes on its extreme values at the end points of the interval over which the function is to be evaluated.

A better approximation, from the viewpoint of utility on the computer, is given by expanding the function f(x) as

$$f(x) = b_0 + \sum_{n=1}^{N} b_n T_n(x) \qquad -1 \leq x \leq 1$$

where $T_n(x)$ is a Tchebyshev polynomial of order n. Such an expansion tends to give a uniform error bound over the entire interval and generally allows an expansion to a lower order than the Taylor series for equivalent accuracy.

For functions that are to be evaluated a large number of times, a greater approximation accuracy (for a given order in the polynomial) can be achieved through the use of rational functions of the form

$$R_{nk}(x) = \frac{\sum_{i=1}^{n} a_i x^i}{\sum_{j=1}^{k} b_j x^j}.$$

There exist sophisticated procedures which evaluate the coefficients, a_i, b_j so as to minimize the maximum error in the approximation over the interval of interest. To save computational time, such techniques are frequently used for the standard library functions provided with the computer such as the sine, cosine, etc. The coefficient evaluation is too complex, however, to be justified in most computer function evaluations.

The errors introduced by the truncation of the polynomials used either directly or in the rational approximations are the first source of error to be considered in implementing a particular function evaluation. This error can generally be made as small as one likes depending on the number of terms which are retained. With single-word-length computer calculations there is obviously no point in carrying a greater relative accuracy than 10^{-t}.

Overlaying the error inherent in the approximating function itself is the accumulated effect of the errors propagated through the arithmetic operations required. These can be estimated through the techniques introduced earlier. Polynomial evaluation, in particular, is susceptible to the growth in relative error caused by subtraction of similar numbers.

Even when care has been taken to minimize the magnification of errors in each required arithmetic operation, the function being evaluated may have characteristics which tend to magnify existing error in the argument. This can be estimated using the first terms of the Taylor series. If we are evaluating f at the approximate argument $u + \Delta u$, then we can say

$$f(u + \Delta u) \simeq f(u) + f'(u)\Delta u + O(\Delta u^2).$$

For arguments where $|f'(u)|$ is much larger than one, the error contained in the argument can be correspondingly magnified.

5. NUMERICAL QUADRATURE

Quadrature, the evaluation of definite integrals, is one of the areas of classical numerical analysis where the computer has had a major effect. Prior to the advent of computer technology, the accurate evaluation of

$$I = \int_a^b f(x)dx$$

posed a formidable problem, impractical in all but a number of simple cases or where the indefinite integral would be evaluated explicitly. Today such evaluations are performed routinely, although there are still problems involving quadrature in multiple dimensions which require unavailable amounts of time even with the fastest computers.

The trapezoidal rule provides one of the simplest quadrature techniques on the computer and is almost a direct extension of the definition of the integral. The interval from a to b is subdivided by a number of points x_i separated by equal steps of length Δx and the integral I is evaluated as

$$I = (\Delta x/2)[f(a) + 2f(x_1) + 2f(x_2) + \cdots + 2f(x_{n+1}) + f(b)].$$

It can be shown that the error from this approximation alone is proportional to $(\Delta x)^3$. This algorithm has the benefit of simplicity but requires the evaluation of the integrand f(x) a total of $(n + 1)$ times. If f(x) is expensive (in terms of computer time) to evaluate, it follows that the same will be true for the trapezoidal rule.

In order to reduce the evaluations required for f(x) without reducing the accuracy of the evaluation, a large number of other quadrature formulas have been derived. The trapezoidal rule is based on approximating the integrand f(x) between x_i and x_{i+1} with a straight line. If a higher-order polynomial approximation is used, then the power of (Δx) appearing in the error term also increases. The most common formula of this type, known as Simpson's Rule, uses a quadratic estimate for f(x) and has an error proportional to $(\Delta x)^4$.

The third major approach usually goes under the name of Gaussian quadrature formulas. In this case the equal intervals Δx are replaced by an unequally spaced sequence of points so as to minimize the error. The determination of the points and the associated coefficients to be used in the approximation for I is difficult so that tables of predefined tabular values are normally used.

Errors in quadrature are likely to arise from a number of sources other than the remainder associated with the integration interval x (or equivalently, with the number of points in the interval at which $\Delta f(x)$ is evaluated). Obviously, there may be error terms associated with the evaluation of f(x) itself as discussed previously. The error terms for all of the quadrature techniques involve the value of a higher derivative of f as well as a power of x. For many functions, this contribution can significantly increase the size of the error term in the evaluation of I unless the integration interval is reduced appropriately.

6. SOLUTION OF EQUATIONS

We have already discussed the common problem of evaluating y when

$$y = f(x).$$

Frequently one is faced with the converse problem of evaluating x given y. When the equation can be solved explicitly for x, this offers no difficulty; however, one is generally not so fortunate.

In summarizing the numerical techniques for finding x, we may assume that $y = 0$ since the problem is easily reformulated so that this is the case. Thus our problem is one of finding the roots of the function f(x).

Almost without exception, the first steps consist of localizing the roots so that iterative procedures, which work only in the neighborhood of the solution, can then be used to find the precise value. If nothing is known about the function f(x) initially, the first step is often the generation of a graph or table of values which will give a gross estimate of where the roots might be.

For general equations, one of the most straightforward techniques for finding a root is the bisection method. In this case we must start with an interval [a, b] where, for example, f(a) is negative and f(b) is positive, and f(x) is known to be continuous between a and b. The function is then evaluated at $x = (a + b)/2$ and the original endpoint (a or b) for which f has the same sign as f(x) is discarded. The root is now known to be in the interval which is half the size of the original. This procedure can be repeated to specify the root to any desired accuracy and therefore actually solve the equation. However, if f(x) is difficult to evaluate (or if the variable x is a multidimensional vector), it often becomes more efficient to apply other techniques once the interval containing the root is sufficiently small.

The most commonly applied procedure is known as the Newton-Raphson algorithm and is based on approximating f(x) with a first-order Taylor series expansion about the most recent estimate for the root. The solution for the root of the resulting equation is taken as the revised estimate and the procedure is repeated until the values converge. The speed of convergence depends on the ratio

$$|f(x)f''(x)|/(f'(x))^2$$

near the root, and the ratio must be less than one for convergence to occur at all.

For some equations, f(x) can be rewritten as

$$x - g(x) = f(x).$$

Then under certain circumstances, the iterative formula $x_n = g(x_{n-1})$ will converge to the root if x_{n-1} is sufficiently close to begin with.

If f(x) is a real polynomial of order greater than four, there are no explicit formulas for the roots in general. However, there is a wide body of classical algebraic theory that can be applied to help localize or define the roots. These range from well-known techniques of factorization to reduce the order of the polynomial once a single root is found to procedures for identifying the number of such roots in an interval based on the changes in size of the coefficients. Even in these cases, however, there is no single method which is completely satisfactory for finding all of the roots of a polynomial.

The errors which arise in the solution of equations by the techniques indicated are relatively easily controlled. If the iterative techniques converge at all, the difference between successive values will bound the error from this source. A more serious problem is likely to arise from the numerical error in evaluating the function. This may result in shifting or even eliminating some roots.

7. MATRIX EQUATIONS

While the techniques of the previous paragraphs are readily extended to finding solutions or roots of simultaneous equations of several variables, systems of linear equations are given their own special category among the classical problems of Numerical Analysis. This is because of the frequency with which such equations arise in solving real-world problems and because of the vast body of relevant mathematical theory. Computers have made a major contribution in extending the scope of numerical problems in this area that can be practically solved today.

In finding the solution of a set of equations of the form

$$A\bar{x} = \bar{b}$$

or

$$\begin{aligned} a_{11}x_1 + a_{12}x_2 + a_{13}x_3 + \cdots + a_{1n}x_n &= b_1 \\ a_{21}x_1 + a_{22}x_2 + a_{23}x_3 + \cdots + a_{2n}x_n &= b_2 \\ a_{n1}x_1 + a_{n2}x_2 + a_{n3}x_3 + \cdots + a_{nn}x_n &= b_n \end{aligned}$$

the most common techniques are based on Gaussian elimination. Successive multiplication of the equations by a constant and subtraction of one from another leaves either a diagonal or upper triangular form for the equations which are then easily solved. Unfortunately, because of the number of subtractions involved, these techniques frequently introduce a great deal of error.

Another approach (Gauss-Seidel) is reminiscent of the successive substitutions used to find the roots of $x - \phi(x)$. The linear equations are rewritten so that the kth equation has $a_{kk}x_k$ isolated on the left side. Starting with an initial estimate, the x_i are introduced on the right side of the resulting equations to produce a new estimate. When the matrix A satisfies a nominal set of conditions, this sequence of operations can be shown to converge to the solution.

A related problem which arises frequently in physical applications involves solving the equations

$$A\bar{x} = \lambda\bar{x}$$

both for $\bar{x}$ and the so-called scalar eigenvalues λ. An entire repertoire of specialized techniques has been developed for solving this problem but is beyond the scope of what can be discussed here.

In general, matrix operations are characterized by large numbers of additions, often of terms of approximately the same magnitude. This makes them particularly susceptible to loss of significance and resulting error. If the matrices involved are large, it is always wise for the analyst to confirm the results of solving any matrix equation by evaluating the original equation with the solution found. In some pathological cases, even this is not enough to expose the existence of large errors.

8. INTERPOLATION, CURVE FITTING, AND NUMERICAL DIFFERENTIATION

During the days of hand calculations, the use of tables was the predominant form of evaluation of functions. As a result, interpolation techniques received intensive study in classical Numerical Analysis. As indicated earlier, high-speed computers have tended to reduce this application, but it is still used often enough to be significant. In any case, changing computer technology offering low-cost and high-density memories may reverse the trend in the future.

Classical interpolation techniques are based on polynomial approximations to the given tabular function. The coefficients are determined by the requirements that the polynomial pass through the tabular values surrounding the point when the new value is desired. The classical formulas, Newton, Stirling, etc., differ essentially in the grouping of the terms in the resulting polynomial or in the requirements on tabular spacing. Errors inherent in the polynomial techniques depend primarily on the distance between tabular values and on the magnitude of the higher-order derivatives of the function being evaluated. Only if it is a polynomial of the same or lesser degree as that used for the interpolation will the results be exact.

As in the other aspects of Numerical Analysis, the arithmetic operations in the interpolation formula may also introduce significant errors.

The computer has made possible significant advances in a more complex but desirable type of interpolation procedure known as spline fitting. Although polynomials are used to estimate the tabular function in this case also, a smooth fit is ensured by forcing the derivatives of the polynomials to be continuous from one interval to the next. The resulting conditions on the coefficients of the estimating polynomials are matrix equations and are too complex for solution except on modern computers.

The polynomial formula used for interpolation is also generally used when numerical differentiation of the underlying tabular function is desired. Each derivative, which lowers the degree of the interpolating polynomial by one, is a less accurate approximation than the previous. As a result, numerical differentiation should be used cautiously and only where absolutely necessary.

Often the tabular values represent physical measurements which contain varying amounts of error. In this case it is more reasonable to evaluate the associated function using a fitted functional form, either known or assumed. The functional forms have coefficients or other parameters to be determined according to some fitting criterion, often to minimize the sum of the squares of the differences between tabular and corresponding fitted values. In most cases, these problems reduce to matrix manipulation problems of the type described previously with all of the associated problems. As a result, it is only with high-speed computers that any but the simplest problems in curve fitting can be attempted.

9. SOLUTION OF ORDINARY AND PARTIAL DIFFERENTIAL EQUATIONS

The behavior of most physical systems is described in terms of a differential equation, ordinary or partial. With the use of the computer it has become possible to extend the evaluation of such equations beyond the small subset which can be solved explicitly. This potential has made this aspect of numerical analysis probably the most intensively studied over the past few decades. Even so, there remain innumerable differential equations arising in the physical sciences which cannot currently be evaluated.

Most of the problems associated with the numerical evaluation of solutions of differential equations can be discussed in terms of the single ordinary differential equation

$$dy/dx = f(x, y); \quad y(x_0) = y_0.$$

Under rather mild conditions on f, each initial condition y_0 determines a unique solution or trajectory y(x). However, in some cases such trajectories tend to diverge from each other very rapidly even if the initial conditions are close together. Under these circumstances, the small errors that are unavoidably introduced, for example in evaluating f, can rapidly grow to the point of invalidating the result. Such differential equations are said to be unstable and cause major difficulty in evaluation.

Almost all numerical integration techniques proceed one step at a time in the dependent variable, of length h. The simplest approach, known as Euler's formula, is given by

$$y(x + h) = y(x) + h\, f(x, y(x)).$$

This is really equivalent to a first-order Taylor series expression and the error at each step is proportional to h^2 and the first derivative of f at some point in the interval from x to x + h.

In practice, the Euler formula is not sufficiently accurate for reasonable values of h. By considering higher-order terms in the series expansion and combining the results, one can derive the so-called Runge-Kutta formulas, which give greater accuracy at the expense of additional function evaluations in the interval. However, even the higher-order formulas reflect the inherent instability in the differential equation if it exists.

A second major category of numerical integration formulas are the so-called predictor-corrector methods. These utilize a polynomial extrapolation for y(x) based on a number of previously evaluated points. The predictor-corrector algorithms tend to be somewhat more stable than the Runge-Kutta methods, although not necessarily more accurate.

The resources of the computer make it practical to determine the step size h adaptively as the integration progresses so as to use almost the largest value compatible with maintaining a given accuracy. This is done by comparing the error terms from two integration schemes of the same order (e.g., a fourth-order Runge-Kutta and Simpson's quadrature rule). Under the assumption that the error is no greater than the sum of both remainder terms, the value for the size of the next step can be derived. Such adaptive techniques are the basis for most numerical integrations performed on computers today.

The solution of partial differential equations (PDE), while drawing on the results from ordinary differential equations, is an immense area of study beyond description here. Specialized approaches are required for the major categories of partial differential equations and entire texts are devoted to the study of each category. The numerical solution of the PDE should not be attempted without a basic understanding of the body of mathematical theory developed in this area.

10. NEW PROBLEMS IN NUMERICAL ANALYSIS

While the classical subdivisions of Numerical Analysis are still very active with many unsolved problems remaining for the researcher, the use of the computer has expanded interest in new areas as well. In most cases, these have arisen through attempts to optimize the behavior of real systems through computer models.

For example, one of the early problems dealt with the optimization of a linear and nonlinear function of many variables x_i which were themselves subject to constraints of the form

$$ak_1x_1 + ak_2x_2 + \cdots + ak_nx_n \leq C_k.$$

The use of additional variables z_r, called slack variables, can be used to express the inequality as

$$ak_1x_1 + ak_2x_2 + \cdots + ak_nx_n + z_k^2 = C_k$$

and reduce the problem to the solutions of simultaneous equations. Similar approaches work for nonlinear optimization with inequality constraints.

Attempts to optimize complex functions of variables taking on discrete values and complex combinatorial problems similar to the "traveling salesman" problem (which attempts to minimize the traveling distance required to visit all points in a complex network) are still unsolved in general. Current research addressing the measurement of problem complexity suggests that success in these areas will always be limited.

The use of the computer in the real-time control of complex systems is forcing the reexamination of many techniques from Numerical Analysis with a view toward making their operation faster. Outgrowths of such studies have resulted in the so-called Fast Fourier Transform and the Kalman Filter, the latter essentially a recursive scheme for doing least squares estimation.

It seems clear that man's pursuit of insight into increasingly complex physical processes, from atomic interactions to weather prediction, will require both faster and larger computing capacity and innovative algorithms in numerical analysis.

SUGGESTED READINGS

Ralston, A., and H. Wilf, eds., *Mathematical Methods for Digital Computers*, 2 vols., Wiley, New York, 1960.

Pennington, R. H., *Introductory Computer Methods and Numerical Analysis*, Macmillan, New York, 1970.

Arden, B., and K. Astill, *Numerical Algorithms*, Addison-Wesley, Reading, Mass., 1970.

Hastings, C., *Approximations for Digital Computers*, Princeton University Press, 1975.

COMPUTER SIMULATION

Mark Franklin

1. SIMULATION: MODELS AND METHODOLOGY

1.1. Introduction. Digital computer based simulation is rapidly becoming the predominant technique used in the analysis of complex systems. The systems and problems tackled by computer simulation span the range from traditional engineering based systems [1], [2] to biological, environmental, urban, and social systems [3], [4]. There are a number of reasons for the wide and increasing use of computer based simulation.

First of all, models of the complex systems with which we are currently concerned are rarely amenable to analytic solution by traditional mathematical methods. Computer based approximation and numerical analysis techniques which lie at the heart of digital simulation, however, can usually be used to "solve" such models. Second, computer simulation languages and facilities have developed to a point where it has become intellectually easier to formulate, program, and obtain solutions to these models. Third, the decreasing cost of computers, due to technological advances, and of computing, due to higher-level languages, have made the simulation approach more economically reasonable.

Finally, over the past few years our society has become increasingly aware of the difficult problems it faces in almost every sphere of activity. Such problems as energy, environmental quality and industrial productivity are complex and interrelated. Engineers, natural scientists, social scientists, and social planners are now coming together in interdisciplinary groups to try to solve these problems. The training of the individuals in such groups often has a quantitative component and their commitment is usually to a quantitative analysis of the problem at hand. This, added to the complexity of the systems they're attempting to deal with, leads directly to the use of simulation methods. The common language of such groups is becoming the language of simulation with the system model being the core to which individuals contribute, on which individuals test hypotheses and argue, and from which policy recommendations are made. This trend will undoubtedly continue since it offers such problem-solving groups a common communications format, a common problem-solving discipline, and a model solution procedure.

The discussion so far has used the terms *system*, *model* and *simulation* in a general, intuitive manner, avoiding the problems of attempting precise definitions. The terms are used in such diverse ways that deciding on reasonable, useful, and yet concise definitions is difficult. Nevertheless, let us proceed.

A SYSTEM may be defined as a set of interdependent elements acting to achieve some implicitly or explicitly defined goal. Thus, for instance, a computer can be considered to be a system whose interacting elements include logical gates and memory units. In this case the goal is explicitly defined by the known properties of the elements and the manner in which they have been organized. On the other hand, in the case of biological processes (e.g., evolution) the goals are often implicitly rather than explicitly defined.

Once a collection of elements is recognized as constituting a system, then the process of describing the system begins. The description itself is referred to as a MODEL of the system. There are clearly vastly different ways of describing or modeling systems and a number of approaches have been taken [5] to classifying these different descriptive modes. Development of such a classification, or taxonomy of models, clarifies how the digital computer can aid in model formulation and solution and also clarifies what types of models are appropriate for investigating different systems.

1.2. Model Classification. The classification scheme presented here follows that of Mihram [6] and is summarized in Table 1. One of the main distinctions to be made is between *material models* and *symbolic models.* A material model represents a spatial transformation from the original physical system of interest (the "real" system) to some other physical system which in some sense is simpler, more easily understood, or more easily manipulated. One type of material model is a direct *replication* of the system in question but on an altered dimensional scale (e.g., a model airplane). Another type of material model is a *quasi-replica* of the real system. Like the replication, it is a physical model in which a spatial transformation of the real system has occurred; however, in this case one or more dimensions have been omitted (e.g., a road map). The final type of material model in Mihram's classification scheme is referred to as an *analogue* model. In this type of model no attempt is made at preserving the physical dimensions of the real system. The primary objective is to preserve behavioral or performance characteristics. Analog computers have been used extensively in this way to model systems governed by sets of differential equations. The analog computer consists of electronic components whose behavior (i.e., variation in voltages and currents with time) corresponds to simple mathematical operations such as addition, multiplication, and integration. These components may be interconnected so that overall behavior of the system represents, for example, the solution to a set of differential equations. The computer, interconnected in a particular manner, represents an analogue model if this set of equations is the same as the set governing the physical system of interest.

While material models attempt to maintain a physical link between the model and the real system, with symbolic models this link is broken. *Descriptive* models are one type of symbolic model in which natural language (e.g., English) is used to represent a system. The symbols in this case are elements of the language, and manipulation of symbols follows the allowed grammatical rules of the language. A botanist's written description of a plant is a descriptive model. Formal models, or *formalizations*, are another type of symbolic model. This type, however, is one in which symbol operations fall within a highly developed mathematical discipline, such as integral calculus or numerical analysis. A differential equation model of a system is representative of this category.

TABLE 1

Model Taxonomy (Reference 6).

	Material			Symbolic		
	Replication	Quasi-replica	Analogue	Descriptive	Simular	Formalization
STATIC Deterministic	Earthen Relief Map	Road Map	Statue of B. Franklin	Ten Commandments	Decision Logic Tables	Ohm's Law
Stochastic	Critical Dosage Test	Weather Map	Die Toss for Russian Roulette	Weather Report	Non-Adaptive, Random, Chess Playing Program	Equilibrium Queue Length
DYNAMIC Deterministic	Model Train Set	Planetarium Show	Analog Computer Circuitry for $\dot{y} = -y$	Constitution of U.S.A	Critical Path Algorithm	Lanchester's Laws
Stochastic	Drosophila Genetic Experiment	CRT Display of Endurance Test	White Noise Generator	Text on Darwinian Evolution	Vehicle-By-Vehicle Transportation Model	Stochastic Differential Equation

Falling between descriptive and formalization models, and often containing elements of both, are *simular* models. Higher-level programming languages, for instance, allow users to represent systems very much as descriptive models; however, internal to the language processor where formal numerical methods are utilized, the models are more of the formalization type. This is one of the reasons that simular models and simulation languages have achieved such widespread use and popularity. That is, on the one hand, the language features available allow individuals to describe systems in a comfortable manner often closely related to natural language descriptions, while, on the other hand, the built-in formal numerical analysis constructs assure, under proper conditions, model solution. In addition, the users of such simulation languages often need not be too deeply concerned with understanding the details of the numerical methods utilized, since these are built into the language processor. This allows users to concentrate on the structure and properties of the system of interest. Finally, due to the symbol-processing capabilities of digital computers, such computers have become the primary tool used for the development and solution of simular models.

The models thus far have been classified in terms of the level of abstraction used to describe the system. Material models of the replication type are the least abstract, while symbolic models of the formalization type are the most abstract. Another dimension of model classification deals with whether the model is *Static* or *Dynamic*, *Deterministic* or *Stochastic*. A static model is one whose behavior does not change with time. Thus, for instance, Ohm's Law is a simple static model of electrical behavior of a circuit in equilibrium. If the circuit is disturbed by, say, introducing a time-varying voltage, then a set of differential equations incorporating Ohm's Law would be a more accurate model, and this would be a dynamic model. If the models described above contain no nonrandom elements, then they would be deterministic models. If, however, there were random elements, then they would be stochastic models. Thus, for instance, if the voltage on the circuit varied in a random fashion, the appropriate model of the system would be a set of stochastic differential equations. Such a set of equations would consititute a formalization model of the stochastic, dynamic variety.

This completes our review of model taxonomy. The remaining discussion centers around simular models, since these are most closely connected with digital computer simulation.

1.3. Simulation Methodology. Consider now the term "simulation" itself. Although it is often used interchangeably with "modeling," "simulation" implies a good deal more. Simulation is a process containing a number of components, one of which is modeling or describing the system of interest. The process can be discussed in terms of a sequence of stages as illustrated in Figure 1.

The first stage is *problem formulation* and *study planning*. Determination of simulation study goals, and plan of attack, is often the most important and difficult part of a simulation. The difficulty arises because there is a close connection between what are es-

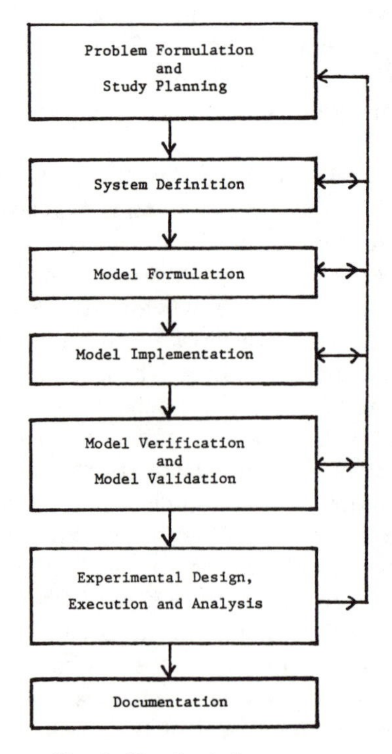

FIG. 1. The simulation process.

tablished as study goals and procedures and how well the system to be studied is itself understood. Indeed, one study goal will often be to obtain a deeper understanding of the system of interest—even here, though an attempt should be made at quantifying what is meant by the goal of achieving a deeper understanding before the simulation study gets under way. Initially, this may entail determining whether certain narrowly defined input/output relationships exist. For instance, if the system to be considered is a hospital outpatient clinic, one might be interested in how patient waiting time varies with the arrival rate of the patients. Once this has been achieved, the problem may be reformulated and expanded to include investigation of various design alternatives. For instance, it might be preferable, from a mean-patient-waiting-time point of view, to schedule patients in different ways depending on their expected resource requirements (i.e., some patients require long physician consultations, others short; some require x-rays, etc.).

Determining the schedule is one aspect of system design and control. The final problem formulation will often relate to *optimizing* the system design. Optimization requires that a measure of system performance be accepted by those individuals involved in the simulation study. Once this measure of performance is decided upon, then system design alternatives can be examined in terms of maximizing or minimizing this performance measure. Preliminary decisions on what this measure should be are part of the problem formulation stage. This can be a nontrivial task, especially when individuals with conflicting personal goals and backgrounds are connected with the study. Often a realistic performance measure can be formulated only later in the simulation process when greater understanding of system operation has been achieved. The outpatient clinic example illustrates how such performance measure differences may occur. The example used outpatient waiting time as the measure of interest. Note, however, that this might not be consistent with a dollar profit measure on clinic operation, or perhaps with a measure based on clinic resource utilization.

Closely connected with establishing study goals is the question of the study plan. A clear plan for achieving the study goals will include time and dollar estimates for the various stages in the study. Stating, in written form, both the goals of the simulation study and the resources required for successful completion is especially important when the study is undertaken under the auspices of a large organization. For instance, in a large corporate environment numerous individuals and groups may have an interest in the development of a corporate financial model. These individuals will often be needed to provide data, to aid in model development, and to provide financial support for the study. Agreement on the goals of the study and a commitment to the general study plan is therefore vital.

The second stage in the simulation process relates to *system definition*. The *entities* or primary objects of interest in the system must first be identified. In the hospital outpatient clinic, for example, the entities might be the patients, doctors, nurses, and X-ray units in the clinic. Associated with these entities are *attributes* which denote various entity properties. Thus, associated with the entity physicians might be the static attributes "quantity" and "specialty," and the dynamic attribute "busy/not busy." The attributes associated with a particular entity (e.g., a particular physician) at a point in time represent the *state* of the entity, while the collected states of all critical entities in the system represent the *state of the system*. Notice that the dynamic attribute "busy/not busy" indicates that an *activity* is in process. Such activities and their interrelationships determine, in part, how the system state will evolve in time. Static or *class* relationships between entities may also be specified (e.g., associated with each X-ray unit are the two nurses required to operate the unit) when needed for system definition.

In addition to entities, attributes, activities, and states, the *system boundaries* must be defined. The system of concern is said to exist in an *environment*. The system boundaries determine those entities in the *environment*, (i.e., outside of the boundaries) which affect system activities but on which the system itself has no effect. Returning to the outpatient clinic example, the weather, for instance, represents an entity in the environment which will affect system operation by altering the arrival rate of patients.

Another aspect of system definition concerns dividing up the system of interest into *subsystems*. Such subsystems each represent an entity or collection of entities which, in some sense, acts as a unit in relationships with other subsystems. The objective in defining subsystems is to simplify specification of system interactions and activities. Fewer subsystems present generally means that fewer attributes, states, and activities are needed in describing the system.

The question of what constitutes a subsystem is closely related to the development of hierarchies of system definitions. Such hierarchies usually represent differing levels of detail which may be used in defining subsystems. An example of this is given in Figure 2. Say that the system of interest is a computer system. At the "highest" level, a very low detail representation of a computer may view the computer as a simple server with given statistical execution properties which acts on incoming customer programs to produce certain program execution results as output. Given a statistical description of the program arrival process, the length associated with the queue of incoming programs, the discipline associated with selecting jobs from the queue (e.g., first come, first served) and a statistical description of the service process, the goal of a simulation study might be to find out what the waiting time is for customer programs (i.e., response time of the system). Such a low detail-level system description could not by itself be used to

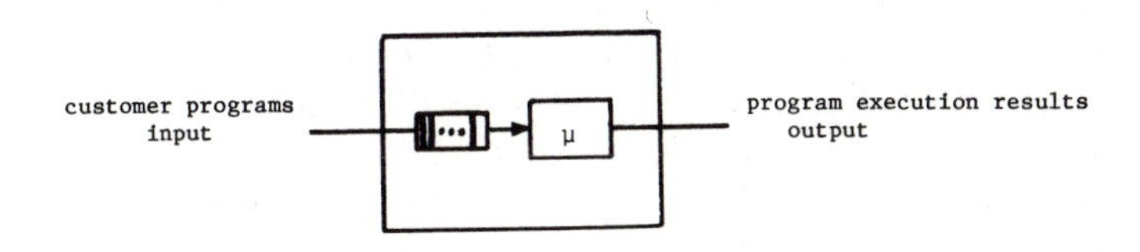

Level 1: Computer System As A Simple Server

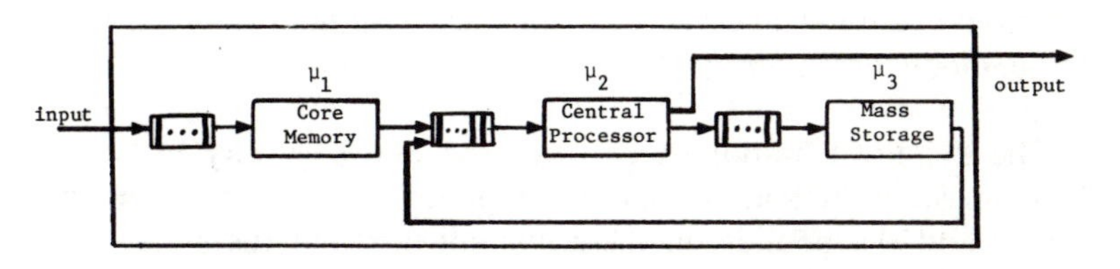

Level 2: Computer System As A Collection of Functional Servers

Level 3: Central Processor As A Collection of Registers and Logical Gates

Level 4: Registers and Logical Gates As A Collection of Electronic Components

Level 5: Electronic Components In Terms of The Basic Physical Laws Governing Their Operation

FIG. 2.

determine the effect of the speed of, say, the mass storage device on response time. To get at this, a lower-level representation is needed.

Level 2 thus considers the computer system as made up of three primary subsystems: central processor, central memory, and mass storage. Notice that other subsystems, such as the system printer, might also be added at this level. The question of what subsystems should be included at each representation level is clearly a matter of judgment. If it was felt that the system printer represented a possible bottleneck or limiting resource in system operation, then it would have to be included. While it will sometimes be necessary to include a subsystem in the model description in order to determine its relevance to system operation, the tendency to construct overly complex models should be avoided. Added subsystems mean added complexity and cost in programming, validating, verifying, and generally understanding what's going on in the model. Initially, it is usually better to err on the side of simplicity rather than complexity.

Pursuing this example a bit further, one can identify at least three more levels of detail. The central processor subsystem, for instance, can be represented in terms of the registers, logical gates, and information transfers between them. This would be an appropriate representational level if the objectives of the simulation study concerned how gate failures affected processor performance. Indeed such studies are often undertaken with a view toward designing fault detection and diagnostic capabilities for computer systems.

A level below the register and gate level would be the circuit and electronic component level, while one level below this would view the components in terms of the basic physical laws governing their operation.

The system definition stage discussed above is very much an analysis stage in that the basic components which make up the model and the bounds on these components and the system are defined. The third stage, the *model formulation* stage, is a synthesis stage, in that the concern here is with the overall structure and interrelationships between the model components. In the model formulation stage, just how the various subsystems and activities affect each other must be defined. Choices must be made, for instance, as to whether a deterministic or stochastic model should be used. Indeed, at this stage the general question of model type (e.g., material, symbolic, etc.) must be resolved. Of course often the model type will be dictated by decisions made in the stages already discussed. Given, say, a dynamic model, these dynamics must be specified. Sometimes this is most easily done by means of logical flow diagrams, while sometimes it is convenient to write down sets of equations which govern these interactions. If a simulation language is available, it may be possible to conveniently describe these relationships directly in the language.

This leads to the fourth stage, model implementation. Clearly, implementation problems and capabilities affect, and are affected by, decisions made in the previous stages. In terms of digital computers, model implementation relates to how the model formulated is mapped into a correct sequence of computer instructions. Simply, how does one produce a computer program corresponding to a given model? Two separate concepts should be identified here. The first relates to the descriptive question of how the model characteristics are represented in the computer's language. The second relates to the numerical techniques used to "solve" the model. This latter question is treated in part in Chapter 7 on Numerical Analysis and is not dealt with here. References to some of the standard works in this field are provided, and additional references will be noted later.

The descriptive problem has been eased considerably in recent years by the wide availability of a host of simulation languages which are often specialized to certain problem areas. For *continuous systems* (i.e., the variables in the system are continuous functions, usually of time) such languages as CSMP, Continuous System Modeling Program [7], [8], [9], and DYNAMO [10], [11] may be conveniently used. For *discrete probabilistic* systems (i.e., the variables in the system are stochastic in nature and change in discrete steps) such languages as GPSS, General Purpose Simulation System [12] [13], GASP II [14], SIMSCRIPT [15], and SIMULA [16], [17] may be used. The GASP IV [18] language is available for those systems which are best represented by a mix of continuous and discrete probabilistic variables. More specialized languages are also available for modeling very specific types of systems. For instance, ECAP, Electronic Circuit Analysis Program [19], is a language which allows for direct modeling of electrical circuits, while ICES STRUDL-II, Integrated Civil Engineering

Systems Structured Design Language [20] may be used to model elastic, statically and dynamically loaded, framed structures. Several of these languages will be examined in some detail in the next two sections of this chapter. It should be noted that while the numerical techniques needed to solve the system are usually built into the code produced by the simulation language compiler, some individuals prefer to code these numerical algorithms directly in a higher-level language such as PL/I or FORTRAN. This generally gives one more control of the detailed implementation questions related to the numerical algorithms and often results in faster executing simulations. On the other hand, this approach typically requires more programming effort. In addition, while the resulting program will indeed be a form of model description, the overall structure and organization of the system will usually be obscured by the detailed level at which coding must proceed. By programming in a simulation language the mapping of the system model into the language commands is often more straightforward, with the program more obviously representing the system of interest. Model documentation and education problems are therefore somewhat eased. This will become clearer in the sections to follow.

Once the model has been programmed on the computer, questions relating to its "correctness" and "goodness" must be dealt with. These questions have been separated into two parts. The first, *model verification*, considers how well the model responses, as now programmed on the computer, correspond to theoretically anticipated results. The questions here concern the basic soundness of the model. Has the program been fully debugged? Is the random number generator working properly? Do simple parameter sensitivity tests perform as expected? Perhaps certain model operating conditions correspond to a system which can be solved analytically (e.g., transform a nonlinear model into a linear one, transform a complex queueing model into a simple single server queue). Under such conditions does the model output correspond to the analytic solution? The objective here is thus to gain confidence in the model's inherent operational characteristics. Often errors will be found at this stage which will cause one to reformulate the model. More often than not programming errors will be discovered which will result in at least partial model reimplementation.

Model validation, the next part, is specifically concerned with how well the implemented model represents reality. Comparisons here are comparisons of model outputs with real collected data. Clearly, this cannot always be done. Often a simulation is used to investigate conditions or systems which do not yet exist, and therefore for which no data is available. If these new situations represent reasonable extensions or modifications of existing systems, then it may make sense to structure the model so that both the existing system and new system are modeled within the same framework. Validation of the submodel representing the existing system can then provide some added credibility for the new system model. When this can't be done, more effort should probably be allocated to model verification. References [21], [22], and [23] deal in part with some of the statistical questions relating to model verification and validation.

Given the preceding stages the modeler is now in a position to begin experimenting with his model. This stage consists of experimental design, execution, and analysis. Although the first stage, problem formulation and study planning, has dictated the type of experiments to be performed, the details must now be set. The large number of variables present in many simulations can lead to excessive computer running times and costs unless some care is taken in this experimental design. Optimization problems, for example, which require multiple runs with parameter variations can be costly even when care is taken [24]. This is especially true with regard to stochastic models where multiple runs or extended runs may be needed to resolve questions of stochastic convergence and statistical validity. This latter problem is dealt with in part in Chapter 8 on Computer Science and Statistics. More information on experimental design can be found in [22], [23], and [25]. Careful experimental design is also necessary if the experimenter is to learn about and be able to analyze the system in an orderly, structured fashion. It is very easy to become overwhelmed with reams of simulation run outputs, each run having, for example, somewhat different parameter settings or initial conditions.

When the analysis has proceeded to the point that the goals initially established have been satisified, the final stage of the simulation process, documentation, is entered. Three types of documentation can be distinguished. The first concentrates on the results of the simulation study. That is: What has been learned about the system of interest? What are the answers to the problems posed in the first stage? The second relates to documenting the simulation program itself so that it can be understood and, if necessary, modified by others. The third concerns the use of the simulation by other parties who perhaps would like to use the simulation to run their own set of experiments. This latter document, in effect, is a user's manual.

This completes the introductory section of this chapter. The remaining sections concentrate on the properties and use of several of the more popular simulation languages.

2. CONTINUOUS SYSTEMS SIMULATION

2.1. Introduction. Continuous systems are those in which the state variables change in a smooth or continuous manner. Most continuous systems simulation is devoted to the solving of systems described by sets of differential equations. Simple differential equation models, such as those with linear, constant coefficients, can be solved without the use of the numerical approximation techniques central to continuous systems simulation. In these cases the use of simulation techniques may still be desirable, however, because of the ease with which such problems can be represented and solved. Once the nonlinearities associated with most real world problems are brought into the differential equation model, it is usually impossible to solve these models without using simulation techniques. It is in this area of complex, nonlinear systems that simulation methods are indispensable.

The remainder of this section on continuous systems simulation contains three parts. The first presents two examples of continuous systems. The second presents a block-oriented simulation language, Block/CSMP, which is well suited for use on a small computer. The third presents an equation-oriented simulation language, 360/CSMP, which is widely used on large IBM computers. These languages are each used to model and solve the example systems.

2.2. Continuous Systems Examples. For the first example consider the simple mechanical system of Figure 3. The system consists of a mass M, suspended from a rigid structure by a spring with stiffness K(X). This is in turn connected to a rigid structure below through a

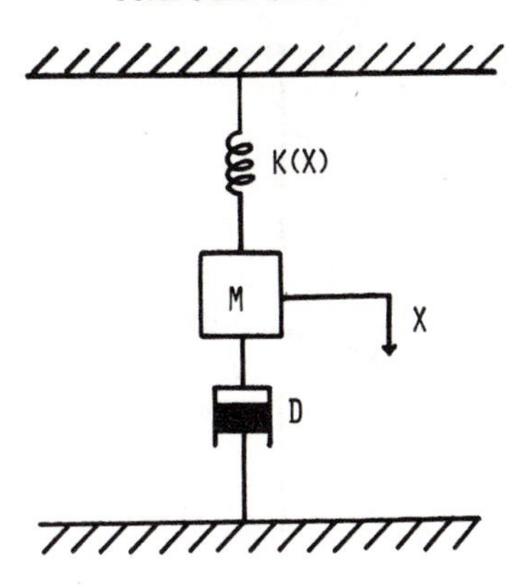

FIG. 3. Simple mechanical system.

dashpot with damping constant D. The object of the simulation is to find out how the mass oscillates with time if the mass is initially pulled down (i.e., the spring is extended) and then released. Such a simple model might represent part of the suspension system of an automobile with the spring corresponding to a suspension spring and the dashpot corresponding to a shock absorber. The objective of the simulation might be to determine the spring/shock absorber combination which produces the "smoothest" ride.

The differential equation which describes the system can be obtained directly by using Newton's Laws. The force needed to accelerate the mass is $M(d^2X/dt^2)$, where d^2X/dt^2 represents the second derivative of the space variable X with respect to time. The force exerted by the spring is K(X). The notation K(X) indicates that this force is a function of the position X. For this problem the function is a nonlinear one which has been empirically obtained and is plotted in Figure 4. The force exerted by the dashpot is proportional to velocity and is given by $D(dx/dt)$. In this system, no other forces act on the mass; hence the nonlinear differential equation governing the system is:

$$M\frac{d^2X}{dt^2} + D\frac{dx}{dt} + K(X) = 0. \tag{1}$$

Note that two initial conditions must be provided in order to solve the system.

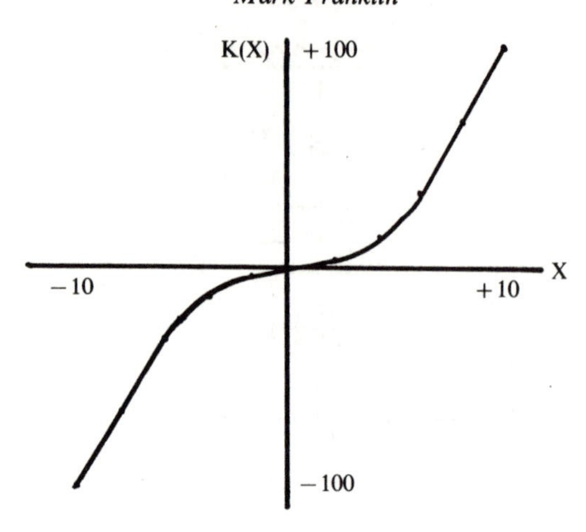

FIG. 4. Spring force K(X) vs. distance X.

It is sometimes convenient to rewrite the system equations in a somewhat different form. First solve for the highest derivative d^2X/dt^2.

$$\frac{d^2X}{dt^2} = -\frac{1}{M}\left(D\frac{dX}{dt} + K(X)\right).$$

Next let the first derivative of X with respect to t be equal to X1DOT and the second derivative be equal to X2DOT. The system described by (1) can now be described as:

$$X(t) = \int_0^t X1DOT(t)\,dt + X(0) \tag{2}$$

$$X1DOT(t) = \int_0^t X2DOT(t)\,dt + X1DOT(0) \tag{3}$$

$$X2DOT(t) = -(1/M)\,(X1DOT(t) \cdot D + K(X(t))) \tag{4}$$

where $\int_0^t$ indicates integration from time 0 to time t, and X(0) and X1DOT(0) represent the initial position and velocity conditions present on the mass. Notice that these equations are fairly obvious. Equation (2), for example, merely says that X at time t is equal to the integral of the derivative of X from 0 to t, plus the initial condition on X. Thus rewriting the equations in this form emphasizes the integrations which govern system operation rather than the differentiations. The primary reason this has traditionally been done is that the integration operation, both when performed electronically on an analog computer, and when done numerically on a digital computer, is a more stable procedure than the differentiation operation. Note that any system of differential equations can be redescribed in the manner given above.

The second continuous systems example is taken from the field of population dynamics. Variations and extensions of this example have been used to study the interactions of different populations with each other and with the environment in which they exist [26]. To begin with consider a single population of size N. On the one hand population growth will be directly related to population size (i.e., the larger the population, the more births and deaths). On the other hand, as the population grows it will exert an increasing impact on its environment. Given a finite or restricted environment, at some point demands on the environment will limit further population growth. Thus the larger the population the lower the expected birth rate. These two ideas can be used to form a simple model of population growth. The resulting differential equation, called the Verhulst-Pearl logistic equation, is given below.

$$\frac{dN}{dt} = aN - bN^2. \tag{5}$$

The first term, aN, represents the rate of increase of a population of size N if there were no resource limitations with "a" being the intrinsic rate of natural increase (i.e., birth rate minus death rate). The second term, bN^2, represents the inhibiting effect of a limited environment, and hence finite resources. The parameters a and b are referred to as the "logistic" parameters.

A simple extension of this model considers two species, with populations N_1 and N_2, competing with each other for the same resources. The equations for this system would be:

$$\begin{aligned}\frac{dN_1}{dt} &= N_1(a_1 - b_{11}N_1 - b_{12}N_2),\\ \frac{dN_2}{dt} &= N_2(a_2 - b_{21}N_1 - b_{22}N_2).\end{aligned} \tag{6}$$

In (6), a_1, b_{11}, and a_2, b_{22} are the logistic parameters for each of the two populations when living alone. The b_{12} and b_{22} parameters indicate the effect of each population on the other. One interesting aspect of this model relates to relative population growth as parameters of the equations are varied. It can be shown [26] that under certain conditions one population can grow while the other becomes extinct, while under other conditions an equilibrium situation can occur with both populations remaining viable. Notice that both of these models represent a continuous approximation to a discrete system. That is, the values of N can only be integer, and can only change by integer amounts in the "real" system while the model allows for fractional values and changes.

To conclude this section let us rewrite equations (6) in the integral form similar to equations (2)–(4).

$$
\begin{aligned}
N_1(t) &= \int_0^t N_1(t)(a_1 - b_{11}N_1(t) - b_{12}N_2(t))\,dt + N_1(0), \\
N_2(t) &= \int_0^t N_2(t)(a_2 - b_{21}N_1(t) - b_{22}N_2(t))\,dt + N_2(0).
\end{aligned}
\tag{7}
$$

$N_1(0)$ and $N_2(0)$ represent the initial sizes of population 1 and 2.

2.3. The Block Continuous System Modeling Program (Block/CSMP). The BLOCK/CSMP simulation language presented here is a successor of the 1130/CSMP language which was originally available on the IBM 1130 computer [8]. Versions of this simulation program have run on numerous minicomputers. The language is easy to use and interactive versions have been developed. These interactive versions allow the user to specify, modify, and run the model while on-line to the computer. The immediate feedback of model solutions from the computer to the user not only speeds up the debugging process but allows the user to experiment with the model in a natural manner letting the user follow interesting leads and insights in a straightforward "hands on" fashion.

Use of BLOCK/CSMP can be divided into the four phases given below:

1. Configuration Phase—The user defines a block diagram representation of the problem.

2. Parameter/Initial Condition Phase—The user specifies model parameter, initial condition, and function generator values.

3. Timing/Output Phase—The user specifies numerical integration algorithm type and step size, output variables to be plotted, and output time step size.

4. Run Phase—The user runs the simulation, obtains the output, and determines what to do next.

For phase 1, the configuration phase, the user must first specify his problem in terms of a block diagram. There are about 30 blocks available to the user in defining his model. The blocks can be broken down into five types: (1) Linear Continuous (e.g., Summer, Integrator); (2) Nonlinear Continuous (e.g., Multiplier, Sine); (3) Nonlinear Discontinuous (e.g., Absolute Value, Limiter); (4) Control and Timing (e.g., Relay, Unit Delay), and (5) Special Functions (e.g., Time, Jitter). A list of ten of these blocks is given in Table 2.

To draw a block diagram for a system initially described by a set of differential equations, it is usually easiest to begin by transforming the equations into an integral form similar to that shown in equations (2)–(4) and (7). Corresponding to each integration operator, an integrator block will be required. For the nonlinear spring example the output of each of the two integrators required will be X(t) and X1DOT(t), respectively. Now X(t) has X1DOT(t) as an input; hence the two integrators are connected as shown below:

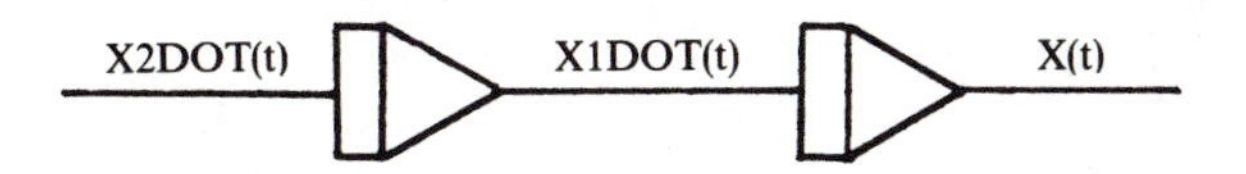

X2DOT(t) is the input to the integrator which is defined by equation (3) and produces X1DOT(t). X2DOT(t) itself, however, is equal to X1DOT(t) and X(t) multiplied by the appropriate parameters and summed together as indicated in equation (4). The function generator and weighted summer blocks in the simulation language can be used for the purpose of implementing equation (4) with the block diagram for the entire problem given in Figure 5. This diagram clearly and visually indicates the interactions and feedback paths which are implicit in the system's differential equations. A similar diagram for the population dynamics problem is also given in Figure 5.

TABLE 2

Block/CSMP Block types.

Block Type	Language Symbol	Flow Diagram Symbol	Operation
Weighted Summer	W	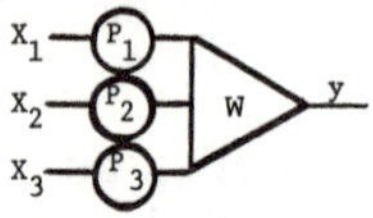	$y = P_1X_1 + P_2X_2 + P_3X_3$
Integrator	I	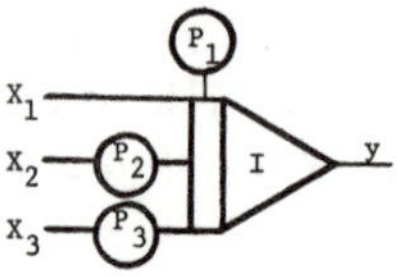	$y = P_1 + \int_0^t (X_1 + P_2X_2 + P_3X_3)\,dt$
Inverter	-		$y = -X_1$
Gain	G	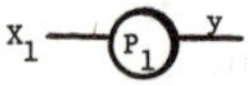	$y = P_1X_1$
Constant	K	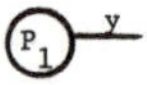	$y = P_1$
Multiplier	X		$y = X_1 \cdot X_2$
Sine	S		$y = \mathrm{Sin}(X_1)$
Function Generator	F		$y = f(X_1)$ P_1 and P_2 define upper and lower bounds, see text)
Limiter	L	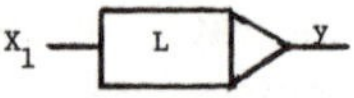	If $X_1 \geq P_1$; $y = P_1$ If $P_1 > X_1 > P_2$; $y = X_1$ If $X_1 < P_2$; $y = P_2$

Given that a block diagram has been established, a unique integer must be assigned to each block in the diagram. The block diagram may now be entered into the computer. In some systems this is done by means of punched cards. Figure 6 indicates how this

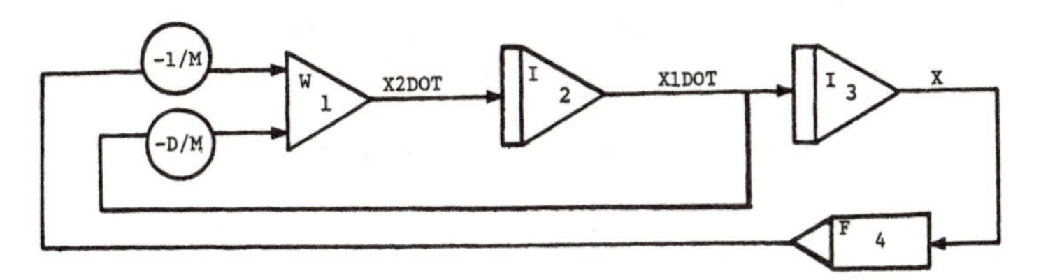

Nonlinear Spring Problem

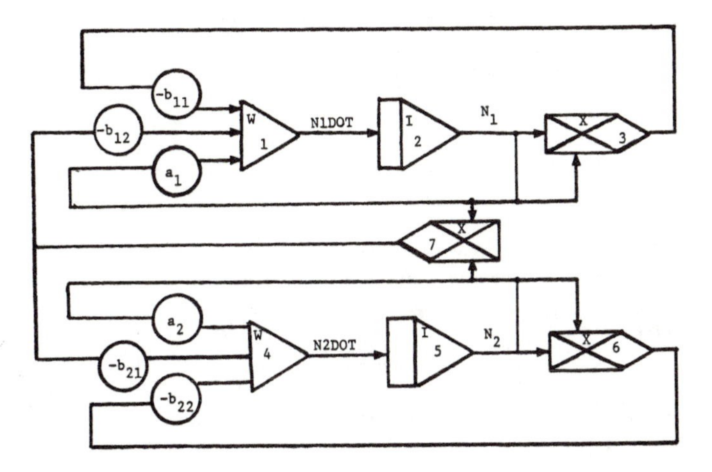

Population Dynamics Problem

FIG. 5. Continuous systems examples (initial conditions omitted).

information is entered in an interactive system where the user has the option of having the system guide him in entering the information. User inputs follow the "?".

The first set of user inputs consists of 1,W,4,2. This says that the block assigned number 1 is a weighted summer, W, whose inputs are the outputs of the blocks assigned numbers 4 and 2. The remaining three configuration specification inputs complete the description of the block diagram to the simulation program. Hitting the carriage return key on the computer input device in response to a ? alerts the program to enter the next phase.

The parameters and initial conditions phase is also illustrated in Figure 6. The first input 3, −10 relates to integrator block 3 and

```
                    CONFIGURATION SPECIFICATIONS

ENTER BLOCK SPECIFICATIONS;
SEPARATE ENTRIES WITH COMMAS
CARRIAGE RETURN ENDS THE OPERATION
BLOCK NUMBER, BLOCK TYPE, INPUT1, INPUT2, INPUT3
?1,W,4,2

?2,I,1

?3,I,2

?4,F,3

?


                 INITIAL CONDITIONS AND PARAMETERS

ENTER: BLOCK NUMBER, IC/PAR1, PAR2, PAR3
SEPARATE ENTRIES WITH COMMAS
CARRIAGE RETURN ENDS THE OPERATION

?3,-10

?1,-0.2,-0.4

?4,10.,-10.

?


                 FUNCTION GENERATORS SPECIFICATIONS

CSMP-11 ALLOWS ONLY 3 FUNCTION GENERATORS
ENTER THE FOLLOWING:
BLOCK NUMBER,   FUNCTION INTERCEPTS #1 TO 4
                FUNCTION INTERCEPTS #5 TO 8
                FUNCTION INTERCEPTS #9 TO 11
SEPARATE ENTRIES WITH COMMAS
CARRIAGE RETURN ENDS THE OPERATION

?4,-100.,-64.,-36.,-16.
     -2.,0.0,2.0,16.
    36.,64.,100.

?


                 TIMING INFORMATION (ENTER DECIMAL POINTS)

INTEGRATION INTERVAL=.1

TOTAL TIME=25
```

FIG. 6. Block/CSMP input statements for the nonlinear spring problem.

specifies the initial conditions on the integrator. The second input defines the values by which the two inputs to the weighted summer are to be multiplied. The third input defines the maximum and minimum values which are to be provided by the function generator. The function generator is defined by 11 equally spaced points. The points vary from -10 to $+10$, and the values associated with each point are given in the function generator specification. These points correspond to the curve of Figure 4. A linear interpolation method is used so that for a particular X input the program provides a K(X) value output.

Timing information related to the integration step size and the total running time for the simulation are now entered. Other information, not shown in Figure 6, relating to the frequency of printing the output, the specific variables to be plotted or recorded, and the integration method to be used must also be entered as part of phase 3. The results of running the nonlinear spring model with the set of conditions indicated in Figure 6 are shown in Figure 7. The left-hand column indicates the time, while the next column indicates the value of the output of block 3, X, at that time. The remainder of the output is a plot of these values. The plot clearly shows the damped oscillatory response of the system with the period of oscillation changing with time due to the presence of the nonlinear spring. Other plots with differing parameter values can be easily obtained. In addition the configuration statements can be saved on magnetic tape for future use.

For completeness the input specifications for the population dynamics example are given in Figure 8. Instructional comments to the user have been omitted in this case. The output, though not provided here, indicates that, with the given parameter values, a stable population mix is obtained and the equilibrium sizes of the two populations are about 52 and 1240, respectively.

2.4. The 360/Continuous System Modeling Program (360/CSMP). The 360/CSMP simulation language [9] is available on many IBM computers. The language is a good deal more powerful than the Block/CSMP language just discussed. One can, for instance, write Fortran statements in as part of the simulation, and provision is made for creation of user subroutines and special functions. The language itself provides a wide variety of simulation-oriented functions (equivalent to the blocks in Block/CSMP) and also contains simple procedures for automatically controlling multiple model runs under varying conditions. Many of the block types explicitly available in Block/CSMP are automatically available as simple Fortran constructs. Note that some of these language differences are due to the fact that Block/CSMP is "interpreted" while 360/CSMP must be compiled using the Fortran com-

```
                              RESULTS PRINT-OUT

 TIME    OUTPUT  3  -0.10000E+02                            0.70000E+01
 0.000 -0.10000E+02 +                                                 I
 0.500 -0.78020E+01 I-----+                                           I
 1.000 -0.32697E+01 I-------------------+                             I
 1.500  0.10539E+01 I-------------------------------+                 I
 2.000  0.44778E+01 I-----------------------------------------+       I
 2.500  0.63310E+01 I-----------------------------------------------+ I
 3.000  0.60805E+01 I----------------------------------------------+  I
 3.500  0.42375E+01 I-----------------------------------------+       I
 4.000  0.18908E+01 I----------------------------------+              I
 4.500 -0.19661E-00 I----------------------------+                    I
 5.000 -0.18965E+01 I-----------------------+                         I
 5.500 -0.31522E+01 I-------------------+                             I
 6.000 -0.37349E+01 I-----------------+                               I
 6.500 -0.35865E+01 I------------------+                              I
 7.000 -0.28861E+01 I--------------------+                            I
 7.500 -0.19509E+01 I----------------------+                          I
 8.000 -0.10647E+01 I-------------------------+                       I
 8.500 -0.29129E-00 I---------------------------+                     I
 9.000  0.35501E-00 I-----------------------------+                   I
 9.500  0.86805E-00 I-------------------------------+                 I
10.000  0.12488E+01 I--------------------------------+                I
10.500  0.15042E+01 I---------------------------------+               I
11.000  0.16453E+01 I---------------------------------+               I
11.500  0.16866E+01 I---------------------------------+               I
12.000  0.16442E+01 I---------------------------------+               I
12.500  0.15353E+01 I---------------------------------+               I
13.000  0.13768E+01 I--------------------------------+                I
13.500  0.11849E+01 I--------------------------------+                I
14.000  0.97435E-00 I-------------------------------+                 I
14.500  0.75800E-00 I------------------------------+                  I
15.000  0.54666E-00 I------------------------------+                  I
15.500  0.34899E-00 I-----------------------------+                   I
16.000  0.17141E-00 I-----------------------------+                   I
16.500  0.18306E-01 I----------------------------+                    I
17.000 -0.10785E-00 I----------------------------+                    I
17.500 -0.20626E-00 I----------------------------+                    I
18.000 -0.27750E-00 I---------------------------+                     I
18.500 -0.32330E-00 I---------------------------+                     I
19.000 -0.34619E-00 I---------------------------+                     I
19.500 -0.34930E-00 I---------------------------+                     I
20.000 -0.33608E-00 I---------------------------+                     I
20.500 -0.31008E-00 I---------------------------+                     I
21.000 -0.27480E-00 I---------------------------+                     I
21.500 -0.23352E-00 I---------------------------+                     I
22.000 -0.18918E-00 I----------------------------+                    I
22.500 -0.14434E-00 I----------------------------+                    I
23.000 -0.10112E-00 I----------------------------+                    I
23.500 -0.61176E-01 I----------------------------+                    I
24.000 -0.25715E-01 I----------------------------+                    I
24.500  0.44753E-02 I----------------------------+                    I
25.000  0.28987E-01 I----------------------------+                    I
```

FIG. 7. Block/CSMP output for nonlinear spring example.

piler. See Chapter 2 on Programming Languages and Systems for a discussion of interpreters and compilers.

A model programmed in 360/CSMP is conceived of as having three distinct sections: Initial, Dynamic, and Terminal. The initial section is used for any computations which should be performed prior to solving the actual system of equations. Values of parameters and initial conditions may be computed in terms of more

```
                    CONFIGURATION SPECIFICATIONS

                  1   W       2       7       3
                  2   I       1       0       0
                  3   X       2       2       0
                  4   W       5       6       7
                  5   I       4       0       0
                  6   X       5       5       0
                  7   X       5       2       0

               INITIAL CONDITIONS AND PARAMETERS

   1    0.8000000         -0.2000000E-03   -0.1060000E-01
   2      300.0000          1.000000          0.0000000
   3    0.0000000         0.0000000
   4    0.6000000         -0.4000000E-03   -0.2000000E-02
   5      300.0000          1.000000          0.0000000
   6    0.0000000         0.0000000
   7    0.0000000         0.0000000


INTEGRATION INTERVAL= 0.10000           TOTAL TIME=  14.000
PRINT INTERVAL= 0.50000
OUTPUT BLOCK=  2,     MIN. & MAX.  = 0.00000          300.00
```

FIG. 8. Block CSMP summary input for the population dynamics example.

basic parameters and data needed for the model may be read in from a peripheral device at this time. Since preliminary calculations may not be necessary for some simulations, this section is optional. Figure 9 contains the 360/CSMP code for each of the two examples presented. Notice that in the initial section variable names are associated with the basic parameters (using the PARAMETER statement) and, for the nonlinear spring problem, the nonlinear function is defined (using the FUNCTION statement).

The dynamic section of the program contains a complete description of the system dynamics. In this section the block diagram description of the problem is a mixture of 360/CSMP and Fortran statements. For instance, in the nonlinear problem the statement X2DOT = −(D*X1DOT + F)/M is an ordinary Fortran statement and corresponds directly to equation (4). The other statements in the dynamic section use the 360/CSMP functions INTGRL and AFGEN. These correspond directly to the integration and function generator blocks in Block/CSMP. The large set of such functions available to the user is specified in [9]. These include, for instance, various logical functions which allow the user to alter the structure of the model dynamically. Note that unlike the Block/CSMP language there is no need to specify block numbers. All interactions are defined by the use of common variable names. These names may be selected by the user for their mnemonic value. Furthermore, the equation form of the input is much

```
              ****CONTINUOUS SYSTEM MODELING****

        ***PROBLEM INPUT STATEMENTS***

***    NONLINEAR SPRING PROBLEM     ***
INITIAL
      PARAMETER M=5. ,D=2.
      FUNCTION  AA=-10.,-100.,-8.,-64.,-6.,-36.,...
                   -4.,-16., -2.,-2., 0.,0., 2.,2., 4.,16.,...
                   6.,36., 8.,64., 10.,100.
DYNAMIC
      X2DOT=-(D*X1DOT+F)/M
      X1DOT=INTGRL(0.,X2DOT)
      X=INTGRL(-10.,X1DOT)
      F=AFGEN(AA,X)
      METHOD ADAMS
      TIMER DELT=.1, FINTIM=25., OUTDEL=.5
      PRTPLT X
      PRINT X,X1DOT,X2DOT,F
      END
      STOP
```

```
              ****CONTINUOUS SYSTEM MODELING PROGRAM****

        ***PROBLEM INPUT STATEMENTS***

***   POPULATION DYNAMICS PROBLEM    ***
INITIAL
      PARAMETER A1=.8, B11=.0106, B12=.0002, N1IC=300.
      PARAMETER A2=.6, B22=.0004, B21=.0020, N2IC=300.
DYNAMIC
      N1DOT=N1*(A1-B11*N1-B12*N2)
      N2DOT=N2*(A2-B21*N1-B22*N2)
      N1=INTGRL(N1IC,N1DOT)
      N2=INTGRL(N2IC,N2DOT)
TERMINAL
      N1DIF=N1-N1IC
      N2DIF=N2-N2IC
      WRITE (6,100) N1DIF,N2DIF
  100 FORMAT (' N1DIF=',F8.3,'  N2DIF=',F8.3)
      METHOD MILNE
      TIMER DELT+.1, FINTIM=14., OUTDEL=.5
      PRTPLT N1,N2
      END
      STOP
```

FIG. 9. 360/CSMP programs for two examples.

closer in appearance to the form of the differential equations making this code fairly easy to read and understand.

The terminal section of the program contains computations which may be desired after completion of each run. For instance, in optimization problems, the terminal section might include the optimization algorithm. This section can initiate rerunning the simulation with altered parameter values. The section is not always needed and has been omitted from the nonlinear spring problem. In the population dynamics problem it is used to calculate and print the difference between initial and final sizes of the two populations.

A number of other control statements will be noticed in the programs. These have the general functions of controlling integration type (eight integration algorithms are available, or the user can supply his own) and step size, and variables to be printed and plotted. The specific functions can easily be inferred from the mnemonic statement names.

3. DISCRETE PROBABILISTIC SYSTEMS SIMULATION

3.1. Introduction. Discrete probabilistic systems are those in which the state variables change at discrete instants of time, and in which some of these changes occur in a stochastic fashion. Queueing systems are an example of this. Typically such systems contain the following elements:

1. Customers or items which arrive into the system. Associated with arrival processes are statistical distributions which describe the probability of different interarrival times and perhaps of various customer types.
2. Resources which in some sense service the customers. Associated with these resources are statistical distributions which describe the probability of different service times.
3. Queues which hold customers waiting to use a busy resource, or waiting for some general system state to occur. Associated with these queues is a capacity which determines how many customers the queue can hold, and a queueing discipline which determines the order in which customers are removed from the queue.
4. System routing paths which determine how customers move through the system of queues and resources. Associated with these paths may be statistical distributions or logical conditions which determine what path a customer will follow.

Note that many of the elements described above may be functions of the system state. For instance, the statistical distributions or routing logic may change as the number of customers in the system changes.

The general queueing system described can be used to represent a wide variety of situations. Customers arriving at a supermarket checkout counter, products being fabricated in a factory, airplanes arriving and departing from an airport, paperwork flowing through a corporation, phone calls being processed at a central switching facility, and patients arriving at an emergency room are a few sample situations which can be modeled as a general queueing system. Other situations such as military armed combat and the stock market do not fit the queueing model very well. They can often be modeled using other discrete event probabilistic simulation methods. The lack of obvious structure in such systems, however, requires that a good deal of attention be devoted to their description. Therefore, given the limitations of a single chapter, the example considered later is of a simple queueing system.

Once the queueing system example has been presented, the following two sections will be devoted to implementing the queueing model in two popular discrete event simulation languages. The first language, GPSS, General Purpose Simulation System, is probably the most widely used language of its type. It is easy to get simple models implemented in the language and it is available on most IBM computers. The version considered is called GPSS/360 (12). The second language, SIMSCRIPT, is extremely flexible with the simulation aspects of the language being embedded in a full-blown general purpose higher-level language. The language is harder to learn initially than GPSS; however, that is perhaps to be expected as the price of flexibility.

A major difference between the languages relates to their general "world view." GPSS is a *transaction or particle-oriented* language in that the focus is on the entities (i.e., customers or items) which move through the system. The simulation follows these entities as they move from one activity (facility) to another. SIMSCRIPT, on the other hand, is an *event-oriented* language in that the focus is on the activities and on the events which define the starting and finishing times of these activities. The simulation in this case follows the progress of the various activities defining the model. This difference will become clearer as the languages themselves are discussed. References [27] and [28] discuss other language differences.

An important aspect of discrete probabilistic simulation concerns the generation of random numbers from various distributions, the testing of such generators, and the general questions of statistical convergence and experimental design. The numerical techniques involved here correspond in importance to the integration algorithm, step size, and stability considerations which form the numerical core of deterministic continuous system simulation. These questions are examined in some depth in references [22], [23], and [29].

3.2. Discrete Probabilistic System Example. The example considered here and illustrated in Figure 10(a) is that of a batch-oriented computer center. Customers arrive at the computer center with mean rate λ (customers/minute). Say the interarrival times of

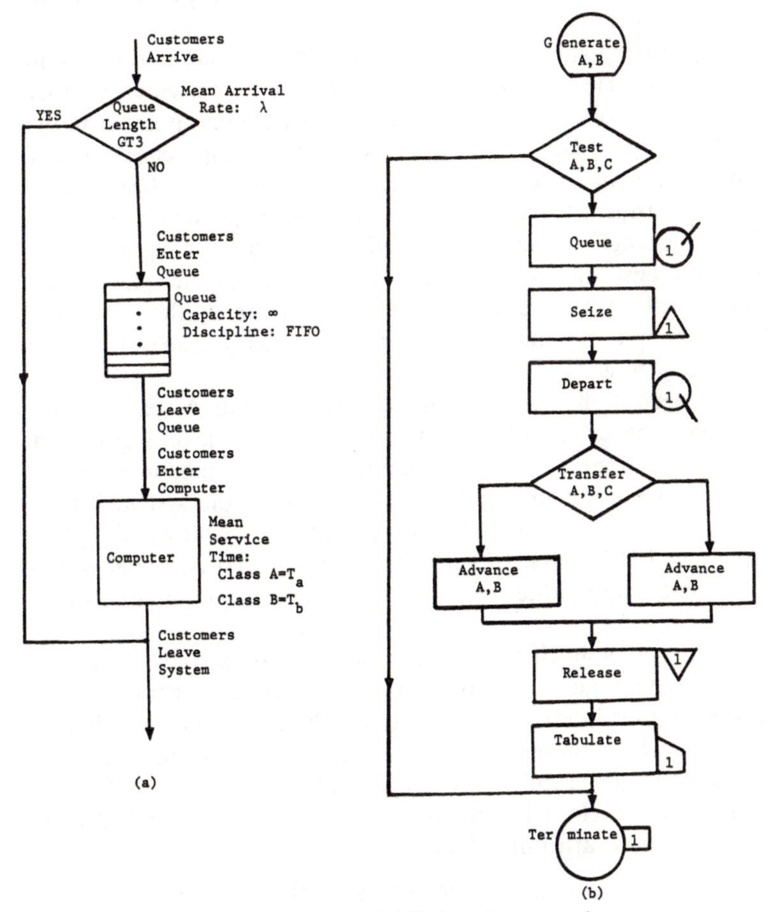

FIG. 10. Discrete probabilistic system example.

customers is exponentially distributed. That is:

$$F_T(t) = \Pr(T \leq t) = 1 - e^{-\lambda t} \qquad t \geq 0 \tag{8}$$

where $\Pr(T \leq t)$ is the probability that the interarrival time T is less than some time t. This is equivalent to saying that the arrival process is Poisson. Note that this assumes that an infinite pool of customers is available.

Associate with each customer a class type so that all customers that arrive are said to be in either class A or class B. This represents an attribute attached to each customer entity. Let XA percent of the customers be of class A.

Each arriving customer is told how many jobs are ahead of him in the queue. The customer, on hearing this, now decides whether or not he will join the queue. If the queue length (i.e., number in the queue plus the number being served by the computer) is greater than some value, the customer leaves, never to be heard from again. He thus engages in a form of queueing system behavior called "balking." If the queue length is less than or equal to this value, the customer joins the queue and once in the queue must remain in the system until his job has been run by the computer. That is, he may not engage in a form of queueing system behavior called "reneging."

The queue itself is taken to be infinite in length. Note, however, that given the strict balking behavior described above, it need not be infinite. Indeed, this type of balking behavior can be modeled by assuming a finite queue with customers lost to the system if they arrive when the queue is filled to capacity. The queueing discipline is taken as FIFO (First In First Out) with the first customer in the queue proceeding to the computer whenever the computer becomes free.

The computer in this model can only handle a single customer job at a time. Once that job enters the computer, it runs to completion without interruption and leaves the computer immediately on ending. The running time or service time associated with the jobs is taken to be uniformly distributed from T − 20 to T + 20. The mean is therefore T(minutes/job) and in this situation is dependent on the class attribute of the job. Thus:

$$T = \begin{cases} T_a \text{ for Class A jobs,} \\ T_b \text{ for Class B jobs,} \end{cases}$$

and the service time distribution is given by:

$$F_R(r) = \begin{cases} 0 \text{ for } r < T - 20, \\ (r - T + 20)/40 \text{ for } T - 20 < r < T + 20, \\ 1 \text{ for } r > T + 20. \end{cases} \tag{9}$$

Assume that the system remains in operation for long periods of time so that start up and shut down effects can be ignored.

Given this system description a number of items might be investigated. From a customer service point of view, mean waiting time (response time) in the system is an important performance measure and might be examined as a function of mean arrival rate, computer service rates, class distribution, and balking point. From a computer-center manager's point of view, computer utilization (i.e., percentage of time the computer is busy) and job throughput (i.e., the rate at which jobs are processed) are important measures which are also functions of the various system parameters. Changes in the system involving the establishment of a priority discipline based on class type rather than the FIFO discipline might also be investigated with a view toward improving, for example, the mean waiting time.

In the sections to follow this system is represented and solved in the GPSS and SIMSCRIPT languages.

3.3. The General Purpose Simulation System (GPSS/360). GPSS is a block-oriented language. The language contains over forty different block types which perform activities commonly found in discrete probabilistic models (e.g., generate an arrival process, place customer in a queue, etc.). These blocks often allow for easy translation from a flow-chart representation of system into GPSS program. Thus a clear correspondence is seen from Figure 10(a), which represents the example system, and Figure 10(b), which is the GPSS block diagram. Short definitions of a few of the more important blocks are given in Figure 11, and a complete definition of all blocks can be found in [12] and [13].

Figure 12 presents a GPSS program which corresponds to the blocks of Figure 10 and solves the example problem. The program begins with a number of comment cards. These are designated by an asterisk in column 1. The first noncomment card is the GPSS control card SIMULATE. This card indicates that an actual simu-

Block Type	Block Symbol	General Function
GENERATE	A,B, C,D,...	Create transactions with mean interarrival time given by A, and distribution specified B.
TERMINATE	A	Removes transactions from the system. Transactions entering the block are eliminated. A is subtracted from termination count value (specified by START command).
SEIZE	A	If facility A is not in use, a transaction entering the SEIZE block will cause it to become "busy". Any other transaction now attempting to enter this facility is halted.
RELEASE	A	Negates the effect of the SEIZE block causing facility A to become "not busy".
QUEUE	A	Arriving transaction is placed in a queue A of unlimited size.
DEPART	A	A transaction is removed from queue A and sent out of the block. A FIFO discipline is used.
Advance	A,B	A transaction entering the block is delayed for a period of time before departing. The mean delay time is A, and distribution is specified by B.
Transfer	A,B,C	A is a number from 0 to 1. An entering transaction will proceed to the block labelled C with probability A, and to block labelled B with probability 1-A.

FIG. 11. Short definitions of some GPSS block types.

lation run is to be made. If this card is omitted, then the GPSS/360 compiler will only scan the input for coding errors.

Following this is a function definition card. EXDIS is a label which the user has associated with the function to be generated. RN1 indicates that the output of random number generator 1 (one of eight available in GPSS/360) is to be used as the independent variable in the function evaluation. That is, every time this function is used, a uniformly distributed random number from 0 to 1 is

```
BLOCK
NUMBER  *LOC    OPERATION  A,B,C,D,E,F,G             COMMENTS
        *
        *       SIMULATION OF A BATCH COMPUTER SYSTEM
        *
                SIMULATE
        *
         EXDIS  FUNCTION   RN1,C24                  NEG. EXPON. DISTRIBUTION
        0,0/.1,.104/.2,.222/.3,.355/.4,.509/.5,.69/.6,.915/.7,1.2/.75,1.38
        .8,1.6/.84,1.83/.88,2.12/.9,2.3/.92,2.52/.94,2.81/.95,2.99/.96,3.2
        .97,3.5/.98,3.9/.99,4.6/.995,5.3/.998,6.2/.999,7/.9998,8
        *
1               GENERATE   60,FN$EXDIS              INTERARRIVALS EXP.DIST.
2               TEST L     Q$LQUEM3MOUT             IF QUQUE LENGTH LT 3 ENTER QU
3               QUEUE      LQUE                     PLACE JOB IN QUEUE
4               SEIZE      COMPU                    USE THE COMPUTER
5               DEPART     LQUE
6               TRANSFER   .8,BBLK,ABLK             80 % CLASS A,20 % CLASS B
7          ABLK ADVANCE    40,20                    CLASS'A JOBS, MEAN 40 SECS.
8               TRANSFER   ,CEND
9          BBLK ADVANCE    60,20                    CLASS B JOBS,MEAN 60 SECS.
10         CEND RELEASE    COMPU                    LEAVE COMPUTER SYSTEM
11              TABULATE   TRNTI                    TABULATE TRANSIT TIMES
12          OUT TERMINATE  1
          TRNTI TABLE      M1,0,20,10
                START      400                      RUN 400 TERMINATIONS
                END
```

FIG. 12. GPSS/360 program for computer system example.

generated for use as the independent variable. Note that this independent variable can be any one of a number of "Standard Numerical Attributes" which are available in GPSS (e.g., queue length, facility, utilization, clock time). The second operand C24 indicates that the function is continuous, and 24 value pairs are to follow which define the function. This particular function is the negative exponential distribution as indicated in the comment at the right side of the statement. The 24 value pairs follow on the next three lines. GPSS performs a linear interpolation between the points when generating a value for a particular RN1. This particular function, as defined, allows one to generate random variates from the exponential distribution. This represents an implementation of the inverse transform method for generating random variates [22], [28].

The function is used in the GENERATE block statement which follows. The GENERATE block creates the transactions which move through the system. In this case these transactions correspond to computer jobs. The first operand indicates that the mean time between creating a transaction is to be 60 time units (seconds in this case). The second operand indicates that the function EXDIS is to be used to determine the distribution associated with this arrival process. This situation therefore results in the generation of transactions with exponential interarrival times. Other capabilities of the GENERATE block include associating attribute values with each generated transaction.

Transactions created by the GENERATE block next enter the TEST block. This block controls the flow of transactions on the basis of an algebraic comparison of two attributes. In this example the L in Test L specifies the Less Than condition. Equal (E), Not Equal (NE), Greater Than (G), Greater Than or Equal to (GE) and Less Than or Equal to (LE) are also possible. Operand A (Q$LQUE) is compared with operand B (3) and, if the comparison condition specified does not hold, then the entering transaction is routed to the block specified in operand C (OUT). If the condition does hold, then the entering transaction is passed through to the next sequential block (QUEUE). In this example operand A specifies the standard numerical attribute queue length for the queue labeled LQUE. Operand B is the constant 3. Thus if the queue length is less than 3, the transaction passes to the next block; otherwise it leaves the system. This, in effect, implements the balking-customer characteristic discussed previously.

The QUEUE block which follows acts as a storage or waiting line for transactions. Such a storage area is necessary if a facility, or piece of equipment, is in use. Operand A (LQUE) specifies a symbolic name for this queue. This name can then be referenced as in the TEST block to obtain various queue parameters. Thus the generated transactions now enter the queue if its length is less than three.

In order to simulate situations in which equipment is used, or in which there exists a single server, GPSS provides a facility entity. Such an entity serves a single transaction at a time and when providing such service is busy (i.e., occupied). Any transactions which may desire to enter at this time are blocked. One way a facility is defined is by the presence of a symbolic name or number in a SEIZE block. COMPU is a symbolic name defining a facility which in this case represents the computer. A transaction, on entering a SEIZE block, causes the facility (COMPU) to become busy and prohibits other transactions from entering until the facility is RELEASEd. The RELEASE block moves the transaction in the defined facility to the next block and causes the facility to now become not busy. Thus the transaction at the head of the queue LQUE continually attempts to enter the SEIZE block and eventually succeeds when the COMPU facility is RELEASEd.

Once the transaction has SEIZEd the facility, it must relinquish its position in the queue. This is done with the DEPART block. Operand A (LQUE) of this block specifies the queue from which the transaction is departing. This causes the number in the queue to decrease by 1 and may therefore affect the direction of subsequent transactions which flow through the TEST L block.

In order to simulate the presence of two classes of jobs, the TRANSFER control block is used. Operand A of this block is a number from 0 to 1. An entering transaction will proceed to the block named in operand C (ABLK) with probability A (.8), and to the block named in operand B (BBLK) with probability 1-A. That is, operand A indicates the proportion of entering transactions which go to the block named in operand C. In the example 80 percent of the jobs are taken to be in class A and 20 percent in class B.

The TRANSFER block moves transactions to one of two ADVANCE blocks. The ADVANCE is used to simulate the passage of time, in this case the execution time of a job. The transaction entering the block is delayed an amount of time specified by operands A and B. The delay time here is a random variate whose value is generated from a uniform distribution with mean A, and minimum and maximum values $A - B$ and $A + B$, respectively. Class A and class B jobs thus represent jobs with different mean execution times, 40 time units for class A and 60 time units for class B.

When the transaction leaves the ABLK ADVANCE block, it enters an unconditional TRANSFER block which moves it to the CEND RELEASE block. This is the same block which is entered by the transactions leaving the BBLK ADVANCE block. The RELEASE block releases the computer facility COMPU. At this point if there is a transaction waiting in queue LQUE, it can SEIZE the computer.

After the RELEASE block, a TABULATE block is entered. This is used to gather certain statistics which are not automatically collected by GPSS. Operand A (TRNTI) points to a TABLE definition statement which must be examined to understand the TABULATE block. The TABLE statement has four operands. The first (M1) is a code which indicates that GPSS should gather statistics on the transit time of transactions entering the TABULATE block. The transit time is the time from transaction creation to entry into the TABULATE block. In this case this represents the time that a job stays in the computer system. The remaining three operands (0, 20, 10) indicate the lower limit, interval size, and number of intervals to be used in the statistics gathering.

The final block, the TERMINATE block, removes transactions from the system, and thus simulates the departure of jobs. The operand value is used in conjuction with a "termination count." The termination count is specified in the operand of the START control card which follows. This count is initialized to the START operand value (400) and is decremented by the TERMINATE operand value (1) every time a transaction enters the TERMINATE block. When the count is equal to zero, the simulation run is ended.

In addition to specifying the termination count, the START control card indicates to GPSS that the set of input cards necessary to execute a simulation has been received and that the execution phase should proceed. The END control card is that last card of the GPSS input deck. Note that, although this example has only a single execution associated with it, multiple simulation runs can be accommodated with altered parameter values or model structure.

The next item to consider is the output produced by the GPSS program discussed above. Although the output produced runs over several pages, it has been condensed and presented in Figure 13. Some of this output represents statistics automatically gathered by GPSS, while some of it (TABLE TRNTI) represents statistics requested by the program through use of the TABULATE block. Not all of these statistics will be discussed here. The first main set of statistics represents a count of transactions which have gone through each block. The block numbers are produced by the GPSS compiler and correspond to the numbers on the left side of Figure 13. Notice that blocks 7 and 9 represent the different number of class A (301) and class B (75) jobs which went through the system. Also, the difference in totals represented by blocks 2 and 3 (403, 379) indicate the number of jobs which turned away from the system because the queue length was greater than two. Thus about 6 percent of the potential customers were lost. The remaining statistics are fairly clear. The computer (Facility COMPU) was busy 68 percent of the time and the average time each transaction used the facility was 44.75 time units. The queue has an average length of .556 and the average transaction spent 36.329 time units in the queue. The table TRNTI indicates the distribution of overall wait-

```
RELATIVE CLOCK          24744  ABSOLUTE CLOCK          24744
BLOCK COUNTS
BLOCK CURRENT   TOTAL    BLOCK CURRENT   TOTAL    BLOCK CURRENT   TOTAL    BLOCK CURRENT   TOTAL
    1       0     403       11       0     376
    2       0     403       12       0     400
    3       3     379
    4       0     376
    5       0     376
    6       0     376
    7       0     301
    8       0     301
    9       0      75
   10       0     376

FACILITY       AVERAGE        NUMBER      AVERAGE       SEIZING      PREEMPTING
             UTILIZATION      ENTRIES     TIME/TRAN    TRANS. NO.    TRANS. NO.
   COMPU         .680           376        44.750

QUEUE      MAXIMUM   AVERAGE    TOTAL     ZERO    PERCENT     AVERAGE     $AVERAGE    TABLE   CURRENT
          CONTENTS  CONTENTS  ENTRIES  ENTRIES    ZEROS    TIME/TRANS   TIME/TRANS  NUMBER  CONTENTS
    LQUE        3      .556      379      137      36.1      36.329       56.896                   3
 $AVERAGE TIME/TRANS = AVERAGE TIME/TRANS EXCLUDING ZERO ENTRIES

TABLE TRNTI
ENTRIES IN TABLE             MEAN ARGUMENT     STANDARD DEVIATION     SUM OF ARGUMENTS
           376                     80.845                 42.375           30398.000     NON-WEIGHTED

         UPPER           OBSERVED        PER CENT CUMULATIVE       CUMULATIVE MULTIPLE     DEVIATION
         LIMIT           FREQUENCY       OF TOTAL PERCENTAGE        REMAINDER  OF MEAN     FROM MEAN
             0                   0            .00         .0            100.0    -.000       -1.907
            20                   3            .79         .7             99.2     .247       -1.435
            40                  56          14.89       15.6             84.3     .494        -.963
            60                 101          26.86       42.5             57.4     .742        -.491
            80                  53          14.09       56.6             43.3     .989        -.019
           100                  49          13.03       69.6             30.3    1.236         .452
           120                  48          12.76       82.4             17.5    1.484         .923
           140                  25           6.64       89.0             10.9    1.731        1.395
           160                  26           6.91       96.0              3.9    1.979        1.867
      OVERFLOW                  15           3.98      100.0               .0
  AVERAGE VALUE OF OVERFLOW                191.19
```

FIG. 13. GPSS example output.

ing times of jobs which enter the computer system with the mean (80.845) and standard deviation (42.375) explicitly noted.

This concludes the discussion of GPSS. The section to follow considers the SIMSCRIPT language which views the world of discrete probabilistic simulation from another perspective.

3.4. The SIMSCRIPT Simulation System. SIMSCRIPT (15) is a general-purpose, higher-level language which has most of the capabilities of FORTRAN in addition to providing list processing and discrete simulation facilities. The version considered here, SIMCRIPT II.5, is documented in [30].

Before considering a SIMSCRIPT implementation of the example problem, a number of concepts central to the SIMSCRIPT word view must be explained. Some of these are similar to ideas presented in Section 1.3. SIMSCRIPT considers systems to be composed of *entities* which have associated with them *attributes.* Entities may be defined in a number of ways. Consider, for instance, the entity called JOB, which has attributes TMEAN and ARR.TI (arrival time). SIMSCRIPT has provision for defining the general entity JOB as follows:

EVERY JOB HAS A TMEAN AND AN ARR.TI

EVERY is a reserved word and has a specified statement format associated with it.

Two types of entities are permitted in SIMSCRIPT. *Temporary entities* are used for those entities which are created or destroyed during the course of the simulation. JOBs may be thought of as entering (CREATE a JOB) and leaving (DESTROY a JOB) the system. By associating them with temporary entities, storage can be dynamically allocated to these JOBs. Assigning JOB as a temporary entity with the TMEAN and ARR.TI attributes is done with the following statements:

TEMPORARY ENTITIES
EVERY JOB HAS A TMEAN AND AN ARR.TI

The other entity type, *permanent entity,* may be used to define those entities which will not be individually created or destroyed during the simulation. Such entities are stored by SIMSCRIPT collectively and are used to represent relatively static objects which remain present in the system for the duration of the simulation. For example:

PERMANENT ENTITIES
EVERY SCHOOL HAS AN ADDRESS

It is usually important to be able to collect entities into groups or *sets* and to provide a means for entering and removing entities from such sets. SIMSCRIPT allows the user to define sets with the DEFINE SET statement, and to enter and remove entities from sets with the FILE and REMOVE statement. For instance, defining the symbol JOB.Q as a set which is in fact a FIFO queue can be done with the statement:

DEFINE JOB.Q AS FIFO SET

In order to link entities with sets, provisions are available for indicating that an entity belongs to a set. If a JOB is to be able to join the JOB.Q, then the following expanded definition of the JOB entity is needed.

TEMPORARY ENTITIES
EVERY JOB HAS A TMEAN, AN ARR.TI
AND MAY BELONG TO A JOB.Q

Sets themselves must be associated with entities in the sense that the set is OWNed by some entity. For instance:

EVERY SCHOOL HAS AN ADDRESS, OWNS SOME
STUDENTS, AND BELONGS TO A SCH.DISTRICT

In this case the entity SCHOOL has an attribute ADDRESS. In addition, the set STUDENTS is owned by SCHOOL, and the SCHOOL itself is a member of the set SCH.DISTRICT. Clearly, complicated set relationships can be established. These relationships, however, provide for great flexibility in modeling complex situations.

One final point in this discussion has to do with SYSTEM attributes and set ownership. The term SYSTEM refers to the overall system being considered. This SYSTEM can itself have attributes and own sets. By having attributes, these attributes now become global variables (i.e., these attributes can be referenced with the same name in different subprograms or subroutines). In addition, certain global pointers are now available for getting at elements of system-owned sets. The set JOB.Q will be owned by the system when defined by the following statements:

THE SYSTEM OWNS A JOB.Q
DEFINE JOB.Q AS A FIFO SET

The logical relationships as defined above are static in nature. To model the dynamics of the system *EVENT ROUTINES* and a process for *SCHEDULING* such routines must be available. Events correspond to transition points between operations or activities in the system modeled. They represent a change of state. Part of the art involved in such simulations is determining the key events which take place in a system. For each of these key events a program or event routine must be written. Event routines have two general functions. First, they perform whatever logical operations or calculations are associated with the event. Second, they determine what future events are to take place given the current system state, and then they schedule these future events.

Scheduling an event effectively means that an event routine name and associated future time of occurrence are placed in a list or stack. Events in this list are ordered by time of occurrence, with the first event being one whose time of occurrence is closest to the current simulation time. Note that as simulation progresses a clock must be maintained by SIMSCRIPT of the current simulation time (variable name TIME.V). Thus, as one goes down the event list, the events listed are scheduled to occur further and further in the future.

SIMSCRIPT and all other discrete event simulation languages provide routines for maintaining the event list, determining which event routine is to be executed next, and transferring control to that routine. A RETURN from an event routine returns control to the "scheduler" which in turn passes control to the routine for the next event to occur. In this way by successively having event routines schedule future events, and by having a "scheduler" routine pass control to the succeeding event routines, the dynamics of the system are modeled.

The events themselves are defined initially by using the EVENT NOTICE statement. For instance:

EVENT NOTICES INCLUDE ISSUE
EVERY END.SERVICE HAS A JNAME

Two events are defined by the statement above, ISSUE and END.SERVICE. Associated with END.SERVICE is an attribute JNAME which, in the example to be discussed, is used to associate a particular JOB with the END.SERVICE event.

Notice that in the discussion above the event routines all occurred due to event scheduling internal to the SIMSCRIPT program (i.e., scheduled by event routines). Such event routines are said to be *endogenous routines.* It is also possible to schedule events external to the system by reading in a list of events from data cards or other input. Such events are referred to as *external* or *exogenous* events.

Consider next the computer example as programmed in SIMSCRIPT II.5 and presented in Figure 14. The program is divided into four sections. The first is the PREAMBLE in which the various entities, attributes, global variables, and set relationships

```
 1  PREAMBLE
 2       NORMALLY, MODE IS INTEGER
 3       THE SYSTEM OWNS A JOB.Q
 4       DEFINE JOB.Q AS FIFO SET
 5       TEMPORARY ENTITIES
 6          EVERY JOB HAS A TMEAN,A ARR.TI AND MAY BELONG TO A JOB.Q
 7       EVENT NOTICES INCLUDE ISSUE
 8          EVERY END.SERVICE HAS A JNAME
 9       DEFINE TOT.JOBS.DONE, UTIL AND JNAME AS VARIABLES
10       DEFINE ARR.TI, SYS.TI AND TMEAN AS REAL VARIABLES
11       TALLY W AS THE AVG OF SYS.TI
12       ACCUMULATE UZ AS THE AVG OF UTIL
13 END                                                 ''END PREAMBLE

 1 MAIN
 2       SCHEDULE AN ISSUE NOW                         ''MAIN ROUTINE
 3       LET UTIL=0  LET TOT.JOBS.DONE=0               ''INITIALIZATION
 4       START SIMULATION                              ''START SIMULATION
 5       END                                           ''END MAIN ROUTINE

 1  EVENT ISSUE                                        ''ARRIVAL OF JOBS
 2       SCHEDULE AN ISSUE AT TIME.V+EXPONENTIAL.F(60.,1)''SET NEXT JOB ARRIVAL
 3       IF (N.JOB.Q GT 2) RETURN                      ''JOB BALKED
 4       ELSE CREATE JOB  LET ARR.TI=TIME.V            ''ASSIGN ATTRIBUTES
 5       IF (UNIFORM.F(0.,1.,1) LT .8) LET TMEAN=40. GO TO T.BUSY
 6       ELSE LET TMEAN=60.
 7       'T.BUSY' IF(UTIL = 1)FILE JOB IN JOB.Q RETURN''COMPUTER BUSY?
 8       ELSE LET UTIL = 1                             ''MAKE COMPUTER BUSY
 9       SCHEDULE AN AND.SERVICE(JOB) AT TIME.V+UNIFORM.F(TMEAN-20.,TMEAN+20.,1)
10          RETURN END                                 ''END ISSUE EVENT

 1  EVENT END.SERVICE(JOB)                             ''END OF COMPUTER USE
 2       LET SYS.TI=TIME.V - ARR.TI                    ''CAL.WAIT TIME
 3       LET TOT.JOBS.DONE=TOT.JOBS.DONE + 1           ''CAL.NUMBER JOBS DONE
 4       DESTROY JOB                                   ''DESTROY JOB
 5       IF (TOT.JOBS.DONE GE 400) PRINT 1 LINE WITH W AND UZ AS FOLLOWS
   MEAN WAIT TIME= ****.****            UTILIZATION= **.****
 6          STOP                                       ''STOP SIMULATION
 7       ELSE IF JOB.Q IS EMPTY LET UTIL=0 RETURN      ''QUEUE EMPTY? YES,RETURN
 8          ELSE REMOVE FIRST JOB FROM JOB.Q           ''NO, REMOVE JOB
 9       SCHEDULE AN END.SERVICE(JOB) AT TIME.V+UNIFORM.F(TMEAN-20.,TMEAN+20.,1)
10          RETURN END                                 ''END END.SERVICE EVENT
```

FIG. 14. SIMSCRIPT II.5 program for computer example.

are defined. Most of the statement types have been considered earlier in this section. Another function of the PREAMBLE is to request that certain statistics be gathered automatically by SIMSCRIPT. In this example the TALLY statement is used to associate the variable name W with the average value of SYS.TI. SYS.TI is a variable whose value is the total waiting time for a job. This W is calculated as a simple average of SYS.TI over all jobs which pass through the system. The ACCUMULATE statement associates the variable name UZ with the average utilization of the computer. Every time UTIL is set to 1 (computer busy) or 0 (computer not busy) the appropriate statistics are gathered and a *time* average of the UTIL busy parameter is maintained.

The second section is the MAIN program. This section schedules the first event to occur (ISSUE) at time NOW which is the current clock time of the simulation. It then initializes several parameters and then transfers control to the event scheduling routine with the command START SIMULATION.

Since an ISSUE event has been scheduled, control will pass to the EVENT ISSUE routine. EVENT ISSUE handles the event of job arrivals. It first schedules itself (i.e., another job arrival) for a time in the future which is exponentially distributed. It then determines if the queue (JOB.Q) has too many jobs in it and, if so, returns control to the scheduler. If not, a JOB is CREATED and its attributes assigned. At this point a test is made to see if the computer is busy. If it is busy (UTIL = 1) the JOB is FILEd in the JOB.Q. If it is not busy, it is made busy, thus modeling the action of the JOB executing on the computer. The end of this execution time, indicated by the END.SERVICE event, is determined and the END.SERVICE event scheduled. The scheduling is done in accordance with the class type for the JOB.

The final routine is the END.SERVICE event routine. The routine first calculates the wait time (SYS.TI) for the JOB that has just finished execution and the number of jobs that have gone through the computer (TOT.JOBS.DONE). The job which has just finished service is DESTROYED at this point, releasing the storage space associated with it. On the basis of TOT.JOBS.DONE the simulation may now be terminated and the gathered statistics printed. If the simulation is not terminated, and there is a JOB available in the JOB.Q, then this JOB begins to execute on the computer and an END.SERVICE event for the JOB is scheduled. Otherwise control is returned to the scheduler. In this example 400 jobs pass through the system before the simulation is terminated. On termination the mean wait time and the computer utilization is printed.

4. SUMMARY

This chapter has considered the general topic of simulation on digital computers. The first section emphasized the broad questions of modeling and simulation methodology. The second section considered two language systems used for modeling of continuous systems, while the third section considered two language systems used predominantly for modeling discrete probabilistic systems. The emphasis in these later two sections has been on presenting some typical, but simple, systems and demonstrating the language system capabilities by modeling these systems. The reader is directed to the references for details on numerical questions related to model solution and for a wider view of these and other simulation systems.

REFERENCES

1. J. S. Rosko, *Digital Simulation of Physical Systems*, Addison-Wesley, Reading, Mass., 1972.
2. G. J. Herskowitz and R. B. Schilling, *Semiconductor Device Modeling for Computer -Aided Design*, McGraw-Hill, New York, 1972.
3. B. C. Patten, ed., *Systems Analysis and Simulation in Ecology*, Academic Press, New York, 1971.
4. D. W. Grooms, "Economic models: A bibliography with abstracts," National Technical Information Service, NTIS COM-74-10103/1GA, Springfield, Va., 1973.
5. K. M. Sayre and F. J. Crosson, eds., *The Modeling Mind: Computers and Intelligence*, Simon and Schuster, New York, 1965.
6. G. A. Mihram, "The modeling process," *IEEE Transactions on Systems, Man, and Cybernetics*, SMC-2, no. 5 (November 1972).
7. F. H. Speckhart and W. L. Green, *A Guide to Using CSMP—Continuous System Modeling Program*, Prentice-Hall, Englewood Cliffs, N.J., 1976.
8. "1130 continuous system modeling program," (H20-0282) IBM Corporation, Data Processing Division, White Plains, N.Y.
9. "System/360 continuous system modeling program-user's manual," (GH20-0367-4) IBM Corporation, Data Processing Division, White Plains, N.Y.
10. A. L. Pugh, *DYNAMO User's Manual*, MIT Press, Cambridge, Mass., 1963.
11. J. W. Forrester, *Industrial Dynamics*, MIT Press, Cambridge, Mass., 1961.

12. "General purpose simulation system/360-user's manual," (GH20-0326-3) IBM Corporation, Data Processing Division, White Plains, N.Y.
13. T. J. Schriber, *Simulation Using GPSS*, Wiley, New York, 1974.
14. A. A. Pritsker and P. J. Kiviat, *Simulation with GASP II*, Prentice-Hall, Englewood Cliffs, N.J., 1969.
15. P. J. Kiviat, R. Villanueva, and H. M. Markowitz, *The Simscript II Programming Language*, Prentice-Hall, Englewood Cliffs, N.J., 1968.
16. O. Dahl and K. Nygaard, "SIMULA—an ALGOL-based simulation language," *Comm. ACM*, **9**, no. 9 (September 1966).
17. UNIVAC, 1106/1108, SIMULA Programmer Reference (UP-7556), 1971.
18. A. A. Pritsker, *The GASP IV Simulation Language*, Wiley, New York, 1974.
19. G. R. Hogsett, "ECAP II," in *Handbook of Circuit Analysis Languages and Techniques*, R. W. Jensen and L. P. McNamee, eds., Prentice-Hall, Englewood Cliffs, N.J., 1976, Chapter 6.
20. *ICES STRUDL-II—Engineering User's Manual*, Structures Division and Civil Engineering Systems Laboratory, Department of Civil Engineering, MIT, Cambridge, Mass., 1968.
21. A. P. Sage and J. U. Melsa, *System Identification*, Academic Press, New York, 1971.
22. G. S. Fishman, *Concepts and Methods in Discrete Event Digital Simulation*, Wiley, New York, 1973.
23. G. A. Mihram, *Simulation : Statistical Foundations and Methodology*, Academic Press, New York, 1972.
24. L. S. Lasdon, *Optimization Theory of Large Systems*, Macmillan, New York, 1970.
25. W. G. Cochran and G. M. Cox, *Experimental Designs*, 2nd ed., Wiley, New York, 1962.
26. E. C. Pielou, *An Introduction to Mathematical Ecology*, Wiley, New York, 1969.
27. H. S. Krasnow, "Simulation Languages," in *The Design of Computer Simulation Experiments*, T. H. Naylor, ed., Duke University Press, Durham, N.C., 1969.
28. H. Maisel and G. Grugnoli, *Simulation of Discrete Stochastic Systems*, Science Research Associates, Chicago, Ill., 1972.
29. D. E. Knuth, *Seminumerical Algorithms*, Art of Computer Programming, vol. 2, Addison-Wesley, Reading, Mass., 1969.
30. *Simscript II 5 Reference Handbook*, Consolidated Analysis Centers, Los Angeles, Calif. 1971.

COMPUTATIONAL TOOLS FOR STATISTICAL DATA ANALYSIS

C. F. Starmer

INTRODUCTION

Discussing the impact of high-speed computing on statistical data analysis is difficult. The issues involved span virtually everything statisticians do and how they do it. In part, general statistical methodology is motivated by our desire to improve both our decision making and our ability to derive new knowledge from experiences. Since experiences are generated in a variety of settings and are based on a number of different substrates, it has been desirable to develop tools for dealing with diversity and heterogeneity. In a typical experiment, a number of substrates are exposed to a stimulus and outcomes are observed. The outcomes reflect the stimulus, the differences among substrates, or both. One task of statistics is to separate that part of the outcome due to the stimulus from that due to substrate heterogeneity. This generally involves studying the interaction between many substrates and the stimulus.

Such studies, traditional producers of sizable amounts of data, steadily have been increasing their size and scope in response to technological advances that have facilitated data acquisition. The process is accelerated even further by the availability of automated computational procedures, so that studies currently considered routine were well beyond contemplation two decades ago. However, the desire to exploit this growth in an effort to improve our observational acuity brings with it a serious risk of overstressing the discriminatory facilities inherent in many statistical techniques. Thus, the introduction of automated large-scale statistical procedures brings into question the way we design observation-gathering ventures and the way we deal with the results. This effect often is obscured by the amount of sheer computational power at the analyst's disposal.

The approach to data analysis taken in this article is motivated by just this availability. As implied above, such a resource is helpful in extending the boundaries associated with both the volume of data to be analyzed and the computational complexity of the analysis. In the past, computation was difficult, and thus the development of statistical methodology frequently was associated with numerical tricks or shortcuts. There appeared statistical cookbooks, comparable to the classic *Joy of Cooking*, that described available computational formulae ("recipes") and example problems to which these formulae were suited. In current textbooks we still find chapters on such topics as linear regression, t tests, one-way analysis of variance (anova), one-way anova with a covariate, two-way anova without interaction, two-way anova with interaction, etc., with accompanying descriptions of computational shortcuts. The volume of material required to present these methods and their associated examples is considerable. As a result there often is only minimal discussion of the limitations and restrictions for valid use of each procedure.

The availability of really inexpensive computing promises to change all this. One can perceive two distinct dimensions of this potential upheaval. First, the removal of computational complexity as a deterrent means that the applicability of a particular statistical technique in a given context can be tested routinely and automatically. Second, the design of an experiment need not be compromised by computational considerations. This metamorphosis, however, is occurring slowly, but it is in progress and we can discuss its implications. Consequently, now that accurate execution of complex computational algorithms is routine, the time has come to change our perspective of data analysis from the computational issues to those issues characterizing the data source itself. In this article we shall concentrate on various models of data sources and we shall investigate how automatic computation can aid in testing hypothesized properties of these models. We shall take as given the notion that these computational facilities significantly extend both the limits on the amount of data analyzed and the "permissible" scope of the analyses to be applied.

DATA SOURCES

Before investigating data models and computational methods, it will be helpful to review data sources as viewed from basic physics and biology. The physical and biological sciences have demonstrated (to my satisfaction) that properties and derived functions reflect internal structure. Properties are derived from the particular configuration of atoms and bonds between them. As a simple example, acetone and propionic aldehyde have the same chemical composition, C_3H_6O, but different structures (Figure 1):

```
        H     O     H
        '     "     '
  H  -  C  -  C  -  C  -  H
        '           '
        H           H              Acetone

        H     H     O
        '     '     "
  H  -  C  -  C  -  C  -  H
        '     '
        H     H                    Propionic Aldehyde
```

FIG. 1. Two-dimensional representation of acetone and propionic aldehyde.

For these two different structures, the following properties are observed:

Acetone	*Propionic Aldehyde*
colorless liquid	colorless liquid
specific gravity = .72	specific gravity = .81
melting point = −94.6	melting point = −81
boiling point = 56.5	boiling point = 49.5
infinitely soluble in H_2O	not infinitely soluble in H_2O

Although the differences are small, this example clearly demonstrates the dependence of properties on internal structure.

Heterogeneity of function is dependent on both the diversity of available building blocks and the diversity of templates used to define objects assembled from such building blocks. Man-made structures seem to enjoy greater homogeneity of function than do living structures. This is due in part to the limited number of templates we use in synthesizing complex objects, and also to the structural uniformity of each building block. With carefully refined materials, we are able to make many uniform copies of complex objects by applying the same template to these materials. For instance, all copies of a particular pocket calculator or food mixer share a high degree of similarity.

In contrast, living objects are replicated by applying a highly variable template to a set of primitive building blocks. The resulting mixture of enzymes, substrates, and membranes form living cells. Biological replication is based on building a new template for each new copy, half of which is derived from each parent template. Such a scheme results in considerable diversity of templates. As an example of the variety of available templates, consider ourselves. Humans have 46 components in their templates, 23 derived from each parent. Thus there are at least 2^{46} possible templates. The magnitude of this number can be contrasted with the current world population of 2^{32} (or 4 billion). In addition to the diversity achieved from nonuniform templates, environmental surroundings contribute to diversity. Each new cell that is created evolves by responding not only to the control that is resident in its internal composition but also to the environment surrounding it. Thus identical cells will evolve into different organisms when exposed to different environments. The growth of our bodies from a single cell certainly exemplifies this.

Outcomes of experiments are determined by both the nature of the stimulus and the nature of the substrate. Thus we need to be able to characterize accurately and then select like stimuli and like substrates. Because of deficiencies in measurement, however, our accuracy of characterization is limited. Our sensory inputs limit the degrees of characterization of cellular aggregates. We characterize structure and associated function of cellular aggregates by properties that can be seen, felt, heard, tasted, and smelled. We quantify these properties by applying yardsticks that are graduated in units of temperature, length, color, sound intensity, pitch, etc. Our indices are limited in their precision. Therefore, we are only able to differentiate among cellular aggregates down to a certain size commensurate with our level of perception. Beyond this level, we cannot detect any further differences without running the risk of classifying unlike substrates as like.

As stated earlier, much of science is "done" by observing the interaction between substrates and stimuli. Presence of an interaction will support one line of reasoning while its lack will support its converse. It is important, therefore, to be able to measure properties accurately so that responses derived from different interactions can be discriminated. The errors associated with our inability to discriminate among outcomes, i.e., classifying unlikes as likes, form the justification for a statistical approach to data analysis.

To fix these ideas, consider a number of patients given one of two blood-pressure-lowering drugs. Our task is to determine whether the drugs are equivalent; hence, we use blood pressure as an indicator of the drug-patient interaction. Measuring blood pressure for each patient, we find a spectrum of pressures. This spectrum arises from the fact that the patients are structurally different from each other, and thus, for each patient, the drug has a different substrate with which to interact. Both the substrate and the drug determine the observed pressures. Part of the spectrum of pressures is due to the heterogeneity among substrates. This portion is usually referred to as "error" or "noise." The rest of the spectrum of pressures is due to the drug-substrate interaction. Our job is to determine, with some level of confidence, which of these two conditions is immediately more significant, i.e., whether the spectrum of pressures is predominantly due to substrate heterogeneity or to the drug.

It is important to remember that, while dealing with collections of data, three sources of variation are operational: one comes from the interaction of an ideal stimulus with an ideal test unit (ideal patient in the example above); another arises from the difference between the ideal stimulus and the real one; while the third stems from the difference between the ideal and actual substrates. As long as we are unable to define likeness or we persist in calling known unlikes as likes, we shall be faced with sources of variation (stimulus and substrate) influencing what we measure. The goal of statistics is to aid in separating outcomes due to differences among substrates from outcomes elicited by the stimulus. Automatic computation can aid significantly in meeting this goal.

HYPOTHESES AND TESTING

Suppose we take a sample of 100 patients who are representative of a particular population, give half of them a new antihypertensive drug, and give the other half a water pill. We measure the blood pressures of each group after one month of treatment and find the spectrum of results shown in Figure 2. Now we repeat the experi-

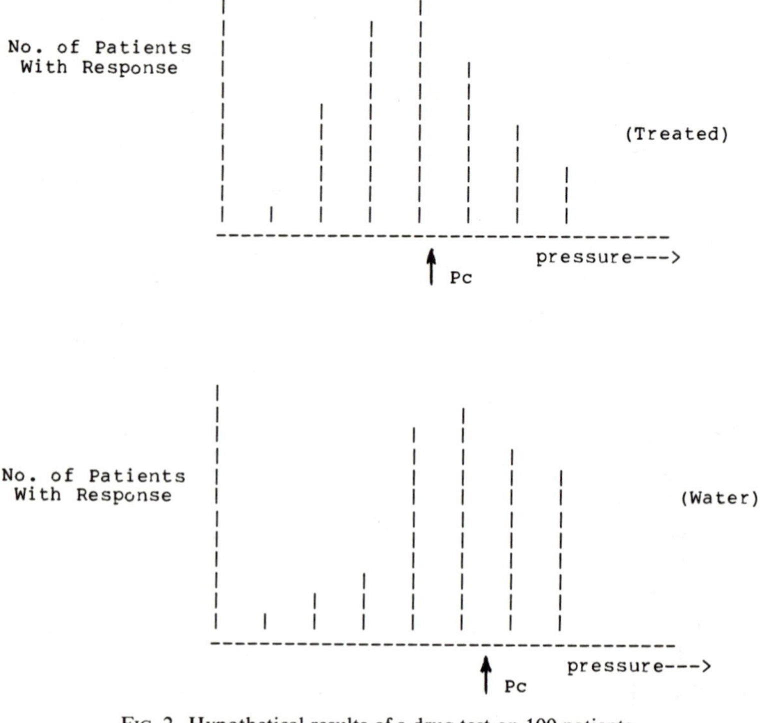

FIG. 2. Hypothetical results of a drug test on 100 patients.

ment on another group of 100 patients and find the results shown in Figure 3. Several observations can be made. At first sight, all the histograms are different. We expect this since we did not study like patients to begin with. The second observation is that the histogram of the treated patients seems to show more patients below a critical pressure, p_c, than the histogram of untreated patients. We would therefore conclude that the drug interaction with patients results in a lowering of blood pressure.

We are worried, though, that there may be sufficient structural differences between patients in the two treated groups that our conclusion is in error (i.e., simply wrong). In order to estimate the variation in blood pressure due to differences in patients, models are used to describe outcome variation as a function of patient

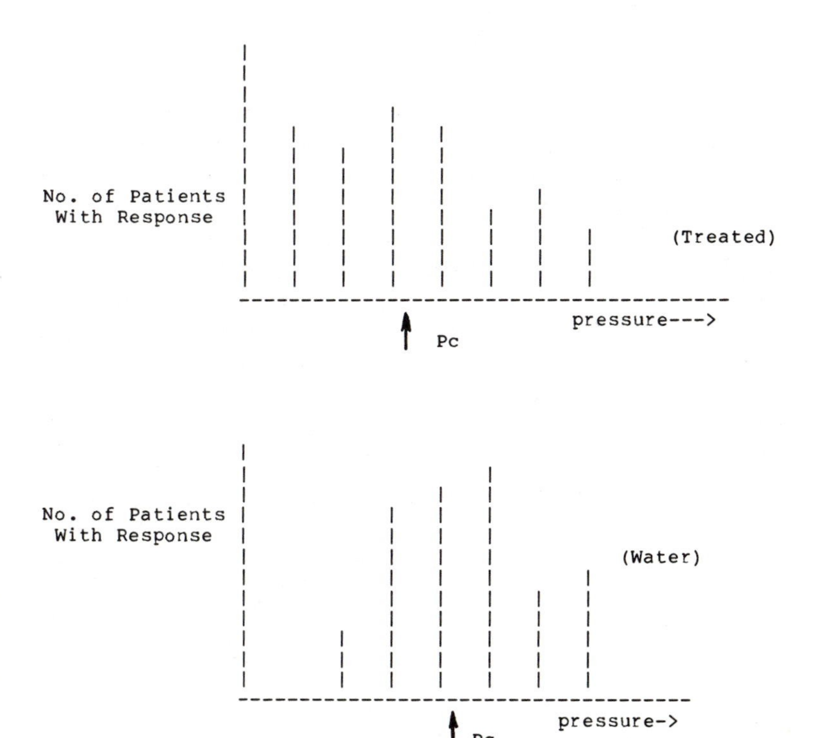

FIG. 3. Experiment for Figure 2 repeated.

differences. These models form part of the theoretical underpinning of statistical data analysis, and their understanding leads to an awareness of the limitations of various procedures.

To begin with, several terms must be defined so that our language parallels that of traditional statistics. A primitive "event" is the outcome of an experiment or some activity, and the "sample space," S, is the set of primitive events representing all possible outcomes of an experiment or activity. A compound event is one expressed as the union of primitive events. A "probability measure" p, then, is a real-valued function defined on a sample space S such that:

$0 = p(e) = 1$ for every event e in the sample space;
$p(S) = 1$;
$p(e_1 \cup e_2 \cdots) = p(e_1) + p(e_2) + \cdots$ for every sequence of disjoint events $e_1, e_2, \ldots$;
two events, a and b, are independent if $p(a \text{ and } b) = p(a)p(b)$.

The probability, $p(e_i)$, of the event e_i usually is perceived as the expected frequency with which event, e_i, occurs. A function whose value is a real number associated with each primitive event in the sample space is called a random variable. A discrete probability distribution is a function associating each possible value that a random variable can take on with the corresponding probability.

We shall simplify the blood pressure experiment in order to demonstrate how probability, used to assess the uncertainty associated with our calling unlikes as likes, is applied in the analysis of experimental results. Instead of recording pressures after taking the medication, we simply record whether or not the pressure dropped in a treated group of 10 patients. We find the following results:

dropped	no change or increased
7	3

Thus, the observed probabilities of a lowered pressure and an unchanged or increased pressure was .7 and .3, respectively. If the study were repeated again and again, we might find observed probabilities of .80 and .20 or .60 and .40, etc. Can we develop a model that might indicate how much variation to expect given a hypothesis or model describing source signals (blood pressure)? Our model might describe various degrees of expected blood pressure change and we could spend a long time evaluating each model. We could circumvent this prolonged evaluation by assuming a model that states the drug did nothing and that therefore any variation in outcomes was due solely to patient differences. Since there is only one hypothesis to evaluate, i.e., no drug effect, the labor spent in testing models is reduced dramatically. This model (no change model), of course, is the familiar "null hypothesis." Mathematically stated it is:

$$H_0 : p = 1/2$$

which states that the probability of a change in blood pressure should be 1/2 if the drug had no effect. On the average, half the patients should experience a decrease while the other half should experience no change or an increase.

Since we are going to deal with quantitative assessment of the hypothesis, we have to assign a numerical value to the various outcome events. Let I be an indicator random variable that has the value 0 when the blood pressure drops and 1 when the blood pressure rises or remains unchanged. Let us define the expected value of a random variable, $\mathscr{E}(X)$. We denote by X_i the value of X associated with the ith event in the sample space. The expected value of X is

$$\mathscr{E}(X) = \sum_i X_i \, p(X_i)$$

where $p(X_i)$ is the probability associated with the event represented by the value X_i. We then define the expected variation of a random variable, X, as

$$\begin{aligned} \mathrm{Var}(X) &= \mathscr{E}(X - \mathscr{E}(X))^2 = \sum_i (X_i - \mathscr{E}(X_i))^2 p(X_i) \\ &= \mathscr{E}(X^2) - \mathscr{E}^2(X). \end{aligned}$$

The expected value of a sum of random variables, Y_i, is

$$\mathscr{E}(\Sigma\, a_i Y_i) = \Sigma\, a_i \mathscr{E}(Y_i).$$

The variance of a sum of independent random variables is

$$\mathrm{Var}(\Sigma\, a_i Y_i) = \Sigma\, a_i^2\, \mathrm{Var}(Y_i).$$

Let p be the probability when the indicator variable is 1. The expected value of our indicator function is then:

$$\mathscr{E}(I) = 1 \cdot p + 0 \cdot (1 - p) = p$$

and its variance is

$$\mathrm{Var}(I) = p - p^2 = p(1 - p).$$

Thus, for our experiment the null hypothesis would produce an expected value of p or 1/2 and a variance of $\frac{1}{2}(1 - 1/2) = 1/4 = .25$. Since we are counting events, we define a new random variable, B, as the number of patients experiencing a blood pressure drop; so

$$B = \sum_{i=1}^{10} I_i .$$

Therefore, $\mathscr{E}(B) = \mathscr{E}(\Sigma\, I) = np = 10(1/2) = 5$

$$\mathrm{Var}(B) = \mathrm{Var}(\Sigma\, I) = np(1 - p) = 10(1/4) = 2.5$$

$$\mathrm{Var}(B/n) = (1/n^2)\, \mathrm{Var}(B) = (1/n)p(1 - p) = pq/n = .25/10$$

$$= .025.$$

The standard deviation is defined as

$$\mathrm{Sd} = \sqrt{\mathrm{Var}},$$

so

$$\mathrm{Sd} = \sqrt{.025} = .16.$$

The issue before us now is whether the fraction of patients experiencing a pressure drop is consistent with observations characterizing 10 patients from a population of untreated patients (the population defined by the null hypothesis). For this we need to know the likelihood of getting results as extreme as the .7 we observed, given a no effect model (an untreated population). This reduces to a problem in combinatorics: Assuming two equally likely outcomes, what are the probabilities associated with all possible partitions of 10 outcomes (0 decrease, 10 no change or increase; 1 decrease, 9 no change or increase, ...). Let i be the number of patients experiencing a decrease in pressure; then the probability of finding i patients out of n, P(i), is

$$P(i) = \frac{n!}{i!(n-i)!} p^i(1-p)^{n-i}.$$

The results are summarized in Table 1.

TABLE 1

Expected pressure drops for patients assuming no drug effect.

Number of patients with a reduced pressure	Number of partitions	Probability		
i = 0	1	1/1024	.000977	
1	10	10/1024	.009765	
2	45	45/1024	.043945	
3	120	120/1024	.117188	
4	210	210/1024	.205078	
5	252	252/1024	.246094	
6	210	210/1024	.205078	
7	120	120/1024	.117188	
8	45	45/1024	.043945	} .171875
9	10	10/1024	.009765	
10	1	1/1024	.000977	

From this discrete probability distribution we see that the likelihood of finding 7 or more patients out of a group of 10 who experienced a drop in blood pressure when untreated was .171875.

It is helpful to recognize that this observation is the result of dealing with a theoretical population of heterogeneous or unlike patients. The model allows us to project the results of experimenting on the "null" or untreated population, and then to compare these results with those obtained from an actual experiment.

For this particular example, we are left with a final decision: Did the drug do something or not? Therefore, we must draw a line somewhere so that observations falling on one side of the line suggest a drug effect and observations falling on the other suggest none. In so doing we know we shall make mistakes from time to time, calling drugs effective that are not (type 1 error) and calling effective drugs ineffective (type 2 error). There is no way to avoid this problem so long as we deal with samples from a heterogeneous group of patients. Wishing to minimize the number of type 1 errors, let us set the threshold for making an error at .05—that is, we wish to be wrong no more than 1 time in 20. This means that to reject the null hypothesis—that the drug is ineffective—we must observe at least 8 patients with a positive response. We only found 7; therefore, we conclude (perhaps in error) that the drug is ineffective.

The role played by sampling in the example cannot be overemphasized. Any time one extracts a sample or subgroup of patients from a heterogeneous group or population, an opportunity is created for faulty interpretation of data. The simultaneous presence of heterogeneity among both the patients and the treatment always presents a dilemma. What is responsible for the observed outcome? Either the heterogeneity or the treatment could be responsible. There are only two guaranteed ways to resolve this conflict. One is to solve the heterogeneity problem by precise matching of patients. This is not attainable with current technology. The other option, however, is attainable: Dispense with sampling and deal with the entire population. Efficient data collection, high speed computing, inexpensive mass storage, and highly effective data organization techniques have made this option a feasible one, whereas 10 years ago it was not. In fact, much of the work on sampling and population estimates came about specifically because of our inability to observe an entire population. With such an approach no longer out of the question, we have a real opportunity to reassess data collection techniques with an eye toward removing misrepresentations due to sampling. Thus the effect of computing and computer science on the very nature of data analysis is a most profound one.

The purpose of this exercise has been to demonstrate some of the underlying concepts central to the experimental method. In performing experiments we study a group of unlike subjects that are called like, and we compare experimental results with a standard or control set. A null hypothesis is proposed because it minimizes the amount of testing necessary to assess a treatment effect. The null hypothesis is stated in terms of an analytical model which provides us with what we would expect from studying a "no treatment" intervention. Such studies are enhanced by exploiting computing resources to examine observed data in their full richness to an extent impossible heretofore. Thus, for example, generation and scrutiny of histograms derived from a "no treatment" population helps set valuable guidelines for accepting or rejecting a hypothesis.

HISTOGRAMS AND THEIR CHARACTERIZATION

In our experiments, we observe a property of the experimental unit or substrate and a sequence of n observations ($t_1, t_2, \ldots, t_n$) is recorded. Under appropriate conditions the spectrum of observed values may be thought of as n samples of a single random variable, x_i, having a given probability distribution. The variation observed by sampling x_i arises from two sources: the heterogeneous nature of the experimental substrates, and the interaction between an intervention or treatment and the substrates.

The essence of hypothesis testing, as we saw earlier, lies in determining whether or not two or more random variables display the same behavior. This behavior is monitored by sampling the random variable. A histogram, the cornerstone of many statistical procedures, is one way of displaying the random variable's behavior.

Histograms are displays of the frequencies of values of the random variable. Thus, for our earlier example, the histogram would appear as shown in Figure 4. We have no difficulty in remembering the features of this particular histogram. However, had we chosen to record intervals of changes in blood pressure instead

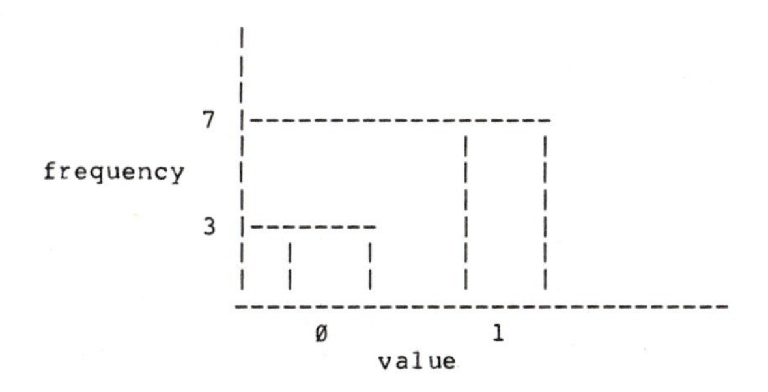

FIG. 4. Histogram of pressure effects (0 = pressure drop; 1 = pressure increase or no change).

of simply the direction of change, our histogram might appear as in Figure 5.

Clearly, as the number of observed values increases, the complexity of the histogram becomes unmanageable. Thus, unless we find a convenient characterization (equation) that describes this frequency-pressure relationship or have access to high-speed computing, we shall have difficulty in comparing this histogram with others. The amount of detail in a histogram increases with the cardinality of the sample space, thereby making manual treatment difficult. Therefore, histograms have given rise to a number of "descriptive" statistics used to represent certain features of the histogram. Hypothesis testing, then, is approached indirectly by comparing these descriptive statistics instead of comparing the histograms from which they were derived.

It is useful to speculate about approaches to histogram comparison, given high-speed computing. Information is lost when using descriptive statistics to characterize histograms. Means and variances, the most popular descriptive terms, characterize only the gross shape of the histogram. Techniques for direct comparison of histograms avoid the information lost through use of these terms. However, such comparison requires more computation such as sorting the data and computing the cumulative frequency function. (The Kolmogorov-Smirnov two-sample test (1) is an example of one such test. This test compares the cumulative frequency distributions arising from different experimental settings.) For large data

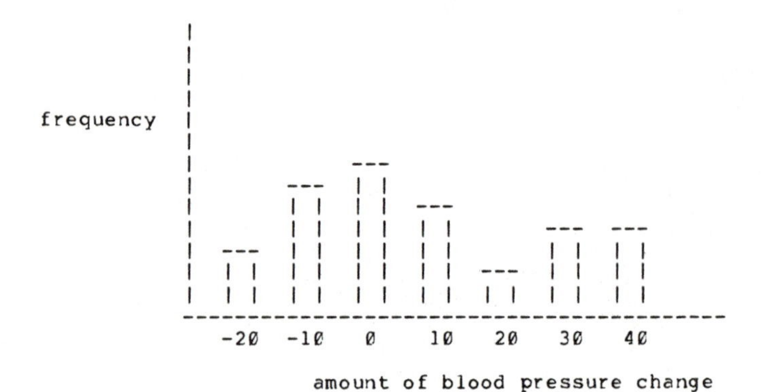

FIG. 5. Histogram of blood pressure changes.

collections (in excess of 50 data points), the manual computational effort required is such to discourage direct techniques. High-speed computing removes this constraint.

The median, mode, skew, kurtosis, etc., are other indices associated with various features of a histogram. Much of hypothesis testing is based on comparing one or more of these features. Because all the information captured in the histogram is rarely reflected in these abstractions, we are essentially adding another noise source to the data, thereby enhancing the likelihood of making a type 1 or type 2 mistake. However, these features do have interesting properties that produce several forgiving tests of hypotheses and allow us to estimate parameters embedded in probability distributions that are associated with many experimental settings.

Earlier, it was stated that an experiment can be viewed as sampling a random variable if certain conditions are met. If sampling is such that

$$\{t_1 = x_1(.), t_2 = x_2(.), \ldots, t_n = x_n(.)\}$$

is an independent class of n observations and the $x_i(.)$ has the same probability distribution for all i, then the set of observations is called a random sample of size n of the variable x(.).

Let $\bar{X}$ be the sample mean. Then,

$$\begin{aligned} \mathscr{E}(\bar{X}) &= \frac{1}{n}\,\mathscr{E}\left(\sum_{i=1}^{n} t_i\right) = \frac{1}{n}\sum_{i=1}^{n}\mathscr{E}(t_i) \\ &= \mathscr{E}(x(.)). \end{aligned}$$

Therefore, the expected value of the sample mean is equal to the expected value of the random variable under investigation. We can use the sample mean to estimate expected values regardless of the probability distribution of x(.).

Recall that $\mathrm{Var}(x) = \sigma^2 = \mathscr{E}[(x^2) - \mathscr{E}(x)]^2$ so that we can estimate $\sigma^2(x)$ with s^2 where

$$\begin{aligned} s^2 &= \frac{\sum_{i=1}^{n} t_i^2}{n} - \left(\frac{\sum_{i=1}^{n} t_i}{n}\right)^2 \\ &= \frac{\sum_{i=1}^{n} t_i^2 - \dfrac{(\sum_{i=1}^{n} t_i)^2}{n}}{n}. \end{aligned}$$

Again, s^2 like $\bar{X}$ can be used reliably to estimate properties of a random variable (and, thus, a histogram) independent of the probability distribution.

To summarize, we assess the interactions between treatments or stimuli and experimental substrates by observing a property of the substrate. This property is represented by a random variable that takes on a different value for each different experimental outcome. The properties of the random variable are monitored by collecting and evaluating a random sample, and the features of the sample are portrayed by a histogram. Hypothesis testing is based on comparing random variables suggested by competing views of an experiment. This comparison is carried out by determining whether the competing random variables are equivalent. Because histograms have more features than we can comfortably handle, they are usually represented by means and variances. The null hypothesis, therefore, is reduced to equivalence of means and variances and the hypothesis test usually is established by comparing these means and variances. It is not outlandish to characterize this approach as being prompted by the necessity of living in a world without computers. The anachronism is clear.

BIAS

When estimating properties of a random variable from a random sample, it is helpful to know whether or not the estimator is biased. An unbiased estimator, t, is one for which its expected value is equal to the parameter, θ, that it estimates.

$$\mathscr{E}(t) = \theta.$$

The sample mean estimates the expected value of a random variable in an unbiased manner. However, s^2, our estimate of σ^2 is biased. Instead of

$$\mathscr{E}(s^2) = \sigma^2,$$

the expected value for s^2 is

$$\mathscr{E}(s^2) = \frac{1}{n}\,\mathscr{E}\left[\sum t_i^2 - \frac{1}{n}\sum t_i \sum t_i\right] = \frac{n-1}{n}\,r^2.$$

Thus, for an unbiased estimator of σ^2 we need to have

$$s^2 = \frac{\sum t_i^2 - \dfrac{\sum t_i \sum t_i}{n}}{n - 1}.$$

This necessity, of course, is well established and it is accommodated routinely in computational procedures.

STANDARDIZED VARIABLES

It is interesting to note that histograms resulting from sums of experimental observations appear almost the same when expressed in the appropriate units. Consider the sum of the points of several dice. An experiment (in this context) consists of a number of throws of n dice. After each throw, the points on the faces are summed and a frequency of occurrence, f_i, of each sum is tabulated. Plotting the relative frequencies of each sum gives the results of Figure 6. As the

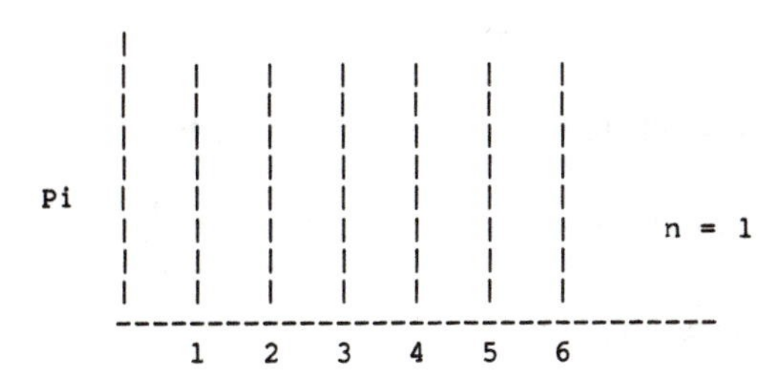

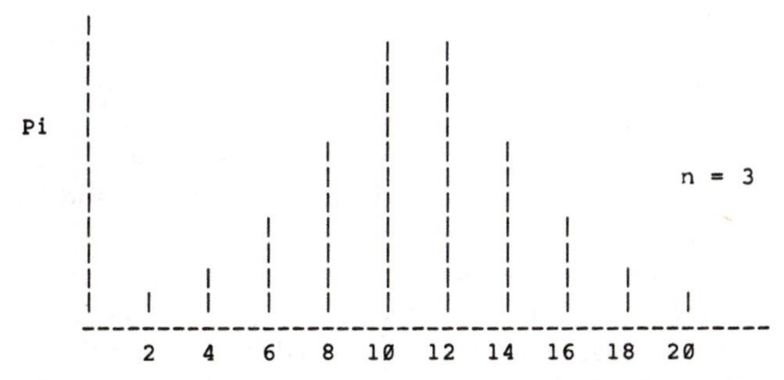

FIG. 6. Histograms of sum of points for n thrown dice, $P_i = f_i/\Sigma f_i$.

number of terms in the sum increases, the histograms get wider and the peak moves to the right. Let S_n be the sum of n dice. Then its expected value is

$$\begin{aligned}\mathscr{E}(S_n) &= \mathscr{E}(S_1 + S_1 + \cdots + S_1)\\ &= n\mathscr{E}(S_1)\\ \mathscr{E}(S_1) &= \sum_{i=1}^{6} i\, p_i = \frac{1}{6}\sum_{i=1}^{6} i = \frac{6 \cdot 7}{12} = \frac{7}{2}\\ \therefore\ \mathscr{E}(S_n) &= \frac{7n}{2}.\end{aligned}$$

The variance associated with S_1 is

$$\begin{aligned}\mathrm{Var}(S_1) &= \mathscr{E}(S_1^2) - \mathscr{E}^2\,(S_1)\\ &= \frac{1}{6}\sum_{i=1}^{6} i^2 - \left(\frac{7}{2}\right)^2\\ &= \frac{35}{12}.\end{aligned}$$

Since the n dice are independent, the variance of S_n is

$$\mathrm{Var}(S_n) = \mathrm{Var}(S_1 + S_1 + \cdots + S_1) = \frac{35n}{12}$$

and the standard deviation, then, is

$$\mathrm{Sd}(S_n) = \sqrt{\frac{35n}{12}}.$$

From these relationships, we see that the mean and variance of the S_n is dependent on the number of terms in the sum.

For purposes of histogram comparison, it is helpful if the histograms are normalized such that the peak amplitude and variance are independent of n. Two parameters need adjusting: the mean and the variance. The mean can be shifted to the origin by subtracting from each sum the expected value of the sum, while the width of the histogram can be normalized by dividing the quantity (sum $- \mathscr{E}$(sum)) by the standard deviation of the sum. Thus we form a new random variable, z, where

$$z = \frac{S_n - \mathscr{E}(S_n)}{Sd(S_n)}.$$

The properties of z can be established by computing its expected value and its variance:

$$\mathscr{E}(z) = \frac{\mathscr{E}(S_n) - \mathscr{E}(\mathscr{E}(S_n))}{Sd(S_n)} = 0,$$

$$Var(z) = Var\,\frac{S_n - \mathscr{E}(S_n)}{Sd(S_n)} = 1.$$

This normalization works for any random variable, not just sums of random variables. Therefore, all random variables of the form

$$z = \frac{X - \mathscr{E}(X)}{Sd(x)}$$

will have expected value = 0 and variance or standard deviation = 1. Without extensive high-speed computational facilities, these adjustments often were considered sufficiently formidable to preclude their use. That deterrent is gone.

When dealing specifically with sums of random variables, their histograms approach the form of a normal curve as n increases. This is generally observed regardless of the underlying probability distribution. A normal curve, describing the behavior of a random variable, x, with mean μ and variance σ^2, is defined (Figure 7) by:

$$N(x; \mu, \sigma^2) = \frac{1}{\sqrt{2\pi\sigma^2}} \exp\left[-\frac{1}{2}\frac{(x-\mu)^2}{\sigma^2}\right].$$

The observation that sums of random variables produce bell-shaped histograms is stated explicitly in the central limit theorem:

Let x_i, i = 1, 2, ..., n, be independently distributed random variables with the same probability distribution having an expected value $\mathscr{E}(x)$ and a finite variance σ^2. Then

$$z = \frac{\sum_{i=1}^{n} \frac{x_i}{n} - \mathscr{E}(x)}{\sqrt{\sigma^2/n}}$$

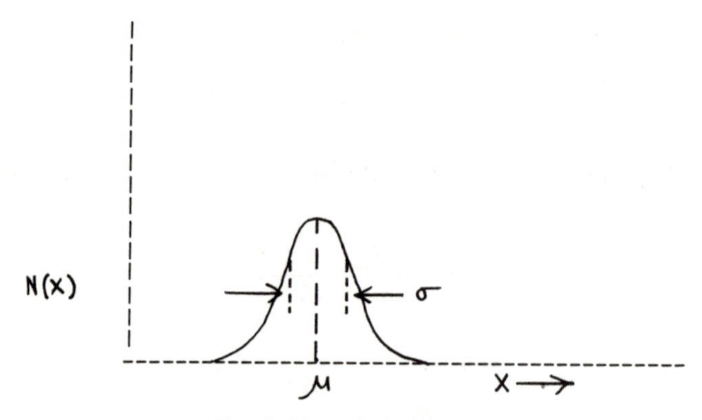

FIG. 7. Normal distribution.

is asymptotically normally distributed with mean of 0 and a variance of 1, N(0, 1).

The importance of this theorem lies in the fact that it provides a very robust (i.e., distribution-free) test for comparing histograms when one considers only the centroid of the histogram as being important. The mean, being a sum of random variables, will be asymptotically normally distributed. A couple of examples here will help demonstrate this point.

EXAMPLE 1. We have two drugs we wish to evaluate: one is a placebo and the other is an antihypertensive drug for lowering blood pressure. Our interest is in observing whether the pressure obtained after administering the drug is lower than it was prior to administering the drug. The results are displayed in Figure 8. N_A patients were studied with drug A while N_B were studied with drug

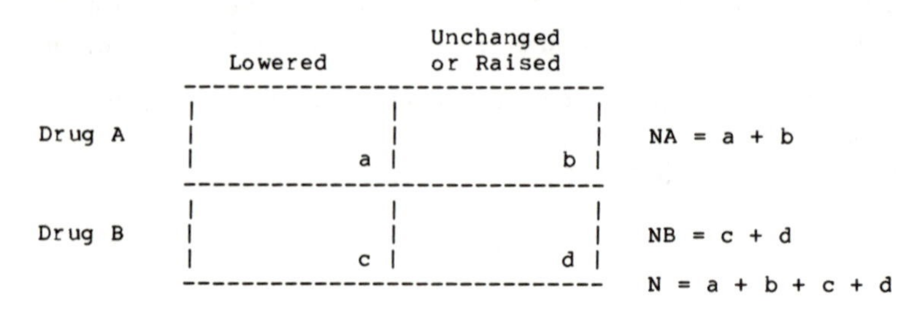

	Lowered	Unchanged or Raised	
Drug A	a	b	NA = a + b
Drug B	c	d	NB = c + d
			N = a + b + c + d

FIG. 8. Results of a drug comparison experiment.

B. The number of patients experiencing a lowering of pressure was a and c, respectively. For this particular experiment, assume that the binary responses (lowered, not lowered) occur with fixed probability (p_A and p_B) for drugs A and B, respectively. The null hypothesis is stated in terms of no drug effect, i.e., $p_A = p_B$. By studying a number of patients we can estimate p_A and p_B with the relative frequencies ($\hat{p}_A$ and $\hat{p}_B$) of the outcomes. Assigning a value of 1 to the outcome "pressure lowered" and 0 to the outcome "pressure not lowered," it is seen that the mean outcome of drug A is the relative frequency of "pressure lowered" for drug A. The same holds for drug B.

Since the estimators are sums of a 0/1 random variable, the central limit theorem holds. Therefore, it is possible to use a z statistic to test the null hypothesis, the latter stated as $p_A = p_B$ or $p_A - p_B = 0$. Let

$$z = \frac{(\hat{p}_A - \hat{p}_B) - \mathscr{E}(p_A - p_B)}{\sqrt{\operatorname{Var}(p_A - p_B)}}$$

which is asymptotically distributed as a N(0, 1). With knowledge of z, one can determine whether or not the observed difference in outcomes is consistent with no drug effect. Continuing with the evaluation of z, it was shown earlier that the variance of a sum of indicator variables (0/1 variables) was $np(1 - p)$. Thus, the variance of a proportion is

$$\frac{p(1-p)}{n}.$$

Because the drugs are independent, we have:

$$\operatorname{Var}(p_A - p_B) = \operatorname{Var}(p_A) + \operatorname{Var}(p_B).$$

By assuming the null hypothesis to be true, we have

$$\operatorname{Var}(p_A) = \frac{\left(\frac{a+c}{N}\right)\left(\frac{b+d}{N}\right)}{N_A}$$

$$\operatorname{Var}(p_B) = \frac{\left(\frac{a+c}{N}\right)\left(\frac{b+d}{N}\right)}{N_B}$$

$$\mathrm{Var}(p_A - p_B) = \left(\frac{a+c}{N}\right)\left(\frac{b+d}{N}\right)\left(\frac{1}{a+b} + \frac{1}{c+d}\right)$$

$$z = \frac{\dfrac{a}{a+b} - \dfrac{c}{c+d}}{\sqrt{\dfrac{a+c}{N}\dfrac{b+d}{N}\left(\dfrac{1}{a+b} + \dfrac{1}{c+d}\right)}}$$

and

$$z^2 = \frac{N(ad - bc)^2}{(a+b)(a+c)(c+d)(b+d)}.$$

Since z is asymptotically distributed as a N(0, 1), we can compute the value beyond which we would say that $\hat{p}_A$ and $\hat{p}_B$ probably come from different populations, i.e., drug A is probably different from drug B. Setting a value of 5% as the type 1 error we shall tolerate,

$$.95 = \int_{-z_c}^{z_c} \frac{1}{\sqrt{2\pi}} e^{-x^2/2}\, dx \qquad |z_c| = 1.96.$$

(Convenient means for evaluating the above already have been built into some hand calculators.) Thus, if $|z| < 1.96$, we accept the hypothesis that drug A = drug B as reflected by blood pressure responses. (For those familiar with statistical tests, note that $z_c^2 = 3.84$ and that z^2 is the same as χ^2 for testing independence in a 2×2 table. The critical value of χ^2 is 3.84 for a 2×2 table.)

EXAMPLE 2. As a second example, we shall study the same two drugs, but instead of simply monitoring change, let us actually measure the amount of change. Let X_i be the pressure in the ith patient. We shall retain the same hypothesis, i.e., response to drug A = response to drug B. The data source model will be a function of the two average responses, μ_A and μ_B. We estimate μ_A from

$$\bar{X}_A = \frac{\sum_{i=1}^{n_1} X_i}{n_A}$$

and μ_B from

$$\bar{X}_B = \frac{\sum_{i=n_A+1}^{n_A+n_B} X_i}{n_B}.$$

The hypothesis we test is that $\mu_A - \mu_B = 0$. From the definition of the z test, we set

$$Z = \frac{(\bar{X}_A - \bar{X}_B)}{Sd(\bar{X}_A - \bar{X}_B)}.$$

The variance of $(\bar{X}_A - \bar{X}_B) = Var(\bar{X}_A) + (Var(\bar{X}_B)$. Since the null hypothesis is assumed true, we have

$$Var(\bar{X}_A) + Var(\bar{X}_B) = Var(X)\left(\frac{1}{N_A} + \frac{1}{N_B}\right)$$

$$Var(X) = \frac{\sum_{i=1}^{N_A} X_i^2 + \sum_{i=N_A+1}^{N_A+N_B} X_i^2 - \frac{(\sum_{i=1}^{N_A} X_i)^2}{N_A} - \frac{(\sum_{i=N_A+1}^{N_A+N_B} X_i)^2}{N_B}}{N_A + N_B - 2}$$

$$Sd(\bar{X}_A - \bar{X}_B) = \left[Var(X)\left(\frac{1}{N_A} + \frac{1}{N_B}\right)\right]^{1/2}.$$

The z statistic, derived in this manner, is equivalent to the t statistic used for comparing the means of two populations. It is interesting to note that the critical value of z for rejecting the null hypothesis at the .05 level of significance is ± 1.96. This is the asymptotic value t approaches as the sample sizes increase.

From these examples one can see how to design a test for many hypotheses when a mean or other linear function of the data represents the information content of a histogram. As a first approximation for developing hypothesis tests, the z statistic is not a bad choice. One is required to specify the underlying data model so that expected means and variances can be evaluated. Critical values for accepting or rejecting the hypothesis are then derived from tables of N(0, 1). The increased analytical facilities provided by the process of standardizing variables certainly are beyond dispute. However, they no longer are something to be wished for. Rather, they are something to be expected.

We have neglected to discuss notions of power, sample size estimation, and type 2 errors. These topics relate directly to the observation that the variance of a mean decreases as n, the number of observations, increases. Many useful exercises exploring these concepts can be developed around the asymptotic approach used above.

RANDOMIZATION TESTS

Although z tests can be used in many common experimental settings, there are many other tests that do not require the invocation of asymptotic properties. Randomization tests, based on enumeration of all possible partitions of the outcomes, constitute such a class of tests. In particular, easy access to computing casts a different light on the practicality of these tests which were developed by Fisher [2]. These tests assume that the observed samples are from a "null" population. By ignoring the original partition (by treatment or intervention) of the observations, and investigating all possible partitions of the data, it is possible to describe a "local" probability distribution attributed to a specific "null" data generator. The original observed data partition is then compared to this distribution and an assessment made as to whether the data are consistent with the "null" data. These tests were little used because of the computational load required to permute the data. High-speed computing, however, makes a certain class of these tests feasible, thereby placing their power and versatility well within reach.

Consider the case of two drugs again. A study of two groups of 5 patients might appear as shown in Figure 9. We assume that the null hypothesis is true, i.e., Drug A responses are no different from Drug B responses. Furthermore, we assume that the proportion of (−) responses to (+) responses reflects the underlying heterogeneity of the experimental substrates (patients) and that the number of patients subjected to each drug was predetermined. The "null" probability distribution, then, is derived by enumerating all data

	Response +	Response −	Total
Drug A	1	4	5
Drug B	3	2	5
	4	6	10

FIG. 9. Outcome of a two-drug test.

configurations possible for fixed marginal totals (5 and 5 in this case).

The number of ways of partitioning 10 patients into two groups of 5 is $\binom{10}{5} = 252$, each grouping being equally likely. The number of ways of partitioning the (+) responses into two groups, one of size a is $\binom{4}{a}$. The number of ways of partitioning the (−) responses into two groups, one of size b is $\binom{6}{b}$. Thus, the number of ways to get a (+) responses and b = 5 − a (−) responses is

$$\binom{4}{a}\binom{6}{b} = \binom{4}{a}\binom{6}{5-a}.$$

Since there are $\binom{10}{5}$ possible partitions of the data, the probability of getting b (−) responses and a (+) responses where b = 5 − a is

$$p(a) = \frac{\binom{4}{a}\binom{6}{5-a}}{\binom{10}{5}}.$$

For a general table as shown in Figure 10 (where a is the smallest integer among a, b, c, and d), we have

$$p(a) = \frac{\binom{a+c}{a}\binom{b+d}{b}}{\binom{a+b+c+d}{a+b}}.$$

Tabulation of the probability distribution then requires adjusting a, b, c, and d such that $a + b = n_A$, $c + d = n_B$, $a + c = l_1$, and

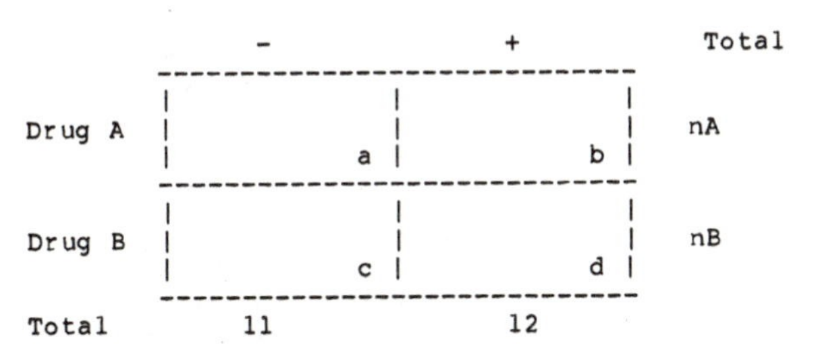

	-	+	Total
Drug A	a	b	nA
Drug B	c	d	nB
Total	11	12	

FIG. 10. Generalization of the two-drug test.

$b + d = l_2$, while a ranges from 0 to $\min(l_1, n_A)$. Thus, the distribution associated with a 2×2 table having marginal totals of 5, 5 and 6, 4 is given in Table 2. The probability of observing 1 or fewer (+) responses with drug A is $.02381 + .23810 = .26190$. The results of our experiment, therefore, are consistent with taking two samples of 5 from the same population, i.e., no drug effect.

Evaluating the probability distribution can be simplified considerably by expanding p(a) in terms of factorials. This leads to an efficient algorithm for assessing the probability of a single table. The additional elements of the probability distribution can then be derived iteratively by multiplying and dividing this quantity by functions of a, b, c, d, and a variable x which is incremented or decremented by one for each iteration [3]. Again, this is discouraging for computational resources limited to manual means, but eminently feasible for integration as part of a computer-based analytical environment.

As a second example we wish to determine whether diet A is different from diet B. We have two groups of patients with n_A of them on diet A, and n_B on diet B. The weight change for each patient is measured based on the weights before and after diet.

The null hypothesis for this particular study is that diet A is equivalent to diet B as reflected by the weight changes. If the diets are equivalent, then the $n_A + n_B$ patients could be considered to have come from the same population. Thus, the association of n_A weight changes with diet A and n_B weight changes with diet B is totally arbitrary. Following the last example, we have $\binom{n_A + n_B}{n_A}$ possible partitions for "labeling" patients with diet A or diet B.

For each partition we compute the average weight changes for

TABLE 2

Probability distribution of a 2×2 table with marginal totals of 5, 5 and 6, 4.

	p(m)	p(i ≤ m)
m = 0	.02381	.02381
= 1	.23810	.26190
= 2	.47619	.73809
= 3	.23809	.97619
= 4	.02381	1.00

patients labeled diet A and diet B. We order the differences of average weight changes and construct a histogram (Figure 11). From this histogram, we determine whether the original weight difference associated with those patients treated with diets A and B is consistent with the null probability distribution.

Actually the complete histogram need not be constructed. We only need the tails of the histogram. The portion required is dependent on the amount of type 1 error we are willing to tolerate. Thus, if we set the type 1 error rate to 5%, then we only need compute the .05 $\binom{n_A+n_B}{n_A}$ most extreme data configurations.

By sorting the data these configurations can easily be found. Let d_i be the ith observed weight change and r_i the ith rank ordered weight change.

$$\text{natural order:} \qquad d_1 d_2 \cdots d_{n_A} d_{n_A+1}\, d_{n_A+2} \cdots d_{n_A+n_B}$$

$$\text{rank order:} \qquad r_1 r_2 \cdots r_{n_A} r_{n_A+1} r_{n_A+2} \cdots r_{n_A+n_B}$$

$$\bar{r}_A = \frac{\sum_{i=1}^{n_A} r_i}{n_A} \quad \bar{r}_B = \frac{\sum_{i=n_A+1}^{n_A+n_B} r_i}{n_B}.$$

The most extreme data configuration leads to a weight change difference of $\bar{r}_A - \bar{r}_B$. Less extreme configurations can be computed by interchanging r_{n_A} with r_{n_A+1}, etc. In this manner the extremes of the histogram can be directly computed. This is a powerful yet simple algorithm whose routine use is made feasible by high-speed computers and techniques developed for them.

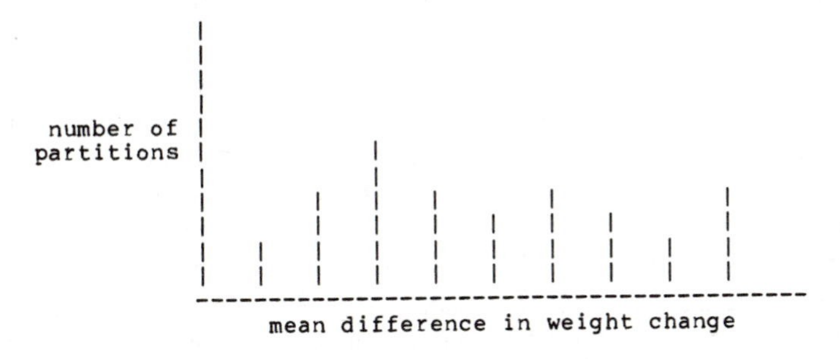

FIG. 11. Histogram for diet experiment.

KOLMOGOROV-SMIRNOV TEST

The Kolmogorov-Smirnov test represents another vehicle for comparing two random variables. The test is based on comparing the observed cumulative frequency distributions associated with the two random variables in question. As such, the test is sensitive to anything that modifies the distribution function, including changes in means and variances. If two samples have been drawn from the same population, then one would expect the sample based cumulative distributions to be fairly close to each other, since the only differences would arise from baseline heterogeneities. When the distance between the two cumulative distributions is too great at any particular point, then the possibility of differences between the two random variables must be considered.

The Kolmogorov-Smirnov test is based on the largest difference between the cumulative distributions associated with two random variables. Let $C(x_i)$ be the observed cumulative frequency for the ith value of random variable x and let $C(y_i)$ be the observed cumulative frequency for the ith value of random variable y. Let n_x be the total number of observations of X and n_y be the total number of y. Then let

$$D_1 = \max_i \left(\frac{C(x_i)}{n_x} - \frac{C(y_i)}{n_y} \right)$$

be the maximum difference in the observed cumulative distributions and

$$D_2 = \max_i \left| \frac{C(x_i)}{n_x} - \frac{C(y_i)}{n_y} \right|$$

be the absolute maximum difference between the two observed cumulative distributions.

This test is not widely used because of the effort required to tabulate the cumulative frequency distribution for large collections of data. However, high-speed computing minimizes the difficulty in sorting the data and hence estimating the cumulative distribution. Like the randomization tests, this test makes no assumptions about underlying probability distributions, and therefore provides an unencumbered view of experimentally derived data. The opportunity to remove such encumbrances constitutes perhaps the most profound effect of computer science on statistical work.

MODELS OF DATA SOURCES

Thus far we have looked at simple data sources where the experimental substrates, though heterogeneous, were considered equivalent. When we can characterize the heterogeneity, then there exists a class of models that can be used to improve our ability to test hypotheses. Let $\mathbf{x}_{(1 \times m)}$ be a $(1 \times m)$ vector of attribute values that describes the properties of a single experimental substrate. Since most experiments compare two or more interventions, treatments, or stimuli, we include these in our definition of properties. The response we observe as an outcome of an experiment, then, is some function of these properties, i.e.,

$$y = f(\mathbf{x}).$$

To a first approximation, the response outcome y can be represented by a linear function of the substrate properties and intervention

$$y = \mathbf{xA}$$

where $\mathbf{A}$ is an $(m \times 1)$ vector of model parameters. We rarely are able to deal with all properties contributing to the heterogeneity of the experiment; so we include a noise source, ϵ. The observed outcome, y, then is expressed as

$$y = \mathbf{xA} + \epsilon.$$

For an experiment involving n independent substrates, we can represent the data in matrix form as

$$\underset{(n \times 1)}{y} = \underset{(n \times m)}{X} \; \underset{(m \times 1)}{A} + \underset{(n \times 1)}{\epsilon}$$

where

$$\underset{(n \times m)}{X} = \begin{pmatrix} x_1 \\ x_2 \\ \vdots \\ x_n \end{pmatrix}.$$

One method of estimating the parameters of the model, A, is to choose the value of A that minimizes the squared difference between the predicted responses, y, and the observed responses, Y, or

$$\min_{A} (\hat{Y} - Y)'(\hat{Y} - Y)$$

$$\min_{A} (Y - XA)'(Y - XA).$$

Thus we can estimate $\hat{A}$ by standard technique:

$$\underset{(m \times 1)}{\hat{A}} = \underset{(m \times m)}{(X'X)^{-1}} \underset{(m \times n)}{X'} \underset{(n \times 1)}{Y}.$$

The expected value of $\hat{A}$ is

$$\begin{aligned}\mathscr{E}(\hat{A}) &= \mathscr{E}[(X'X)^{-1}X'Y] = (X'X)^{-1}X'\mathscr{E}(Y) \\ &= (X'X)^{-1}X'XA = A\end{aligned}$$

and its variance is

$$\begin{aligned}\text{Var}(\hat{A}) &= \text{Var}[(X'X)^{-1}X'\hat{Y}] = (X'X)^{-1}X'X(X'X)^{-1}\ \text{Var}(\hat{Y}) \\ &= (X'X)^{-1}\ \text{Var}(\hat{Y}).\end{aligned}$$

For experiments where the variance is independent of the substrate used, we can let

$$\text{Var}(\hat{Y}) = \sigma^2 \begin{pmatrix} 1 & & & & \\ & 1 & & & \\ & & 1 & & \\ & & & \ddots & \\ & & & & 1 \end{pmatrix}.$$

Note that the estimate of A, the model parameter vector, is based on a linear combination of the outcome or response variables. Thus, each model parameter should have an asymptotic normal distribution and we can test hypotheses using z statistics.

HYPOTHESIS TESTING

We frequently test ideas about the a_i's with hypotheses of the form $a_i = 0$ or $a_i = a_j$, $i \neq j$. A general class of these hypotheses can be described by a matrix product of the form

$$\underset{(s \times m)}{C}\underset{(m \times 1)}{A} = \underset{(s \times 1)}{0}$$

where C is an arbitrary matrix of rank $s \leq m$. Suppose the experimental model is

$$y = a_1 x_1 + a_2 x_2$$

and the null hypothesis H_0: $a_1 = a_2$. Then,

$$C = (1 \quad -1).$$

To test such a general class of hypotheses, we need to know what contribution the restrictions make to the fit between the data and the model. This can be assessed by performing a constrained optimization of

$$\min(Y - XA)'(Y - XA)$$

$$CA = 0.$$

From this we find the minimum to be

$$(Y - X\hat{A})'(Y - X\hat{A}) + \hat{A}'C'[C(X'X)^{-1}c']^{-1}C\hat{A}$$

where $\hat{A} = (X'X)^{-1} X'Y$, the unrestricted estimate of the parameter vector. The term

$$(Y - X\hat{A})'(Y - X\hat{A})$$

is the squared error due to the unrestricted model while $\hat{A}'C'[C(X'X)^{-1}C']^{-1} C\hat{A}$ is the additional error due to the restrictions.

Let

$$S_H = \hat{A}'C'[C(X'X)^{-1}C']^{-1}C\hat{A}$$

and

$$S_E = (Y - X\hat{A})'(Y - X\hat{A}) = Y'Y - Y'X\hat{A}.$$

Then, a ratio of average errors can be formed for testing hypotheses. Thus, to test the hypothesis $CA = 0$,

$$F = \frac{S_H/\text{rank}\ (C)}{S_E/(n - m)}$$

where F is the variance ratio with rank (C) and $n - m$ degrees of freedom. Clearly, high values of F would suggest the null hypothesis does not hold while low values of F (<4) are consistent with the null hypothesis. Note that we have made no serious distribution assumptions since estimates of parameters, A, are linear combinations of identically distributed random variables.

IMPLEMENTATION

Access to computing considerably simplifies the management of this general linear model. We shall look at the computational requirements necessary for having such a general model at our fingertips. Expanding the general model $\hat{Y} = X\hat{A}$, we obtain

$$\begin{pmatrix} \hat{y}_1 \\ \hat{y}_2 \\ \vdots \\ \hat{y}_n \end{pmatrix} = \begin{pmatrix} x_{11} & x_{12} & \cdots & x_{1m} \\ x_{21} & x_{22} & \cdots & x_{2m} \\ \vdots & & & \\ x_{n1} & x_{n2} & \cdots & x_{nm} \end{pmatrix} \begin{pmatrix} a_1 \\ a_2 \\ \vdots \\ a_m \end{pmatrix}$$

where each row represents data associated with a single substrate. The parameters of the model are estimated from

$$\underset{(m \times 1)}{\hat{A}} = \underset{(m \times m)}{(X'X)^{-1}} \underset{(m \times 1)}{X'Y}.$$

Expanding X'X we find

$$\underset{(m \times m)}{X'X} = \begin{pmatrix} \sum_{i=1}^n x_{i1}^2 & \sum_{i=1}^n x_{i1} x_{i2} & \cdots & \sum_{i=1}^n x_{i1} x_{im} \\ \sum_{i=1}^n x_{i1} x_{i2} & \sum_{i=1}^n x_{i2}^2 & \cdots & \\ \vdots & & & \\ \sum_{i=1}^n x_{i1} x_{im} & \sum_{i=1}^n x_{i2} x_{im} & \cdots & \sum_{i=1}^n x_{im}^2 \end{pmatrix}$$

and expanding X'Y and Y'Y, we find

$$\underset{(m \times 1)}{X'Y} = \begin{pmatrix} \sum_{i=1}^n x_{i1} y_i \\ \sum_{i=1}^n x_{i2} y_i \\ \vdots \\ \sum_{i=1}^n x_{im} y_i \end{pmatrix} \quad Y'Y = \sum_{i=1}^n y_i^2 .$$

These two matrices can be accumulated incrementally. Starting with $X'X = 0$ and $X'Y = 0$, each term of each sum is computed based on an ith substrates characterization $(y_i, x_{i1}, x_{i2}, \ldots, x_{im})$. Each element of the matrices is then updated. By exploiting these data-structuring techniques, one can analyze models independent of n, the number of substrates studied without requiring additional computer memory for the data. The necessary computations for hypothesis testing are simple functions of these three matrices.

To minimize the programming necessary to support the general linear model computations, it is helpful to have processing modules for performing the following tasks:

1. inversion of an $(m \times m)$ matrix $A = B^{-1}$,
2. pre/post multiplication of a matrix with another matrix

$$\underset{(s \times s)}{C} = \underset{(s \times m)}{A'} \underset{(m \times m)}{B} \underset{(m \times s)}{A}$$

3. matrix multiplication $\underset{(m \times s)}{C} = \underset{(m \times p)}{B} \underset{(p \times s)}{A}$.

The following sequence outlines the computations necessary to evaluate a linear model:

1. Initialize Y′Y, Y′X, X′X to zero.
2. Accumulate Y′Y, Y′X, X′X for n data vectors.
3. Estimate model parameter from $\underset{(m \times 1)}{A} = \underset{(m \times m)}{(X'X)^{-1}} \underset{(m \times 1)}{X'Y}$.
4. For each hypothesis, get contrast matrix C and form

$$S = \underset{(1 \times m)}{A'} \underset{(m \times s)}{C'} \left[\underset{(s \times m)}{C} \underset{(m \times m)}{(X'X)^{-1}} \underset{(m \times s)}{C'} \right]^{-1} \underset{(s \times m)}{C} \underset{(m \times 1)}{A}$$

(note: two applications of pre/post multiply)

$$S = \underset{(1 \times 1)}{Y'Y} - \underset{(1 \times m)}{Y'X} \underset{(m \times 1)}{A}$$

$$F = \frac{S_H/\text{rank }(C)}{S_E/(n - m)}.$$

The resulting significance level for $F(\text{rank}(C), n - m)$ can be looked up in tables or computed directly. Here again, standard computer science techniques make these implementations relatively straightforward.

EXAMPLES

Consider the drug problem. We have two groups of patients of size n_A and n_B. Group A is treated with drug A; group B is treated with drug B. The null hypothesis states that drug A is equivalent to drug B. We measure the blood pressure to assess the drug-patient interaction. One model for this problem is

$$y_i = x_{i1} a_1 + x_{i2} a_2$$

where the independent variables x_{i1} and x_{i2} are 0/1 variable, a_1 is the model pressure associated with drug A, a_2 is the model pressure associated with drug B, and y_i is the blood pressure response of the ith patient. Let

$$x_{i1} = 1 \text{ if patient is given drug A, else } 0;$$

$$x_{i2} = 1 \text{ if patient is given drug B, else } 0.$$

The data from an experiment then will appear as

$$\begin{pmatrix} y \\ y \\ y \\ \vdots \\ y \\ y \\ y \\ \vdots \\ y \end{pmatrix} = \begin{pmatrix} 1 & 0 \\ 1 & 0 \\ 1 & 0 \\ \vdots & \vdots \\ 1 & 0 \\ 0 & 1 \\ 0 & 1 \\ \vdots & \vdots \\ 0 & 1 \end{pmatrix} \begin{pmatrix} a_1 \\ a_2 \end{pmatrix}.$$

To test the equivalence of the two drugs as reflected by the average responses, let

$$C = (1 \quad -1).$$

The $F(1, n_A + n_B - 2)$ is equivalent to the square of the t statistic obtained from a standard Student's t test. Next, let us assume that the blood pressure response to the drug is dependent on the original blood pressure. (People with normal pressures will have a small response while people with high pressures will have a larger response.) Define a new model,

$$\hat{y}_i = x_{i1} a_1 + x_{i2} a_2 + x_{i3} a_3 + x_{i4} a_4$$

where x_{i1} and x_{i2} are 0/1 variables depending on the drug used, x_{i3} is the initial blood pressure of patients treated with drug A, and x_{i4}

is the initial blood pressure of patients treated with drug B. Graphically, this assumed behavior would appear as in Figure 12. The X matrix now appears as

$$\begin{pmatrix} 1 & 0 & p_1 & 0 \\ 1 & 0 & p_2 & \cdot \\ 1 & 0 & \cdot & \cdot \\ 1 & 0 & \cdot & \cdot \\ 1 & 0 & \cdot & \cdot \\ 1 & 0 & p_{n_A} & 0 \\ & & & \\ 0 & 1 & 0 & p_{n_A+1} \\ 0 & 1 & \cdot & p_{n_A+2} \\ 0 & 1 & \cdot & \cdot \\ 0 & \cdot & \cdot & \cdot \\ 0 & \cdot & \cdot & \cdot \\ 0 & 1 & 0 & p_{n_A+n_B} \end{pmatrix}$$

There are several ways of stating null hypotheses for this study. One way is to state that the two lines are equivalent to

$$C = \begin{pmatrix} 1 & -1 & 0 & 0 \\ 0 & 0 & 1 & -1 \end{pmatrix}$$

where row one tests equivalence of intercepts while row two tests equivalence of the slopes.

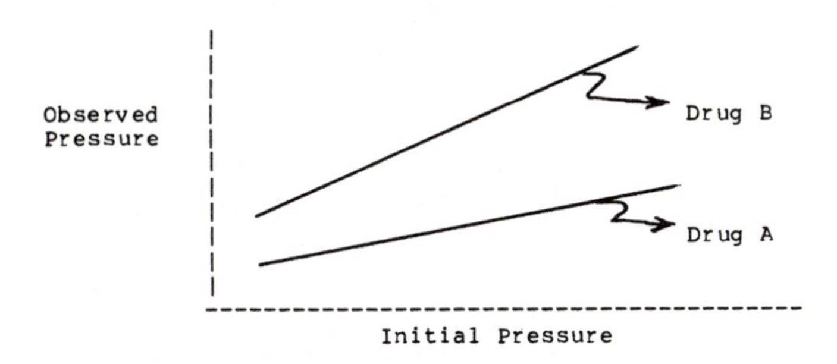

FIG. 12. Treatment experiment with blood pressure response sensitive to initial blood pressure.

As a final example consider simultaneous drugs, A and B, where we are testing three different doses of drug A and two different doses of drug B. A schematic view would appear as

```
              A1          A2          A3
        -------------------------------------
        |           |           |           |
   B1   |           |           |           |
        |         1 |         2 |         3 |
        -------------------------------------
        |           |           |           |
   B2   |           |           |           |
        |         4 |         5 |         6 |
        -------------------------------------
```

Thus, patients are treated with various combination doses of drugs A and B. An appropriate model here would be

$$y_i = x_{i1}a_1 + x_{i2}a_2 + x_{i3}a_3 + x_{i4}a_4 + x_{i5}a_5 + x_{i6}a_6$$

where a_i is the model parameter for a particular dose combination. Thus, a_1 is the parameter for A_1B_1, a_6 is the parameter for A_3B_2; x_{ij} is 1 for treatment combination j and 0 for other treatment combinations. Graphically we can visualize this experiment as shown in Figure 13. There are three hypotheses of primary interest:

1. Is there any interaction between drug A and drug B?
2. If there is no interaction, is there a dose effect with drug A?
3. If there is no interaction, is there a dose effect with drug B?

The interaction hypothesis tests whether the presence of one drug affects the response of another drug in a variable manner. Thus, we

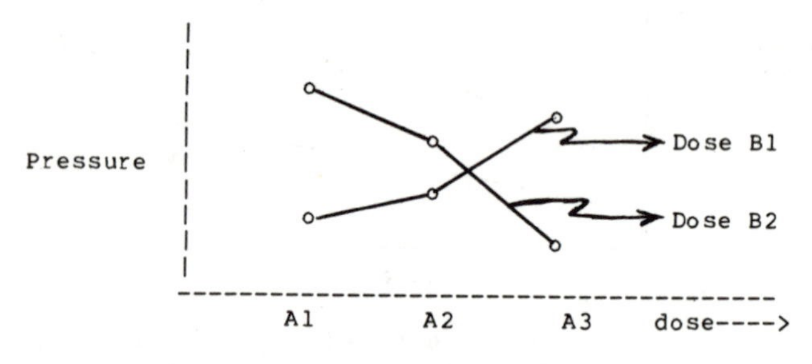

FIG. 13. Simultaneous dosage experiment.

are interested in whether the Dose A profile is parallel to the Dose B profile. The test for parallelism is

$$A_1B_1 - A_1B_2 = A_3B_1 - A_3B_2$$

$$A_2B_1 - A_2B_2 = A_3B_1 - A_3B_2.$$

Therefore,

$$C = \begin{pmatrix} 1 & 0 & -1 & -1 & 0 & 1 \\ 0 & 1 & -1 & 0 & -1 & 1 \end{pmatrix}.$$

The test of a dose effect with drug A is derived by summing over the B doses such that

$$\text{dose 1 effect} = A_1B_1 + A_1B_2$$

$$\text{dose 2 effect} = A_2B_1 + A_2B_2$$

$$\text{dose 3 effect} = A_3B_1 + A_3B_2.$$

The dose effect is assessed by testing

$$\text{dose 1 effect} - \text{dose 3 effect} = 0$$

$$\text{dose 2 effect} - \text{dose 3 effect} = 0$$

or

$$C = \begin{pmatrix} 1 & 0 & -1 & 1 & 0 & -1 \\ 0 & 1 & -1 & 0 & 1 & -1 \end{pmatrix}.$$

Similarly, for the dose effect with drug B,

$$C = (1 \quad 1 \quad 1 \quad -1 \quad -1 \quad -1).$$

These examples demonstrate the simplicity of hypothesis testing when one starts with a data source model. We have not been rigorous in the development of the probability distributions associated with testing these models. However, these tests are quite robust, and in practice with moderate-sized samples, the central limit theorem saves the day [4].

OTHER APPLICATIONS OF LINEAR MODELS

Linear models have found very wide use in the analysis of experimental data. The fact that a Taylor expansion can be used to linearize a nonlinear model has been particularly useful in dealing with biological data. As stated earlier, the assumption of an experimental error, ϵ, that is independent of the dependent or response variable, is necessary for useful parameter estimation. However, there are a number of settings where the experimental error is not independent of the response variable.

One such case arises when dealing with frequency or counting data. Data obtained from radiation detectors has an underlying Poisson distribution. For this distribution, the variance, σ^2, is equal to the mean, μ. Thus, if one acquires a count of 10,000 over a one-minute period, the variance is 10,000 and the standard deviation is 100 counts/min. However, if one acquires 1,000 counts over the same period, the standard deviation is approximately 32 counts/min. Clearly, the lower the count, the greater the percentage error. The percentage error for the 10,000 counts is only $\pm 1\%$ while the percentage error for the 1,000 counts is $\pm 3.2\%$.

Another case arises from the use of an algebraic transformation to linearize a nonlinear model. For instance, the gamma function is used frequently to characterize indicator dilution data. Let C(t) be the concentration of an indicator at time t. The time dependent behavior of C is expressed as

$$C(t) = Kt^{\alpha}e^{-t/\beta}$$

where k, α, and β are model parameters. By taking the natural logarithm of both sides of this equation we obtain

$$\ln C(t) = K + \alpha \ln t - \frac{t}{\beta}$$

which is a linear model with independent variables 1, ln t, and t, and model parameters k, α, and $1/\beta$.

Taking logarithms, however, modifies the error term, ϵ. Initially, it is assumed that ϵ is independent of C(t). The variance introduced by the transformation is proportional to $1/C^2(t)$, thereby violating the constant variance assumption.

Estimation of model parameters from either of these two examples by minimizing

$$(Y - XA)'(Y - XA)$$

results in unreasonable estimates. Large deviations between the data and the predicted data will be weighted more than small deviations. To adjust for this effect, a weighted minimization of the form

$$(Y - XA)'W(Y - XA)$$

is used where

$$W_{(n \times n)} = \begin{pmatrix} \frac{1}{w_1} & & & \mathbf{0} \\ & \frac{1}{w_2} & & \\ & & \ddots & \\ \mathbf{0} & & & \frac{1}{w_n} \end{pmatrix}.$$

Each individual weight, w_i, is an estimate of the noise level or variance associated with the ith data point. Thus, a large deviation between the model and the data

$$y_i - x_i A$$

will be scaled appropriately. Parameters are estimated in the standard way. Thus,

$$\min_A (Y - XA)'W(Y - XA)$$

yields

$$\hat{A} = (X'WX)^{-1}X'WY.$$

Hypotheses of the form $CA = 0$ are tested by forming the statistic

$$F = \frac{S_H/\text{rank}(C)}{S_E/(n - m)}$$

where

$$S_H = A'C'[C(X'WX)^{-1}C']^{-1}CA$$

and

$$S_E = Y'WY - Y'WXA.$$

Observe that the differences in $\hat{A}$, S_H and S_E, between the weighted model and the unweighted model are restricted to $X'WX$ and

X′WY. Thus, if X′WX and X′WY are accumulated instead of X′X and X′Y, the computer program for supporting the unweighted model becomes a program for supporting a weighted model.

The weighted model has found considerable application when dealing with categorical or discrete valued observations. Grizzle et al. [5] showed that many problems in categorical data analysis were simply weighted regression problems. The Grizzle approach has led to a unification of categorical data analysis procedures just as the unweighted linear model has unified the analysis of continuous data. These are major strides toward improved effectiveness of statistical analyses, and their widespread introduction is linked inexorably with the computerization of the associated algorithms.

DISCUSSION

Data analysis is a context-sensitive activity. For the first time, with high-speed computing and inexpensive bulk storage, we are able realistically to match each experiment with an appropriate model. Therefore, in addition to carrying out an experimental protocol, the investigator can also develop a model relating experimental outcomes with the properties of each substrate. Manual computing resources limit the complexity of the investigator's model; worse, it can force inappropriate oversimplifications that obscure important effects. High-speed computing, however, removes this constraint and allows the investigator to account for more properties contributing to substrate heterogeneity. A model that accurately describes substrate properties improves our ability to separate treatment effects from substrate effects.

High-speed computing facilities and techniques also provide new avenues for data analysis. Histograms, depicting the behavior of random variables, can be readily created and displayed, providing the investigator with a rich, graphical representation. Since histogram generation is no longer difficult, comparison of random variables can be carried out by visual inspection of their histograms. Procedures for histogram comparison (Kolmogorov-Smirnov) can now be routinely performed for large data sets, since the sorting required to generate the histogram is only a minor issue, treatable as a single conceptual operation with which the analyst need not be burdened.

High-speed computing also provides new life for old methods. Randomization of permutation tests, first suggested by Fisher, requires the computation of a probability distribution for each data set that is derived from an experimental investigation. The probability distribution is derived by first assuming the null hypothesis true, and then computing a measure of the treatment effect for each permutation of the experimental data. This poses little difficulty when only a few observations are in the data set. However, manual methods bog down for, say, 20 or more observations. High-speed computing provides a practical means for preparing each permutation. These randomization tests seem ideally suited for today's computers and provide a class of "exact" tests that are free of assumptions about underlying probability distributions.

Computing opens new options for data analysis. These options suggest avenues for statistical research which will significantly aid the scientific investigator. Similarly, statisticians and their colleagues are taking on problems that develop larger and larger amounts of data. The resulting data-management activities provide new directions for research in computing hardware and software. Computing and statistics are irreversibly bound together. The evolution of computing hardware can no more ignore the area of data analysis than can the evolution of statistical methodology ignore the tools that computer science provides.

REFERENCES

1. S. Siegel, *Nonparametric Statistics for the Behavioral Sciences*, McGraw-Hill, New York, 1956.
2. R. A. Fisher, *Statistical methods for research workers*, 11th ed. (rev.), Hafner, New York, 1950.
3. W. H. Robertson, "Programming Fisher's exact method of comparing two percentages," *Technometrics*, **2** (1960), 103–107.
4. D. F. Morrison, *Multivariate Statistical Methods*, McGraw-Hill, New York, 1967.
5. J. E. Grizzle, C. F. Starmer and G. G. Koch, "Analysis of categorical data linear models," *Biometrics*, **25** (1969), 489–504.

INDEX